# Archaeology, Soil- and Life-Sciences Applied to Enclosures and Fields

Proceedings of the Session
'From microprobe to spatial analysis – Enclosed and buried surfaces as key sources in Archaeology and Pedology'

European Association of Archaeologists 12th Annual Meeting
Krakow, Poland, 19th to 24th September 2006

Edited by

Kai Fechner
Yannick Devos
Mathias Leopold
Jörg Völkel

Institut national de recherches archéologiques préventives
Inrap

CReA

TUM
Technische Universität München

ROOTS

BAR International Series 2222
2011

Published in 2016 by
BAR Publishing, Oxford

BAR International Series 2222

*Archaeology, Soil- and Life-Sciences Applied to Enclosures and Fields*

ISBN 978 1 4073 0780 0



BAR Publishing is the trading name of British Archaeological Reports (Oxford) Ltd. British Archaeological Reports was first incorporated in 1974 to publish the BAR Series, International and British. In 1992 Hadrian Books Ltd became part of the BAR group. This volume was originally published by Archaeopress in conjunction with British Archaeological Reports (Oxford) Ltd / Hadrian Books Ltd, the Series principal publisher, in 2011. This present volume is published by BAR Publishing, 2016.

Printed in England

BAR
PUBLISHING

BAR titles are available from:

BAR Publishing
122 Banbury Rd, Oxford, OX2 7BP, UK
EMAIL info@barpublishing.com
PHONE +44 (0)1865 310431
FAX +44 (0)1865 316916
www.barpublishing.com

# Table of Contents

# Foreword

In order to study spatial and functional organisation in the past, ancient enclosed areas such as enclosures, houses and courtyards offer an important source of information, whether they were limited by ramparts, walls, hedges, posts or ditches (e.g. Marion & Blancquaert, 2000; Wieland, 1995). In recent years, the understanding of the genesis and organisation of such ancient enclosed areas has been improved by the increased application of multi-proxy research programs.

In the same perspective, buried surfaces, either protected by walls, ramparts, burial mounds or urban waste deposits, are another valuable source of information for archaeologists, geoarchaeologists and pedologists. Excellent overviews on the study of such buried surfaces can be found in Goldberg and Macphail (2006) and Holliday (2004). Both research subjects are partly based on soil and life sciences in order to provide significant means for interpretation and objective comparisons between the case studies.

During the session 'From microprobe to spatial analysis – Enclosed and buried surfaces as key sources in Archaeology and Pedology' the potential of different methods to study these ancient buried and enclosed surfaces have been presented. Contributions of archaeopedology, micromorphology, phytolith studies, palynology, geophysics and geochemistry to this understanding have been discussed.

This first volume contains part of the papers presented during this session, in particular those focussing on archaeology, soil and life sciences applied to the study of **ancient enclosures and fields**. The second volume will focus on archaeology, soil and life sciences applied to the study of **ancient houses, gardens and courtyards**.

The first two contributions are respectively discussing the potentials and limitations of geoarchaeology to the study of enclosed surfaces (Leopold), and the application of phosphorus cartography (Devos, et al.(a)).

The other papers published in these proceedings are either case studies dealing with the study of enclosures and fields (Fercoq de Leslay, et al.; Raab & Völkel; Rouppert, et al.), or regional syntheses on the study of enclosures (Fechner, et al.; Malrain) and cultivated soils (Devos, et al. (b)).

# References


Goldberg, P. and Macphail, R., 2006. *Practical and theoretical geoarchaeology*. Blackwell Publishing, Malden.

Holliday V.T. 2004. *Soils and Archaeological Research.* Oxford University Press, Oxford.

Marion St. and Blancquaert, G., Ed., 2000. Les installations agricoles de l'âge du fer en France septentrionale. Etudes d'Histoire et d'Archéologie 6.

Wieland G. 1995. Die spätkeltischen Viereckschanzen in Süddeutschland- Kultanlagen oder Rechteckhöfe ? In: Haffner A. (ed.). Heiligtümer und Opferkulte der Kelten. Archäologie in Deutschland. Sonderheft 1995 (Theiss, Stuttgart) : 85-99.

## Reading Committee and Acknowledgements

We are particularly grateful to the reading committee for their thorough and critical corrections. The papers published in these proceedings have been reviewed by :

- Prof. Roger Langohr, soil scientist, Ghent University, Belgium
- Dr. Geertrui Louwagie, soil scientist and soil chemist, European Commission, Joint Research Centre, Institute for Prospective Technological Studies, Sevilla, Spain
- Prof. Günther Mossbauer, archaeologist, Universität Osnabrück, Germany
- Prof. Peter W. Birkland, geomorphologist, University of Colorado at Boulder, USA
- Dr. Cécilia Courbot, archaeologist, Université de Lille III, France
- Dr. Matthias Leopold, soil scientist, Technische Universität München, Germany
- Kai Fechner, archaeologist and soil scientist, Institut national de recherches archéologiques préventives (INRAP), France
- Yannick Devos, archaeologist and soil scientist, Centre de Recherches Archéologiques, Université Libre de Bruxelles, Belgium

Furthermore, we would like to thank the institutions and persons who have supported the organisation of the session and the editing of the proceedings: Institut national de recherches archéologiques preventives (INRAP), in particular **Nathan Schlanger**, **Hervé Guy** and **Laurent Sauvage**, the 'Centre de recherches archéologiques' (Université Libre de Bruxelles), the Brussels Capital Region, and in particular the Direction of Monuments and Sites, the Research Team in Archaeo- and Palaeo-Sciences (ROOTS), the organising committee of the 12th Annual Meeting of the European Association of Archaeologists in Cracow. We are particular grateful to Danko Josić, who agreed to host these papers in the British Archaeological Reports.

We would like to dedicate this volume to our colleague **Pascal Querel** for his enthusiastic implication in the reading committee before his unfortunate passing away.

Kai Fechner, Yannick Devos, Matthias Leopold and Jörg Völkel

# Chapter 1
# Buried Soils and Surfaces, a Key Archive for Geoarchaeology

*Matthias Leopold*

*Special Field for Geomorphology and Soil Science,*
*Research Department for Ecology and Ecosystem Sciences,*
*Center of Life and Food Sciences Weihenstephan,*
*Technische Universitaet Muenchen, D-85350 Freising, Germany,*
*matthias.leopold@wzw.tum.de*

## Introduction

The cultural landscape in Europe is the product of many different human and natural activities and processes. Settlement, clearing and farming over thousands of years have constantly transformed the landscape as well as the bio- and geocomponents like flora and fauna, relief, hydrology, soils and others. As geoarchaeology tries to reconstruct certain stages of this cultural landscape development, geoarchives of the various time periods are crucial. Soil science combined with colluvium stratigraphy are such geoarchives, because colluvial sediments have a large distribution. The existence of the geoarchive colluvium means that in places the original soil has been removed by erosion. Therefore there is a lack of *in situ* soils for certain time periods throughout Europe. Enclosed (proto)historical areas which are protected against erosion offer the unique and rare chance to gather information about the stage of soil development at certain times and allow further reconstructions and calculations of human impact outside the enclosures. Furthermore, earth ramps, walls, burial mounds preserve the original soils effectively over thousands of years.

This paper describes the theoretical background of the strength and weaknesses in using buried soils and surfaces. The theory is applied to field studies of two different Celtic ramparts. These impressive earthworks buried and protected the old pre-human disturbance surfaces and soils, which helps in the reconstruction of the soil formation conditions and provides environmental data for the Archaeologists for the Celtic time period.

## Theoretical background of using buried soils and surfaces

Soils and their associated surfaces are part of a geosystem, which is also the space of human activity. Farming, clearing, settling, etc. are all somehow linked with soils, and archaeology tries to reconstruct the kind and shape of human activity over different time periods (Holliday, 2004: 290 ff., Goldberg and Macphail 2006: 193 ff.). Thus, the reconstruction of soils and surfaces can play an important role in archaeological questions especially in setting up a stratigraphic chronology.

Soils can be defined as four dimensional natural bodies which are the syngenetic product out of different soil forming factors. It is generalised in the following function (1):

$$S = f(SPM, C, R, O, T, HI) \qquad (1)$$

S soil, SPM soil parent material, C climate, R relief (topographic setting),O organism (fauna, flora), HI human impact, T time (after Dokutschajew 1883, modified; see also Jenny 1980)

These factors control the direction and the intensity of different pedogenic processes which are summed up as transformation, translocation and turbation processes. As the factors change from place to place slightly or significantly, the resulting pedogenic processes either get amplified, or weakened, or they don't take place at all. However, the processes lead to a distinct combination of layers (horizons) of mineral and/or organic constituents of variable thicknesses, which differ from the soil parent materials in their morphological, physical, chemical, and mineralogical properties, and their biological characteristics (Joffe 1949). The result of this is the soil profile.

For geoarchaeological work it is important to keep in mind that factors which control pedogenic processes not only change from place to place but also during time (for discussion see Holiday 2004). This means that different soil forming factor combinations occur at different times of human history. It can be easily imagined that changing climate, changed human impact like farming, clearing or settling which changed the topography due to heavy upslope soil erosion and downslope accumulation of sediments. Soil scientists must be aware of all of these changes because opening and describing a soil profile nowadays is

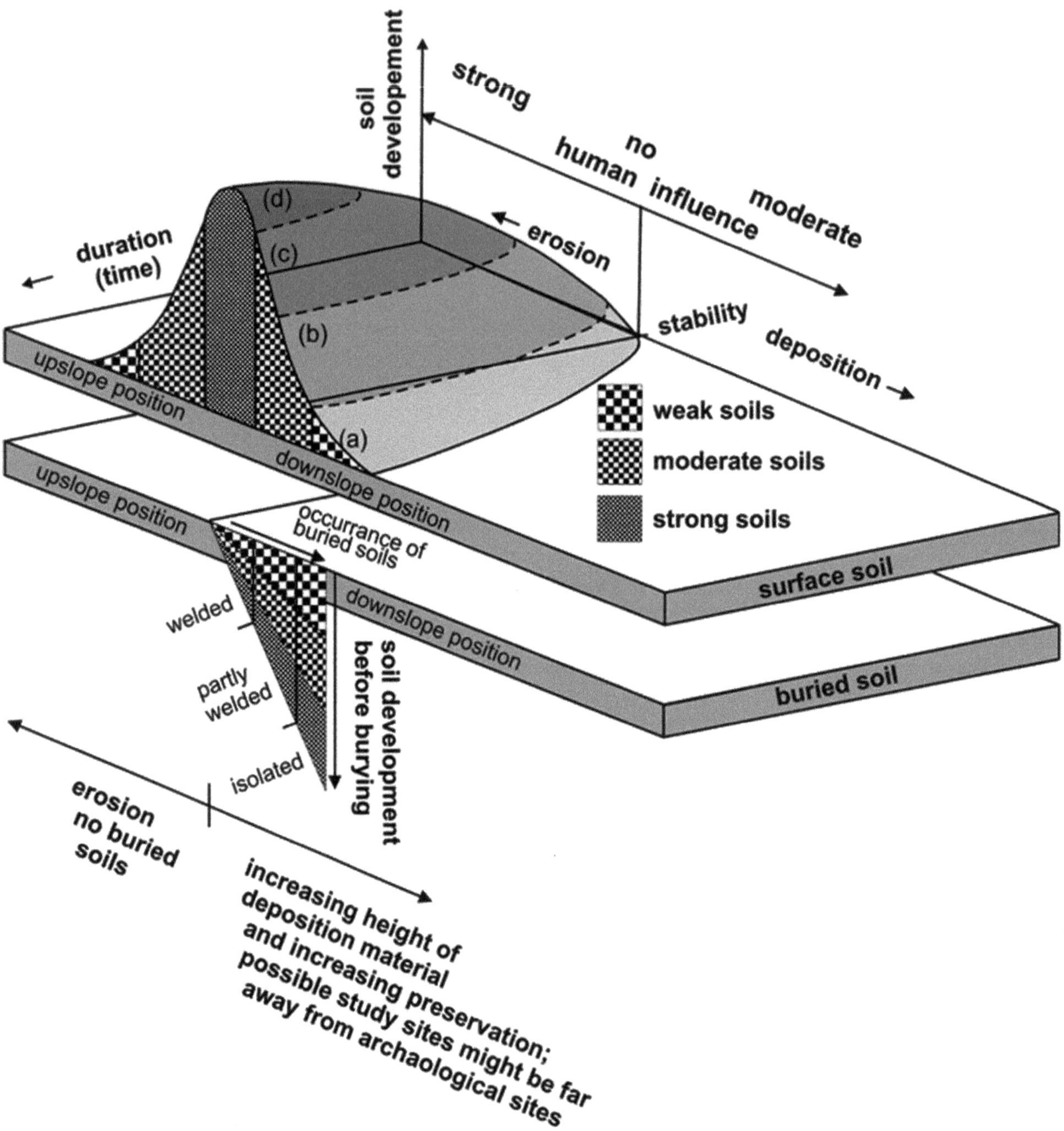


*Figure 1. Scheme showing the intensity of soil development on the surface in relation towards time and landscape stability in contrast to the possibility of the occurrence of buried soils. (d)-(a) show levels of soil development if soils remain at the surface for a long time (a), if they get truncated more and more (c – a at the erosion side / upslope position) or if they get buried to various depth and frequency (c-a at the deposition side / downslope position). Depending on the time and intensity of the soil development before a soil gets buried and depending on the thickness of deposition material, week, moderate or strong soils can form welded, partly welded or isolated buried soils. The chance to find buried soils rises with the topographic position towards the downslope area of a hill. Image from Birkeland 1999, modified.*

describing a syngenetic natural product that was probably altered over time. Figure 1, which was originally presented for alluvial plains, presents a scheme of the intensity of soil development against time as modified by erosion or deposition which, in Middle Europe, can be partly replaced by the intensity of human influence (see also Holiday 2004: 27). However, the original scheme (upper part of fig. 1) only shows part of the complexity. This is mainly because diagnostic soil systems are based on diagnostic features (e.g. the US-soil classification system). The diagnostic features that are observed today are the result of a continuous building up of a certain soil feature over time. Diagnostic features are just the product of soil genetic processes but it is the process itself that has to be in the focus of investigation. Genetic soil classification systems like for example the French, the German (Ad Hoc Boden 2006) or partly the WRB (IUSS 2006) offer the advantage to include processes and time in a soil description which

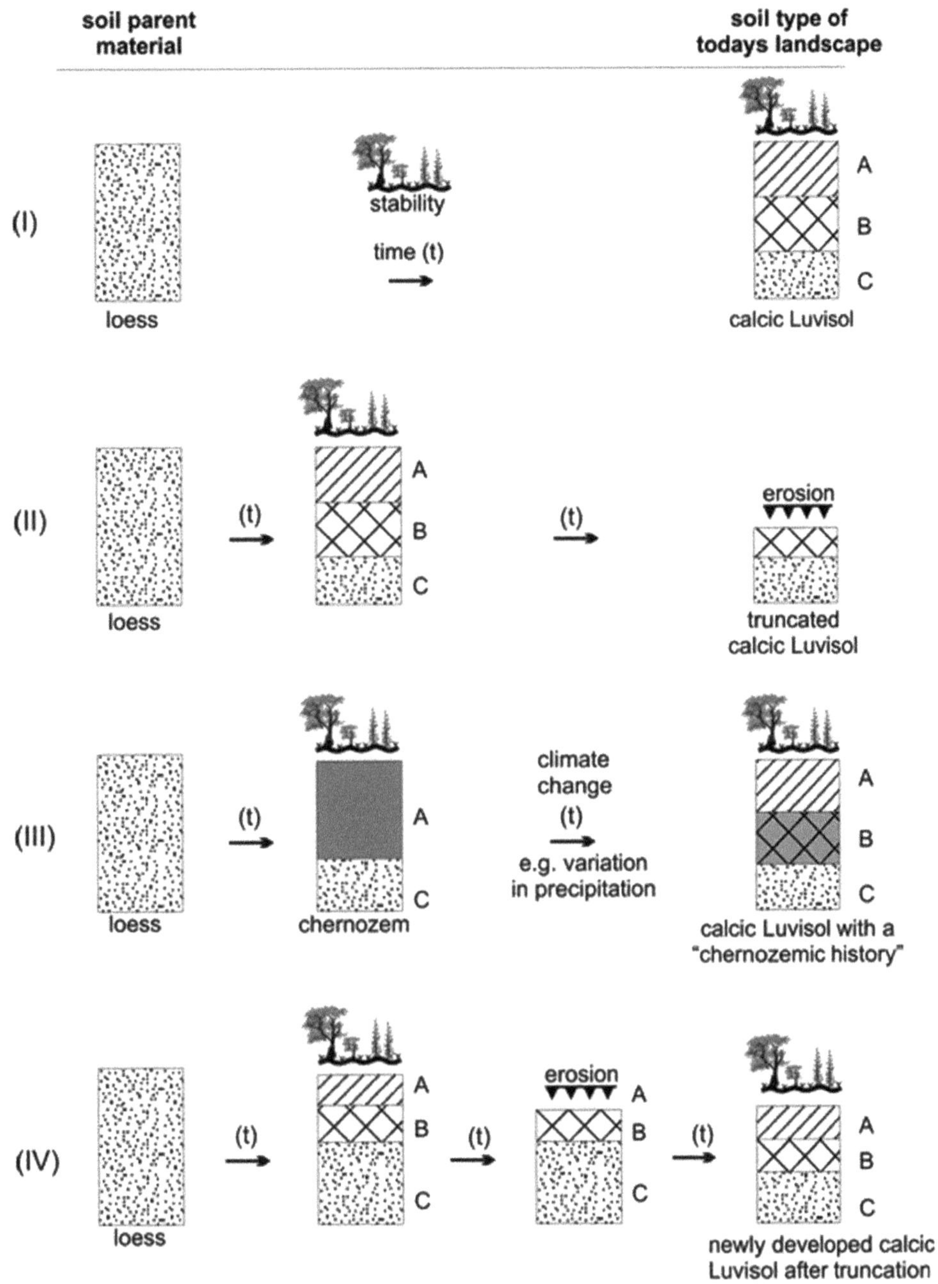


*Figure 2. Different possibilities of soil development under changing soil formation factors and time. Diagrammed are four main development possibilities, but many combinations and other minor important constellations exist, all showing the complexity of the final product, soil (A, B, C indicate the main soil horizons).*

can help to avoid misinterpretations. A good example of the advantage to use soil genetic descriptions rather than diagnostic features is given by Wieder and Gvirtzmann (1999). In old settlement areas however, excavations show that soil development can be a non-continuous process. It can be disrupted for some thousand years and re-established if physicochemical parameters and/or land use change. The soil development can be set back to the level of a younger soil (level d-a in fig. 1). Soils with a climax development (level d) can get eroded so that only the base remains, the result being old in situ sediments (parent material) with a young soil at level a on the erosion side. On the other side, ongoing deposition preserves the original soil now forming buried soils, which primarily belonged to one of the levels a-d depending on the age and land use history. A buried soil is the result. As soil development is an ongoing process, it also takes place in the deposited material which now covers the buried soil described above. To sum up, soil development can occur at different times, with or without interruption due to changes in the soil forming factors. This leads to four basic constellations which we can expect when investigating a modern soil profile (fig. 2).

(I) A soil develops under constant soil forming factors, or possible minor changes are well known (e.g. climate variability). Pedogenic processes have lead to distinct soil

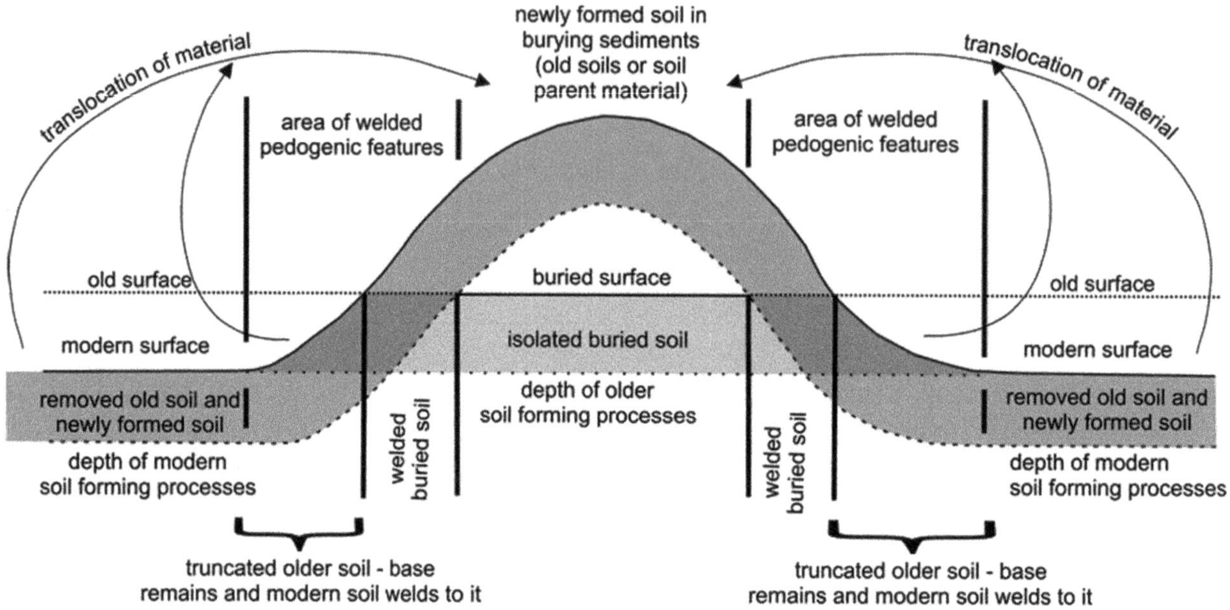


*Figure 3. Overview of the position of buried soils and surfaces within an user-defined earthwork. Depending on the thickness of the overlying material, the buried soil contains different information which can be read out.*

horizons and their physical and chemical properties can be described and measured. As soils develop under certain rules and in distinct stages, data collected from modern soil allow a reconstruction of the different soil developing stages. (II) A soil formed under known soil forming factors as in (I) but has been partly eroded due to human impact (farming). If the truncation is shallow or the erosion process is modern, there is a good chance, that the different stages of a soil development can be re-constructed. (III) A soil formed under constant soil forming factors for a certain time followed by a change of some factors (e.g. climate variability). Pedogenic features of the modern soil now exhibit remnants of the older phase and features of the modern phase. If the change of factors lead to two distinct types of soil, the separation and therefore the reconstruction of the different stages of soil development might still be possible (e.g. the transformation of a Chernosem into a calcic Luvisol). However, there are other combinations of factors where a reconstruction would be difficult. (IV) A soil has formed under constantly changing factors. Additionally, the soil has been partly truncated or totally eroded probably more than once. There might be hardly any material left with which to reconstruct certain soil development stages. Beside these four basic constellations other combinations are possible and repeated erosion with following reforestation and soil development leads to complex polygenetic soil profiles (see also Vreeken 1975).

The above constellation no. (I) is rarely found throughout the Central European landscape due to the long settlement history which reaches back at least several thousand years. Combinations of no. (II) and (III) or (III) and (IV) are most common in Europe, forming complex polygenetic soils. Their use as an archive in geoarchaeological studies is a great challenge for soil scientists and often the achievable results are of diverse quality.

However there is a possible solution to achieve higher data quality. The existence of buried soils and surfaces under historic and prehistoric objects or natural accumulations could help to gather soil data of certain time periods of the past (see Vreeken 1975, Goldberg and Macphail 2006: 248 ff.). In this context the thickness of the burying material is important (fig. 3 and fig. 1). Deeply buried soils can be isolated from subsequent soil forming processes and therefore they may retain many features from the time they were formed. They are defined as isolated buried soils (Schaetzl and Sorenson 1987). In contrast thinly buried soils can be still influenced by modern soil forming processes, like the translocation of clay particles or the solution and precipitation of carbonate or others (eg. Leopold and Völkel 2006b). In the latter case the modern soil and the buried soil form a welded soil (Olson and Ruhe 1980). If a soil is a relict soil and can be traced to a welded soil or to an isolated soil one can back out soil process information for different length of time. A good discussion about the different use of the several terms is given by Holliday (2004: 77-79). However, buried soils offer the possibility to both reconstruct ancient surfaces and ancient soil formation factors. Depending on the position (isolated or not, see fig. 3) buried soils contain different information. The age of the burying object (e.g. a wall or rampart) sets a time of which the buried soil stores certain data. With that soil science can provide data which is needed for erosion history, the reconstruction of single weather and/or climate conditions or of the intensity of human impacts before a certain time in history. Much of this information helps archaeologists to

| reconstruction of: | surface | water table | erosion | pedology | human activity | pedo-ecology | chronology (relative) | chronology (absolute) |
|---|---|---|---|---|---|---|---|---|
| buried surface | xxx | x | xx | x | xx | - | xxx | - |
| buried soil | x | xx | xx | xxx | xx | xxx | xx | x |

| | |
|---|---|
| xxx | very valuable |
| xx | valuable |
| x | partly valuable |
| - | not valuable |

*Table 1. Overview of some possible uses of buried soils and surfaces and their benefit to certain aspects of pedology and geoarchaeology.*

draw a more precise picture of the history at an excavation site and the surrounding areas.

## The use of buried soils and surfaces in geoarchaeological contexts

The following paragraph refers to cases discussed during the session "From microprobe to spatial analysis – Enclosed and buried surfaces as key sources in Archaeology and Pedology" that use buried soils and surfaces for their respective studies. The role of the examples given is to point out the diversity of the potentials of using buried soils and surfaces in geoarchaeological studies.

Other case-studies attest of Neolithic cultivating surfaces conserved on sites in the Moselle valley, in France and Luxembourg (Baes and Fechner, in preparation). The involved techniques are pedology and micromorphology.

An inventory of traces of gardens and fields of Northern France, Belgium and Luxembourg, using the same identification techniques, is proposed by Fechner et al. (in preparation).

A particular case is the one of Ribemont-sur-Ancre (Fercoq et al., this volume), where a former cultivated surface is transformed by a sacral site of the Late Iron Age. Both the impact by the deposition of human corpses and by a new man-made vegetation cover is documented. In order to do so, a multi-proxy approach has included malacology, palynology, pedology and micromorphology.

The buried surfaces studied by the article of David and Fechner (in preparation) are mostly comprised in medieval and modern upfills in Northern and Central France, including the so-called dark earths. A buried surface of the Late Bronze Age is also mentioned. The methods are mostly geophysical mapping, crossed in some cases with phosphorus tests and analyses.

A medieval buried surface conserved in a former house brings us back to the Moselle valley (Blaising & Gérard, in preparation). Here, some phosphorus tests, pedology and micromorphology have been applied, but the article mostly documents the archaeological results ad implications.

Additionally, the examples from different medieval sites in Brussels illustrate proto-urban cultivation areas, using a multi-proxy approach that includes among others, pedology, micromorphology, phytolith and charcoal analyses (Devos et al., this volume).

Finally, Raab and Völkel (this volume) describe the use of buried soils and a Celtic rampart to reconstruct stages in soil development, especially in clay mineral transformation an acidification. Celtic ramparts and its buried soils and surfaces have been used in Germany for the reconstruction of soil erosion rates during the period of the Latène age (Leopold 2003, Leopold and Völkel 2005, 2007a, b). Furthermore the Celtic rampart of the well known Oppidum at Manching preserved a well developed buried soil which was used to reconstruct hydrological conditions during the time of wall construction (Leopold & Völkel 2005).

## Conclusion

Buried soils and surfaces are very valuable geoarchives as shown by studies in the last decades (e.g. Litt et al. 1987; Saile 1993; Holliday 2004). Also experimental Archaeology uses soils and buried soils for their studies about soil development on archaeological objects (e.g. Breuning-Madson et al. 2001, Goldberg and Macphail 2006), however, published examples are few.

Above the theory description and some examples show how the information in geoarchives is used in archaeological reconstructions. Table 1 sums up some of the information a buried soil or surface can provide. As indicated the value of buried soils and surfaces for the reconstruction of different aspects relevant for Geoarchaeology is variable. Buried surfaces are very valuable to reconstruct the level of a former surface, or to establish a relative chronology at a site. It is also valuable to determine heights of erosion and to estimate the influence of humans at a site before the time of burying. In contrast isolated buried soils are very valuable to get information about the pedology before the time of burying. Welded buried soils can provide very valuable pedo-ecological data of the time after the burying. In general buried soils give valuable information about old water tables, stages of erosion, human influence on the land, and provide data to establish an absolute chronology. Both buried soils and surfaces can help to reconstruct important details of a geosystem.

Discussed above, a soil is a constantly changing and developing four-dimensional natural body, so many processes can have happened before, during, and after the time of burial by human activity. Archaeologists need to be aware of that. Some of the pedogenic processes are rather special and complicated so we recommend that archaeologists involve soil scientists as collaborators at archaeological excavations to get as much information as possible out of the buried soils and surfaces geoarchive.

## References


Ad hoc Boden (2005[5]). Bodenkundliche Kartieranleitung. Bundesanstalt für Geowissenschaften und Rohstoffe, Hannover, 392 p.

Blaising, J.M., and Gérard, F., in preparation. The interest of the archaeological study of « *akkerbergs* » for recognizing Early Medieval rural settlements in Lorraine (France, VIIIe-XXe centuries). In: Archaeology, soil- and life sciences applied to houses, gardens and courtyards. Second volume of the Proceedings of the Session 'From microprobe to spatial analysis – Enclosed and buried surfaces as key sources in Archaeology and Pedology' organised at the European Association of Archaeologists, 12th Annual Meeting, Krakow-Poland. 19th to 24th September 2006.(= British Archaeological Reports, International Series).

Birkeland, P.W., 1999. *Soils and Geomorphology*, New York.

Breuning-Madsen, H., Holst, M.K., and Rasmussen, M., 2001. The Chemical Environment in a Burial Mound Shortly after Construction – An Archaeological-Pedological Experiment, *Journal of Archaeological Science,* 28 : 691-697.

David, C., and Fechner, K., in preparation. Man-made « dark earth » and palaeosurface horizons in France: case studies based on geophysical mapping and other soil approaches. In: Archaeology, soil- and life sciences applied to houses, gardens and courtyards. Second volume of the Proceedings of the Session 'From microprobe to spatial analysis – Enclosed and buried surfaces as key sources in Archaeology and Pedology' organised at the European Association of Archaeologists, 12th Annual Meeting, Krakow-Poland. 19th to 24th September 2006.(= British Archaeological Reports, International Series).

Devos, Y., Vrydaghs, L., Fechner, K., Laurent, C., Degraeve, A., and Modrie, S., this volume. *Buried Anthropic Soils in the Centre of Brussels (Belgium): Looking for Fields in a (Proto-) urban Context.*

Dokutschajew, W.W., 1883. *Russkij Chernozem*, St. Petersburg.

Eheim, A., and Völkel, J., 1994. Vergleich des Tonmineralbestandes und ihrer pedogenen Transformation in Böden unterschiedlichen Alters anhand eines kletischen Bauwerks. In: Kohler, E. (ed.): *Berichte der deutschen Ton- und Tonmineralgruppe e.V.*, DTTG 1994: 55-65.

Fechner, K., Baes, R., Langohr, R., Louwagie, G., Mikkelsen, J., Slachmuylder, J.-L., and Devos, Y., in preparation. Pre-modern *in situ* traces of agriculture in High and Middle Belgium, Luxembourg and Northern France. A status quaestionis. In: Archaeology, soil- and life sciences applied to houses, gardens and courtyards. Second volume of the Proceedings of the Session 'From microprobe to spatial analysis – Enclosed and buried surfaces as key sources in Archaeology and Pedology' organised at the European Association of Archaeologists, 12th Annual Meeting, Krakow-Poland. 19th to 24th September 2006.(= British Archaeological Reports, International Series).

Fercoq du Leslay, G., Court-Picon, M., Loicq, S., Fechner, K., this volume. Environmental and archaeological studies in sanctuaries, functional and spatial interpretation: the case of Ribemont-sur-Ancre (Somme).

Goldberg, P., and Macphail, R., 2006. Practical and theoretical geoarchaeology. Blackwell 455 p.

Holliday, V.T., 2004. *Soils and Archaeological Research.* Oxford University Press, Oxford, UK.

IUSS Working Group WRB (2006). World reference base for soil resources 2006. Worlds soil resources reports, 103, FAO Rom, 128 p.

Jenny, H., 1980. The soil resource. Origin and behaviour. Ecological. Studies 37, Springer Verlag, Berlin, pp 377.

Joffe, J.S., 1949. *Pedology*, Pedology Publication, New Brunswick, NJ.

Leopold, M., 2003. *Multivariate Analyse von Geoarchiven zur Rekonstruktion eisenzeitlicher Landnutzung im Umfeld der spätlatènezeitlichen Viereckschanze von Poign, Lkr. Regensburg,* Regensburger Beiträge zur Bodenkunde, Landschaftsökologie und Quartärforschung, 2. URL: http://www.opus-bayern.de/uni-regensburg/volltexte/2003/233

Leopold, M., and Völkel, J., 2007a. Reconstruction of the palaeoenvironment and the old cultural surface of a Bronze Age Settlement – An example for a multi-methodological interaction of Soil Science and Archaeology in Southern Germany, *Geodinamica Acta. In press*

Leopold, M., and Völkel, J., 2007b. Quantifying prehistoric soil erosion - the discussion of different methods by the example of a Celtic square enclosure in southern Germany, *Geoarchaeology,* 22[8]: 873-889.

Leopold, M., and Völkel, J., 2006a. The Celtic Rampart of the Oppidum at Manching. An example for a multi-methodological interaction of Soil Science and Archaeology in Southern Germany, *Zeitschrift für Geomorphologie* N.F., Suppl. 142: 135-148.

Leopold, M., and Völkel, J., 2006b. Colluvium: Definition, differentiation and possible suitability for reconstructing

Holocene climate data, *Quaternary International* 162-163: 133-140.

Leopold, M., and Völkel, J., 2005. Methodological approach and case study for the reconstruction of a (pre) historic land use model, *Zeitschrift für Geomorphologie* N.F., Suppl. 139: 173-188.

Litt, T., Kohl, G., Göndorf, J. and Jäger, K.-D., 1987. Zur Datierung begrabener Böden in holozänen Ablagerungsfolgen, Jahresschrift für mitteldeutsche Vorgeschichte, 70: 177-189.

Raab, A., and Völkel, J., (this volume): Holocene soil development and pedogenic clay mineral transformation in Luvisols with different development ages at the Viereckschanze of Poign (Central Bavaria, Germany).

Ruhe, R.V., and Olson, C.G., 1980. Soil welding, Soil Science, 130, 132-139.

Saile, T., 1993. Holozäner Bodenabtrag im Bereich einer Bandkeramischen Siedlung am Rande des Reinheimer Beckens bei Wembach (Hessen), Archäologisches Korrespondenzblatt, 23: 187-196.

Schaetzl, R.J., and Sorenson, C.J., 1987. The concept of "buried" versus "isolated" paleosoils: examples from northeastern Kansas. Soil Science, 143: 426-435.

Vreeken, W.J., (1975). Principal kinds of chronosequences and their significance in soil history. European Journal of Soil Science, 26: 378-394.

Wieder, M., and Gvirtzman, G., 1999. Micromorphological indications on the nature of the Late Quaternary Paleosols in the southern coastal plain of Israel. Catena 35 : 219-237.

# CHAPTER 2
# THE APPLICATION OF PHOSPHORUS CARTOGRAPHY TO ARCHAEOLOGICAL SITES AND STRUCTURES: A STATE OF THE ART AND PROPOSAL OF A PROTOCOL APPLICABLE FOR BELGIUM, LUXEMBURG AND NORTHERN FRANCE

*Yannick Devos (1), Kai Fechner (2, formally 1/3), Jari Hinsch Mikkelsen (3, 4)*

(1) Centre de Recherches Archéologiques, Université Libre de Bruxelles.
Avenue F.D. Roosevelt, 50, CP850, B-1050 Brussels, Belgium.
(2) INRAP Nord-Picardie ZI de la Pilaterie,
11 rue des Champs, 59 650 Villeneuve d'Ascq (France),
(3) ROOTS (Research Team in Archaeo- and Palaeo- Sciences).
Av. H. de Brouckère, 82 B-1160 Brussels, Belgium.
(4) Research Institute for Nature and Forest (INBO).
Kliniekstraat 25, B-1070 Brussels, Belgium.

Abstract: In the last few decades, there has been an exponential increase in the application of phosphorus analyses for archaeological purposes, often combined with other chemical and geophysical analyses. The growing realisation that knowledge on anthropogenic accumulation of phosphorous in the soil can provide important information have lead to an increasing insight in relative and absolute phosphorus distribution patterns in time and space. This knowledge, scattered in numerous publications and reports, has been reviewed in this paper, demonstrating the importance of a thorough taphonomic study. Finally, recommendations for further development of the application and interpretation of phosphorus analysis for archaeological cartography are proposed, more specifically, for Belgium, Luxemburg and Northern France.

Keywords: Phosphorus cartography, sampling, taphonomy, geoarchaeology

## Introduction

Present day archaeological research covers a whole field of partner sciences aiming to reveal ancient human presence, to reconstruct ancient activities and to understand human interaction with the broader environment. Geoarchaeology takes an important position amongst the sciences implicated in this interdisciplinary research. Geoarchaeologists not only deal with palaeoenvironmental reconstruction; they also investigate traces – other than the classical artefacts – left by human activities. In identifying and characterizing anthropogenic traces phosphorus analysis has proven to be a powerful tool (Provan, 1971; Proudfoot, 1976; Bakkevig, 1980).

Human activities like manuring, harvesting, storage, preparation, cooking, consumption, but also burial activities and stabling can cause high concentrations of phosphorus to accumulate in the soil (Proudfoot, 1976; Sjöberg, 1976; Edwards, 1983; Lippi, 1988; Derry, et al., 1999; Parnell, et al., 2002; Sarris, et al., 2004; Terry, et al., 2000; Terry, et al., 2004; Louwagie, et al., 2005; Holliday and Gartner, 2007). Once the phosphorus enters the soil, it gets easily immobilized, has a very low solubility and therefore remains traceable for a relatively long period of time (Edwards, 1983; Lillios, 1992; Crowther, 1997; Leonardi, et al., 1999; Wells, et al., 2000; Terry, et al., 2000; Fernández, et al., 2002; Terry, et al., 2004). Nevertheless, particular human activities (e.g. cleaning, sweeping and cultivation without the application of manure) can accelerate phosphorus depletion (Proudfoot, 1976; Sandor, et al., 1990; Leonardi, et al., 1999; Ayala, et al., 2004).

Since the late 1920s and the early 1930s – with the pioneering research of the Swedish chemist and archaeologist Olof Arrhenius (1931) – phosphorus analyses have been applied as a tracking tool to find human occupation sites. With time, methods have been developed that allow to determine the presence and extent of archaeological sites, to investigate the use of internal space of the sites and their individual structures and to determine the presence of specific activities (e.g. liming with phosphorus-rich material). Several excellent reviews on the general history of phosphorus analysis in archaeology have been published over the years (e.g. Proudfoot, 1976; Sjöberg, 1976; Kyhlberg, 1993; Crowther,

1997; Parnell, et al., 2002; Holliday and Gartner, 2007). In the following this public knowledge is applied in a review of the phosphorus knowledge specific for the environmental settings characterising Belgium, Luxembourg.

Phosphorus analysis is widely applied in archaeological research, as demonstrated by studies within very different fields of research covering extended time periods, such as:

- Pre-Hispanic conquest sites in South-America (Lippi, 1988; Dunning, 1993; Smyth, *et al.*, 1995; Terry, *et al.*, 2000; Wells, *et al.*, 2000; Fernández, *et al.*, 2002; Parnell, *et al.*, 2002, Terry, *et al.*, 2004; Wells, 2004),
- Pre-Dynastic sites in Egypt (Hassan, 1981),
- prehistoric and medieval sites in Scandinavia (Prøsch-Danielsen and Simonsen, 1988; Kyhlberg, 1993; Smith, *et al.*, 2001; Persson, 2005; Mikkelsen, *et al.*, 2006),
- ancient fields and manuring practices (Berlin, *et al.*, 1977; Edwards, 1983; Sandor, *et al.*, 1990; Lillios, 1992; Dunning, 1993; Mikkelsen and Langohr, 1996; Simpson, 1997; Simpson, *et al.*, 1998; Leonardi, *et al.*, 1999; Louwagie and Langohr, 1999, Macphail and Cruise, 2001; Spek, *et al.*, 2003; Devos, *et al.*, this volume),
- stabling/herding practices (Cruise and Macphail, 2000; Macphail, *et al.*, 2000; Mikkelsen, *et al.*, 2003; Macphail and Linderholm, 2004b),
- the evolution of tells (Davidson, 1973; Ottaway and Matthews, 1988),
- potential burial places (Solecki, 1951; Crowther, 1997; Farswan and Nautiyal, 1997),
- (post-)medieval gardens (Foss, 1992; Miller and Gleason, 1994; Foss, *et al.*, 1995; Persson, 2005; Vissac, 2005),
- "Dark-earth" layers in Europe (Macphail, *et al.*, 2000; Alexandrovskaya and Panova, 2003; Macphail, 2003a; Macphail and Linderholm, 2004a) and the Amazon (da Costa and Kern, 1999; Lima, *et al.*, 2002),
- the spatial organisation of Celtic rectangular ditched structures (e.g. Viereckschanzen) (Fechner, *et al.*, in preparation).

Applications of phosphorus analysis in archaeology can be classified into three main groups:

- determination of the presence of archaeological sites during surveys (Sjöberg, 1976; Hassan, 1981; Cavanagh, *et al.*, 1988b; Lippi, 1988; Crowther, 1997; Bolender and Steinberg, 1998; Parnell, *et al.*, 2002),
- delineation of the use of space within a site or one particular structure (Sjöberg, 1976; Conway, 1983; Cavanagh, *et al.*, 1988b; Quine, 1995; Smith, *et al.*, 2001; Parnell, *et al.*, 2002; Macphail, 2003b, Milek, 1999; Middleton, 2004; Sarris, *et al.*, 2004; Sullivan and Kealhofer, 2004; Terry, *et al.*, 2000; Fernández, *et al.*, 2002; Terry, *et al.*, 2004; Fechner, *et al.*, in preparation),
- investigation of past agricultural activities (Berlin, *et al.*, 1977; Edwards, 1983; Prøsch-Danielsen and Simonsen, 1988; Dunning, 1993; Simpson, 1997; Simpson, *et al.*, 1998; Leonardi, *et al.*, 1999; Louwagie and Langohr, 1999, Macphail, *et al.*, 2000; Dupouey, *et al.*, 2002; Parnell, *et al.*, 2002; Dercon, *et al.*, 2003; Spek, *et al.*, 2003; Guttmann, *et al.*, 2005; Devos, *et al.*, 2007; Devos, *et al.*, this volume).

Although the applications, as demonstrated, are multiple and applied on very distinct geographical areas, some problems remain unsolved.

One problem concerns the lack of a precise protocol, which takes into account possible post-occupational contamination.

The aim of the present article is to delineate the potentials and restrictions of phosphorus analyses for archaeological purposes and to present a protocol for phosphorus mapping in Belgium, Luxembourg and North France.

## Terminology

The element P can take many forms, and is in the literature found under several names. Some of the most commonly used ones are phosphorus, phosphate, phosphor, and P. Also different fractions of P are applied with names, such as Ptot, Porg, Pinorg or PE. The wide range of names and terms can cause some confusion, and should be clarified (table 1).

## Parameters influencing soil phosphorus concentration

Phosphorus studies in archaeological contexts often make

*Table 1. overview of the most common terms applied in phosphorus analysis.*

| *Term* | *Explanation* |
|---|---|
| P | Most simple form, refers to the element as found in the periodical system |
| Phosphorus | Refers to the element P |
| Phosphate | $PO_4^{3-}$ |
| Phosphorus oxide | $P_2O_5$, often the content of P analysed in the laboratory is calculated into $P_2O_5$, which is a tradition that originates from the chemical fertilizer industry, which always expressed the content of nutrients as oxides |
| Ptot | The total amount of phosphorus in inorganic (mineral) and in organic form |
| Porg | Organic phosphorus |
| Pinorg, Pi, PE | Inorganic phosphorus |
| Pavailable, Pa | Plant available or extractable phosphorus |
| P non-available, Pna | Plant non-available or non-extractable phosphorus |

the assumption that phosphorus accumulations remain stable in the soil for long periods (Proudfoot, 1976; Sjöberg, 1976; Fernández, *et al.*, 2002; Parnell, *et al.*, 2002). However, some changes over time can affect this stability. The most obvious being the actual erosion of the phosphorus-rich layer, but also direct and indirect factors such as climate change or changes in soil humidity, can play a role.

Understanding these changes is essential for the interpretation of phosphorus data. It allows determining whether or not it is appropriate to perform phosphorus analysis on an archaeological site or structure and it can avoid producing erroneous data. For example, analysing the phosphorus concentration in an archaeological layer without associating it to the background phosphorus level/content (the pre-occupational phase), is of little use.

To understand the complexity of phosphorus accumulation and storage in the soil, an insight into the responsible parameters is required.

Furthermore, one has to take into account that there are three potential periods during which phosphorus can accumulate in the soil.

1. The pre-occupational phase of the site, the (natural) background.
2. The accumulation deposited concomitant with the occupation of the site, the period of interest for archaeological investigations.
3. Post-occupational accumulations, for example the use of fertilizers during more recent agricultural activities.

Obviously, multiple occupations or visits to one site complicate this basic sequence. For instance the reoccupation of a site after its abandonment can lead to new inputs of phosphorus. These accumulations are not related to the previous occupation phase and may therefore be considered as post-occupational considering the scope of the research. When one wants to study the phosphorus enhancement due to this new occupation, one has to consider the phosphorus enhancement due to the previous occupation as "background noise".

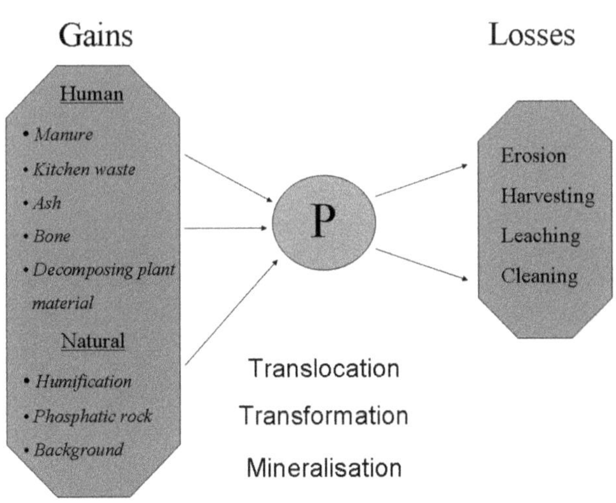


*Figure 1. Flow chart with the major in- and outputs of phosphorus and processes affecting the phosphorus concentration.*

In order to produce quality data, it is crucial to keep in mind the importance of differentiating these phases already during the field work by performing a taphonomic study and appropriate sampling.

## 1. Parent material

### *1.1. Mineralogy*

The soil parent material often contains a certain amount of phosphorus (Proudfoot, 1976; Lindsay, *et al.*, 1989), referred to as the background or native phosphorus. Several authors have noticed that, according to type and mineralogy of the soil parent material, the background concentrations vary (Bakkevig, 1980; Leonardi, *et al.*, 1999).

Background data measured on archaeological sites on some of the more important parent materials for Northern France, Belgium and Luxembourg can be found in table 2. Marls, often containing a considerable fraction of shell fragments, which are made up of different forms of the phosphate mineral apatite, coated foraminiferal tests, phosphatized macrofossil fragments, faecal pellets, phosphatized intraclasts and vertebrate remains tend to have high background values (Jarvis, 1992; Jarvis, 2006). Peats also show high phosphorus levels, due to the presence of phosphorus-rich plant remains. Quartz rich sandy soils, with only a small fraction of weatherable minerals that potentially can release phosphorus will have a very low background level. The variation in the mineral composition of parent materials makes it essential to measure this phosphorus background value.

A complicating element, is site variability of parent materials. Therefore a good field recording of the lateral and vertical variability of parent materials should be performed, e.g. a Neolithic house of Dampierre-le-Château (F) built on two different parent materials, namely marl and limestone (Fechner, *et al.*, in press), presenting a different phosphorus background value.

### *1.2. Physical characteristics*

#### 1.2.1. Texture

Texture is an important factor for phosphorus fixation in the soil. Due to the large specific surface area of clay compared to silt and sand, clay particles are able to adsorb cations (White, 1997: 31-32; Sjöberg, 1976; White, 1997: 31-32; Holliday and Gartner, 2007). This does not imply that sandy soils are unable to store phosphorus, but this rather happens under the form of precipitates (e.g. neo-formed minerals), or associated with organic matter.

To deal with textural variation within a site and its different

*Table 2. different background values of parent material (C-horizon and base of B horizon) from several sites in Belgium, France and Luxembourg. Analyses have been carried out following Mikkelsen and Langohr, 1996.*

| ***Parent material*** | ***Site*** | ***P-background (mg/kg)*** | | | ***Reference*** |
|---|---|---|---|---|---|
| | | Ptot | Pi | Porg | |
| Loess | Ormeignies (B) | 880 | 820 | 70 | Collective, in press. |
| | Fexhe-le-Haut-Clocher (B) | 1030 | 840-870 | 150-190 | Fechner, 2007. |
| | Aubechies (B) | 1170 | 1060 | 100 | Mikkelsen and Langohr, 1996 |
| | Lieusaint (F) | 420-640 | 320-430 | 100-210 | Fechner and Devos, 2006. |
| | Remerschen (Lux) | 1190-1290 | 740-1070 | 230-450 | Baes and Fechner, in preparation. |
| Limestone | Bussy-Lettrée, Vatry (F) | 920-950 | 550-620 | 290-360 | Fechner, 2005. |
| | Champfleury (F) | 840-920 | 560-570 | 280-340 | Fechner and Slachmuylder, 2009. |
| Marl | Altwies (Lux), contact loess-marl | 660-780 | 370-540 | 240-290 | Baes and Fechner, 2003 |
| | Dampierre-le-Château (B) | 1090 | 580 | 510 | Fechner, *et al.*, 2007. |
| Tertiary clay sands | Hélecine (B), contact loess-tertiary clay sands | 960 | 240 | 720 | Unpublished |
| | Savigny (F) | 180 | 70 | 110 | Fechner, this volume; Fechner, *et al.*, in preparation. |
| Coversands | Ekeren, 99/P1 (B) | 110 | 50 | 60 | Mikkelsen, *et al.*, 2002 |
| | Braine l'Alleud "Bosquet" (B) | 80-180 | 60-160 | 20 | Fechner, *et al.*, in press, in preparation. |
| Brusselian sand | Hôtel de Lalaing, Brussels (B) | 600 | 500 | 100 | Devos, 2003. |
| Peat | Rue du Midi, Brussels (B) | 2360 | 930 | 1530 | Devos, 2002b. |

ability to store phosphorus, Crowther (1997) suggests to interpret the phosphorus concentration on a fractional basis (clay, silt), although care should be taken, as some phosphate nodules or crystals can be quite large, even sand size (figure 2).

### 1.2.2. Soil humidity (drainage)

In the past it was believed that the impact of phosphorus leaching through percolation water is negligible because phosphorus tends to form extremely complex and insoluble compounds (Sjöberg, 1976). Recently it has been shown that the situation is somewhat more complex. When water percolation through the subsoil occurs rapidly (e.g. after heavy rainfall or snow melt) the formation of these insoluble compounds becomes more difficult, because of the shorter contact time between the soil particles and the percolating water carrying the phosphorus. Obviously, this

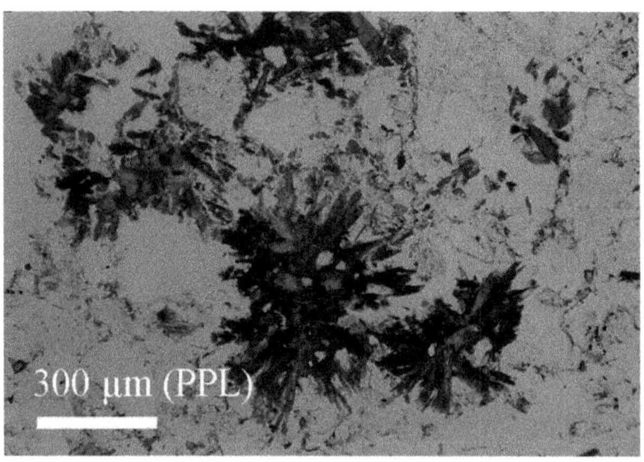


*Figure 2. Sand sized vivianite (iron-phosphate) crystal (plain polarized light).*

results in elevated phosphorus leaching (Djodjic, *et al.*, 2004). Leaching of soil phosphorus appears to be more important in heavily fertilized soils (Djodjic *et al.*, 2004; see also ch. 6).

In most soil profiles we observe a distribution pattern where organic phosphorus decreases with depth while the inorganic fraction increases relatively. Two schoolbook examples where the content of organic phosphorus drops with depth and relatively inorganic phosphorus becomes more increasingly more important are the plaggen soils excavated along the high speed train track between Antwerp and the Belgian Dutch border (Mikkelsen, *et al.*, 2002). Other profiles have a general trend, which is similar but may have as an example a Bh horizon (subsurface horizon enriched by organic matter). In such a horizon the contents of phosphor and especially organic phosphorous will obviously increase.

This general pattern might be disrupted by the presence of a fluctuating water table, especially when the water causes a temporary oxygen poor environment.

On archaeological sites things may complicate. For instance, if a site becomes buried below a few meters of sediment, this implies on the one hand that a former surface layer rich in organic phosphorus with respect to drainage becomes a subsoil, and on the other hand that the groundwater table will rise. This can cause post-burial redistribution of the phosphorus.

## *1.3. Chemical characteristics*

### 1.3.1. Soil pH

The solubility of phosphorus depends not only on the type

of minerals present in the soil and on the supply of oxygen, but also on the pH of the soil.

Several authors indicate a pH (analysed in a water solution) of about 6 to 7 for mineral soils as being the range where phosphorus is most mobile (Eidt, 1973; Lindsay, *et al.*, 1989; Crowther, 1997; Heron, 2001; Tyler and Olson, 2001; Holliday and Gartner, 2007). Nevertheless, early Neolithic examples from the loess region in Belgium show that even at this pH range the movement of phosphorus has been rather limited through time (Fechner *et al.*, in press).

According to Heron (2001) phosphorus in organic soils is most mobile at pH 5.5. At lower pH values the phosphate-anion reacts with iron and aluminium compounds and at higher pH values with calcium-compounds (Sjöberg, 1976; White, 1978). Below pH 5 and above pH 8, phosphorus is thus most tightly fixed in the soil (Brady and Weil, 1999).

As a consequence, changes in pH may induce transformations in the mineralogy of the phosphorus; e.g. calcium phosphates will gradually transform into iron- or aluminium phosphates with decreasing pH. As such, it may be helpful to know the evolution of the pH during and after the deposition of the phosphorus. Apatite will in most cases be an indication of the accumulation of bone but it may also merely reflect an increase of the pH to values above 7.

#### 1.3.2. Cation exchange capacity (CEC)

The CEC concerns the ability of the soil to fix cations in the soil. The CEC is directly affecting the adsorption of newly mineralized phosphorus (Stevenson, 1986; Schlezinger and Howes, 2000).

Soil particles and especially clay and organic matter have negative charges. As the cations have positive charges (such as $K^+$, $Mg^{2+}$, $Ca^{2+}$, $P^{4+}$) they are easily attracted towards these negative loaded surfaces until they eventually are introduced into the biological cycle of the soil (e.g. root uptake).

The higher the CEC, the more charges present in the soil, and thus the easier the phosphorus will get fixed in the soil.

Important resources contributing to a high CEC are different types of clay (especially montmorillonite and vermiculite (White, 1997: 33; Gobat, *et al.*, 1998) and organic matter. If a soil has a low content of clay and organic matter the storage capacity is limited whereby phosphorus or other cations are easily leached out with the soil water unless they get fixed in other ways.

#### 1.3.3. Organic matter

The addition of organic matter to a soil can affect the forms, interactions and redistribution of phosphorus (Holliday and Gartner, 2007). In a direct way, input of organic matter in the form of decomposing residue and litter- containing important amounts of nutrients like phosphorus- in the topsoil (e.g. by manuring), will therefore result in the accumulation of (organic) soil phosphorus. Indirectly such decaying organic matter stimulates microbiological activity (see ch. 1.4) that will in turn affect the phosphorus mineralization and immobilization (Salas, *et al.*, 2003; Holliday and Gartner, 2007).

Finally, the decomposition of the organic matter (humification) will produce organic colloids of high specific surface and high CEC (values often even higher than for clay) (White, 1997). These organic colloids can, together with the clay (see ch. 1.3.2) contribute to the fixation of phosphorus in the soil profile.

### *1.4. Microbiological activity*

Microbiological activity (fungi, bacteria, etc.) plays an important role in the humification of organic matter in the soil profile. In function of this humification, organic phosphorus becomes available for mineralization into inorganic forms of phosphorus. This may result in a stronger immobilisation of the element (Cross and Schlezinger, 1995; Salas, *et al.*, 2003; Holliday and Gartner, 2007).

## 2. Soil forming processes

Although phosphates are assumed to remain more or less immobile in soils - a number of ways how to immobilize phosphorus have been explained above - it appears that significant changes can occur during pedogenesis (Letkeman, *et al.*, 1996).

### *2.1. Oxido-reduction*

Changes in redox (e.g. changes in the content of available oxygen) can have an impact on the mobility of phosphorus.

Reducing conditions imply that the soil experiences low oxygen content. It has been established that under such conditions phosphorus mobility increases (Lindsay, *et al.*, 1989; Scalenghe, *et al.*, 2002).

Phosphorus bound to iron oxides (iron phosphate complex) is very insoluble under well-aerated conditions. If on the other hand iron in these complexes is reduced from ferric ($Fe^{+++}$) to ferrous ($Fe^{++}$) iron under prolonged oxygen poor (anaerobic) conditions, the iron-phosphate complex becomes soluble and phosphorus is mobilized (White, 1997).

Water-saturated soils are therefore particularly problematic for phosphorus cartography, as part of the phosphorus present in the soil profile can become mobile and can migrate horizontally and vertically, creating spurious patterns. This problem has previously been recognised by Eidt (1973), but according to him the human affected soils should still produce much stronger phosphorus readings than non archaeological soils. Nevertheless, investigations on a waterlogged site in the centre of Brussels, showed phosphorus concentrations that were higher outside than inside the archaeological structure (Devos, 2002a), a phenomenon, which was also observed in the dolina/sinkhole in the French Ardennes (Louwagie and Langohr, 1999; Laurelut and Louwagie 2001) and in a roman stable house in northern Belgium (Mikkelsen, *et al.*, 2002; Mikkelsen, *et al.*, 2003;Langohr, *et al.*, 2004). These examples show that the complete change of phosphorus pattern is possible in some cases because of local waterlogging.

Changes in redox status can also occur when a soil gets compacted. Soil compaction (e.g. roads, postholes, walking areas) will lead to a reduced porosity, making it easier to attain reducing soil conditions. Under such reducing conditions phosphorus becomes mobilized. This is often short distance redistribution, because the phosphorus will immobilize as soon as it reaches oxidising conditions.

### 2.2. Clay translocation/ clay migration

Under certain conditions, including a pH in water of about 5.5-6.5, clay particles are able to migrate within the soil, usually downwards. This process can have an important impact on the textural differences between horizons, with typically a clay enriched horizon (Bt-horizon) underlying a clay impoverished horizon (E-horizon).

Phosphorus released by weathering from P-bearing minerals and plant residues tend to concentrate primarily in the clay fraction (Black, 1968). Clay migration can therefore influence the concentration and even the presence/ absence of phosphorus in relation to the clay migration pattern. Goldberg and Macphail (2006: 347) also mention clay translocation as an important factor in downwards phosphorus transport.

### 2.3. Podzolisation

Podzolisation is a soil process that affects the phosphorus distribution pattern (Proudfoot, 1976; Bakkevig, 1980; Crowther, 1997; Goldberg and Macphail, 2006: 69). Breakdown of organic matter with formation of organic acids is an important part of the podzolisation process. These acids are able to mobilise sesquioxides of iron, aluminium and manganese. However phosphorus from the organic matter becomes equally mobile. The fixation of phosphorus is also enhanced where these organic substances and sesquioxides accumulate.

Bakkevig (1980) suggests that the iron pan formed in some podzol soils will cause a more complicated pattern of water displacement. This would in turn cause a wider horizontal distribution of the phosphorus. Own research also suggests that the phosphorus distribution within soils can be affected by podzolisation (e.g. Fechner, *et al.*, in preparation: Bs(h)-horizon of Braine l'Alleud).

### 2.4. Mineralization

The total amount of phosphorus (Ptot) present in soils may be divided into two broad categories, inorganic (Pi) and organic (Po). The relative proportions of the phosphorus in these two forms vary widely. Furthermore, the two forms of phosphorus are continuously interchanged within the dynamic P-cycle of the soil (figure 1; Lindsay, *et al.*, 1989).

The velocity of mineralization is still debated. Boer and Exaltus (1994, referring to Mattingley and Williams, 1962 and Gardiner and Walsh, 1966) report that 70% of the organic phosphorus has been mineralized for a Roman villa in England and a prehistoric burial mound in Ireland. Ottaway (1984) on the other hand, could not detect any phosphorus mineralization on 1600 to 1800 year old archaeological deposits. Cross and Schlezinger (1995) mention a series of investigations where the mineralization of phosphorus depends on the phosphorus availability.

The ratio of the organic and the inorganic phosphorus is sometimes used to determine ancient human activities (Macphail, *et al.*, 2000; Goldberg and Macphail, 2007: 347), e.g. long house locations gave low Ptot/ Pi values (high degree of mineralization), which can be clearly differentiated from the high Ptot/Pi values of the neighbouring arable fields (Macphail, *et al.*, 2000 referring to Engelmark and Linderholm, 1996; Macphail and Linderholm, 2004).

Although it is often assumed that inorganic phosphorus is stable in the soil profile, Schlezinger and Howes (2000) report that newly mineralized inorganic phosphorus, while generally retained in the soil, may be subject to some vertical translocation in the soil.

### 2.5. Bioturbation

Bioturbation may mix and redistribute soil particles, nutrients including phosphorus etc., both vertically and horizontally. We can distinguish two main types of bioturbation.

The floral perturbations are caused by treefalls (windthrows), or by roots. Indeed the uptake of nutrients by plant roots can be seen as a type of floral perturbation, as it causes fluxes in the phosphorus concentrations (Proudfoot, 1976; Leonardi, *et al.*, 1999).

The faunal perturbations are mainly due to digging animals like earthworms, moles, badgers, rabbits. The digging of galleries causes a mixing of the soil and can give spurious patterns. This phenomenon was recognised by Crowther (1997) in his study on vertical movement of phosphorus. In fact bioturbations cause a homogenisation of the soil. This means that a phosphorus rich layer can get mixed with a horizon much poorer in phosphorus, giving an average result. The archaeological chemical pattern will fade away through time in soils with a high degree of bioturbation as a consequence.

## 3. Erosion/sedimentation

Erosion can be a very important factor when dealing with phosphorus cartography. When parts of the ancient phosphorus enriched surface are eroded, it becomes more difficult to interpret the data/site (figure 3; Sjöberg, 1976). Although erosion can strongly influence the results of a survey, little attention has been paid to its impact on the results of phosphorus analyses (Crowther, 1997). While erosion may cause direct changes in the phosphorus distribution, sedimentation will rather cover ancient occupation surfaces enhancing preservation possibilities of the phosphorus (Parnell, *et al.*, 2002).

Crowther (1997) gives an example of a site (Caer Cadwgan, Wales, Great Britain) where phosphorus analysis only provided very limited results. The potential phosphorus rich layers had been exposed to erosion and subsequent colluviation. As a consequence, the eroded areas had lost part of the phosphorus enriched soil and areas buried with colluvium had been enriched due to the relatively high content of phosphorus in the colluvial material.

It is very important to identify when erosion has taken place in relation to the ancient occupation. Only post-occupational erosion has to be considered, as only this type of erosion will disturb the phosphorus record on past human activities. However, erosion concomitant with the occupation period can inform about particular events or activities, whereas erosion prior to the occupation will have no impact on human induced phosphorus concentrations. The latter type can cause differences in the background values though.

## 4. Precipitation

Earlier it was explained that drainage can play a rather

*Figure 3a*

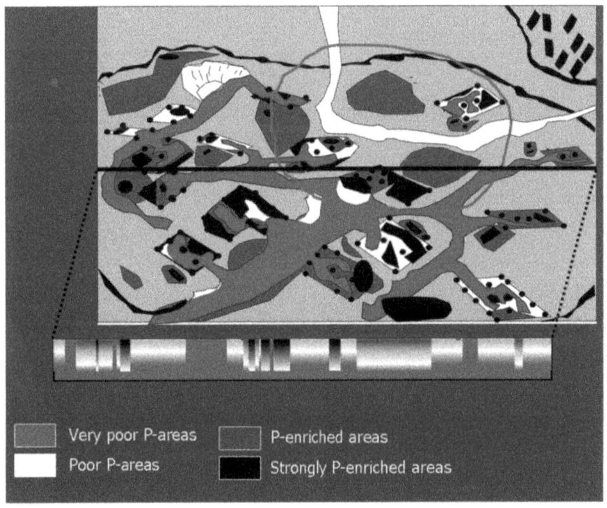


*Figure 3c*

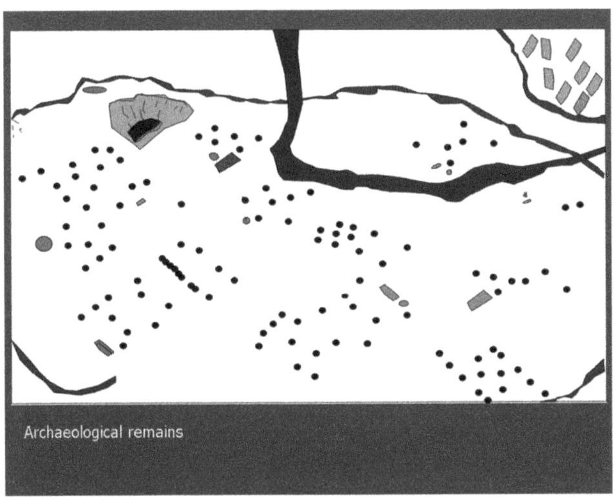


*Figure 3b*

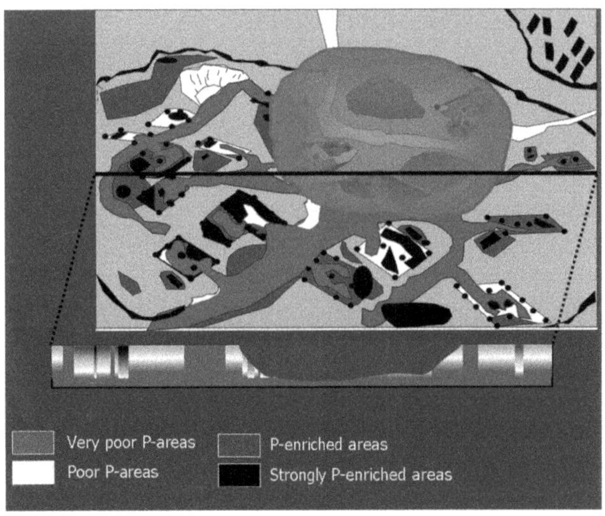


*Figure 3d*

*Figure 3. theoretical example showing the potential impact of erosion. A. site during occupation; B. remains uncovered during excavation; C. theoretical P-map, with transect showing the vertical distribution of the phosphorus; D. effect of erosion on the P-map.*

important role in the redistribution and even leaching of phosphorus in the soil (see ch. 1.2.2). This is of course closely linked to the water infiltration (water breaking through the soil surface) and percolation (the water movement in the soil) and thus also to the magnitude and intensity of precipitation (e.g. rainwater and snow).

When dealing with archaeological structures that are protected from rainfall, like houses, or shelters with a roof, the phosphorus concentrations inside the structure cannot be leached as long as it is occupied or the cover remains intact. In contrast open air structures like for example pits may suffer from leaching.

Parnell, *et al.* (2002) found some phosphorus leaching patterns around a Maya house (Piedras Negras, Guatemala), which they interpreted as ancient roof drip lines where leaching of phosphorus was enhanced due to the local high concentration of water infiltrating the soil. This illustrates how even a local change in precipitation distribution can have an important impact on the phosphorus concentration and distribution pattern.

## 5. Soil cover

The vegetation cover (e.g. crops, grasses, shrubs) is responsible for absorption of part of the phosphorus (Proudfoot, 1976).

Experiments carried out by Leinweber, *et al.* (1999) showed a good correlation between type of land use and the degree of phosphorus leaching or loss.

Sealing of the soil by impermeable covers (modern buildings, asphalt etc.) will also prevent any further soil evolution and therefore also any further movement of phosphorus in the soil, given that the soil environment remains oxygen rich.

## 6. Application of fertilizers

Interpreting phosphorus cartographic results in an archaeological context complicates when the soil is under influence of modern fertilizers.

According to Sjöberg (1976) modern inputs of phosphorus may only slightly change the phosphorus pattern of archaeological interest. Nevertheless, he advises to record their absence or presence. Woods (1977) also states that modern agricultural fertilization does not present problems of interpretation. Eidt (1973) on the other hand recognises the potential problem of the application of modern fertilizers. He therefore performs his analyses on samples taken below the actual plough layer, as he expects that the actual fertilizers will not move significantly downwards. Leonardi, *et al.* (1999) referring to Gurney (1985), also indicate that the fixation of modern fertilizers tends to happen in the upper 30 cm of topsoil only.

According to Delgado and Torrent (2000) it is difficult to quantify the contribution of modern fertilizer to the amount of total phosphorus in the soil. Nevertheless, they found that a substantial portion of total phosphorus in calcareous soils where modern fertilizer had been applied seemed to correlate to this recent input.

Due to excessive phosphorus fertilization through decades, many European agricultural soils have higher phosphorus content. This build up of phosphorus leads in turn to an increased risk of phosphorus leaching (Djodjic, *et al.*, 2004).

The application of fertilizers has also an indirect impact on the phosphorus concentrations in the soil. Through input of fertilizers the biological activity will increase in the soil and therefore enhance the degree of bioturbation (Louwagie, *et al.*, 2005: 82).

## 7. Time

Long-lasting and/or intensive human activities will leave increased levels of phosphorus in the soil (Wells, *et al.*, 2000).

The non-chronological nature of soil chemical input has often been used as the major criticism in regard to the use of soil-chemical surveying in archaeology (Linderholm, 2003; Linderholm, 2007). Therefore it is important to cross-check phosphorus data with other available data.

In cases of complex utilisation of an area, from prehistoric to contemporary time, phosphate analysis can provide ambiguous information (Linderholm and Lundberg, 1994). In those cases, especially urban contexts where we often encounter extremely complex occupation history, the phosphorus data cannot be used, unless they are confronted with other data sources such as archaeological, historical and palaeoenvironmental recordings. One such example is the site of Hôtel de Lalaing-Hoogstraeten (Brussels) where peaks of phosphorus from the 13th and 16th century are cross-checked with the concentrations of heavy metals (Devos, unpublished data).

# Critical literature study on the field and laboratory methodology for phosphorus cartography on archaeological sites and structures

*Field study*

Although most of the articles commenting on the application of phosphorus analysis for archaeological purposes acknowledge the potential influence of circumstantial factors on phosphorus concentrations in the soil (see above), discussions on the field study remain scarce. Nevertheless, the pioneering study of Proudfoot on the site of Goodland clearly demonstrated the importance of recording soil variability (Proudfoot, 1976). He concluded that differences in the phosphorus content often were linked to the heterogeneity of the parent material.

Several authors (Bakkevig, 1980; Crowther, 1997; Sampietro and Vattuone, 2005) discuss the importance of including the vertical distribution of phosphorus within the study of horizontal phosphorus patterns. Bakkevig (1980) and Schlezinger and Howes (2000) remind that an important part of the field study should be concentrated on the search for off-site reference profiles. These profiles should reflect the phosphorus background in the soil at the moment of occupation. Linderholm (2007) acknowledges the importance of careful field work and reminds the importance of understanding the soil formation processes when carrying out phosphorus analyses.

## Sampling

On regional surveys where the aim is to uncover the presence of potential archaeological sites the sampling is often performed following a grid system (Kyhlberg, 1993).

Basically, two types of grid are used. Firstly, the Cartesian grid system, where the lines are perpendicular to each other, and where the cross-points make up the sampling points. Secondly, the isometric system, which is based on equilateral triangles (Sjöberg, 1976). Although this method is more complicated than the Cartesian grid system, it is advantageous given the same number of samples for a 10.5% increase in coverage (Sjöberg, 1976).

Linderholm (sd.) proposes a probability sampling strategy to reduce the data body and the costs. This means that visible features, landscape configuration and soil characteristics are decisive for sampling. This strategy has one obvious problem: survey will only be conducted on places where sites are anticipated and thus depends on existing knowledge.

About the distance between the sampling points several proposals have been developed. Sjöberg (1976) for example proposes 25 m, Bjelajac, *et al.* (1996) applied 20 m intervals. It is obvious that, the smaller the distance between two samples, the more accurate the phosphorus map will be. The intervals between the sampling points will depend on the reasons behind the investigations (Cavanagh, *et al.*, 1988a).

Linderholm (2007) recalls the importance of sampling different soil horizons.

In general, sampling of a site or a particular archaeological structure happens also according to a grid system.

According to Fechner, *et al.* (2003), it can be advantageous when dealing with house structures where the plan is more or less known, to sample along the axe of the house.

Several authors acknowledge the importance of collecting off-site data to be able to detect enrichments of phosphorus (Bakkevig, 1980; Wells, *et al.*, 2000; Fernández, *et al.*, 2002). Wells, *et al.* (2000) specify that the application of statistical tests will not overcome this problem without background data.

The distance between the samples should, according to Sjöberg (1976), not exceed a 1x1 m grid. This distance has also been used by Conway (1983) when sampling a hut floor.

There is also a lot of discussion at which depth samples should be taken. Sjöberg (1976), Eidt (1973) and Bjelajac, *et al.* (1996) suggest to sample below the actual plough zone for soils under agriculture. Staüble and Lüning (1999) sampled in the clay enriched horizon (Bt), rather than the ancient surface soil. This is an approach with serious limitations though. Firstly, it implies that the migration is posterior to the phosphorus accumulation, which is by far not always the case (e.g. Van Vliet, 1990; Van Vliet, *et al.*, 1992; Semmel, 1995). Secondly, through this approach it is virtually impossible to estimate the phosphorus concentration present at the ancient surface and it is impossible to know how much has migrated downwards at each location of the mapping area. Thirdly, the phosphorus distribution is far more complex than just a simple clay-phosphorus relationship.

Zimmermann (1998) proposes to sample always at the same depth but according to us this seems not the best solution, as the actual topography might not always correspond to the ancient one (see Ch. 3).

When sampling vertical profiles Sjöberg (1976) proposes to take samples for each 10 cm. da Costa and Kern (1999) sampled each soil horizon when studying Amazonian Black Earth. In our opinion this seems to be a more reliable approach taking into account soil processes that might affect phosphorus distribution within the soil.

Several authors (Bakkevig, 1980; Conway, 1983; Middleton and Price, 1996; Entwistle, *et al.*, 2000; Terry, *et al.*, 2004) acknowledge the necessity of sampling outside the structure itself.

For their study on ancient Anthrosols in Piedras Negras (Guatemala) Wells, *et al.* (2000) sampled apart from the grid system also the areas where they remarked a huge accumulation of shards.

## Laboratory methods

A major problem, which arises when dealing with phosphorus analyses, is the almost endless list of available methods.

Basically following series of methods have been applied:

- A widely applied tool for the easy, fast and cheap determination of phosphorus during non-destructive surveys of landscape figures is the 'field spot test', which is, strictly seen, not a laboratory method. The method has been developed by Gundlach who adapted the method of Feigl to field conditions (Eidt, 1973). This method was again modified since the nitric acid, recommended by Gundlach gave poorer results than the hydrochloric acid (Eidt, 1973). The field spot test provides relative results concerning phosphorus concentrations in soils (Miller and Gleason, 1994). A main criticism on the application of this method is that the vague class limits, basically an optical interpretation on the intensity and extension of the developed blue colour, make it rather difficult to compare results obtained by different researchers. Moreover it is very important that all tests from the same site are executed and compared at once. As the data are not linear the results cannot be submitted to statistical treatments (Eidt, 1973; Kyhlberg, 1993). Recently, Bjelajac, *et al.* (1996) made an attempt to quantify the results of the field spot test. They concluded that this type of method is only of value when making rough estimates during fieldwork, but in no case to map the space use within an archaeological site. A similar conclusion was made earlier by Sjöberg (1976). Tests are also useful as a first estimate on the phosphorus content and distribution, before eventually genuine laboratory analyses are requested (Fechner *et al.*, in preparation: Merbes-le-Château).

- A first group of proper laboratory methods measures the total phosphorus concentration in the soil (Ptot). The advantage of this analysis is that it provides a measure of the overall phosphorus concentration, irrespective of the stage of mineralization (Crowther, 1997). The Ptot-content will include any phosphorus added in the past, provided the adsorbing agents are not themselves leached from the soil (Conway, 1983). This is also one of the major drawbacks of the method, as it does not differentiate at all between recent inputs and 'archaeological' relevant phosphorus accumulations. Within this method there still exist some variants according to the acids applied to extract the phosphorus. It should be taken into account that the application of a different type of acid will inevitably lead to different results concerning the concentration of Ptot. In fact, it would be more correct to talk about P(nearly)tot, as most methods, except total analyses involving hydrogen-fluoride, are unable to extract all the phosphorus chemically bound in the mineral structures. One type of acid that has been largely applied is fluoric acid. This type of acid is able to completely dissolve the silicate structures, which is the basic building stones of many minerals - such as quarts - making up bulk of the upper part of the continental plates. However, this kind of phosphorus is of little interest, as it has nothing to do with phosphorus accumulated in the soil due to human interference (Conway, 1983; Middleton and Price, 1996). Other acids used are hydrochloric acid (Middleton and Price, 1996) and citric acid (Macphail, *et al.*, 2000). Nitric acid ($HNO_3$) has the disadvantage that it may become neutralised by calcareous soils before phosphorus is released (Eidt, 1973). In order to counter this effect so-called double acid digestion procedures can be applied (Parnell, *et al.*, 2002). In short: when applying a different acid as extracting agent, different portions of the phosphorus are measured. Moreover, the result will depend on the alkalinity or acidity of the soil, as part of the acid used as extracting agent can be neutralised in alkaline soils, which may affect the measured phosphorus concentration.
- A second group of analyses measures the concentration of organic (Porg) and inorganic phosphorus (Pi).
  Inorganic P is usually determined by weak acid extraction or by salt solutions. In Umea, Sweden a 2% citric acid is applied as the extractant (Macphail, *et al.*, 2000). This method is very effective in podzolic soils (Macphail, *et al.*, 2000). In more calcareous soils the method becomes less effective, because the weak acids can become neutralised and less phosphorus will be released (Macphail, *et al.*, 2000).
  The analysis of Pi has the advantage that once phosphorus becomes unavailable to plants, it tends to remain in place as long as the sediments remain *in situ* (Miller and Gleason, 1994).
  Basically, two methods are applied to determine the amount of organic phosphorus in the soil. One method is to extract the organic matter from the soil by acid and alkali extractions. Afterwards the amount of inorganic phosphorus is measured. The amount of organic phosphorus is determined by calculating the difference between the total phosphorus and inorganic phosphorus content (Proudfoot, 1976).
  Another method is to destroy the organic matter by ignition or by an oxidising agent (e.g. peroxide). By destroying the organic matter the organic phosphorus is transferred into inorganic phosphorus. The inorganic part is subject to extraction with a suitable acid solution before (measuring the inorganic fraction only) and after the combustion (measuring the inorganic plus the organic fraction in the meantime due to the combustion transferred into inorganic phosphorus) (Proudfoot, 1976; Macphail, *et al.*, 2000).

The ratio Ptot/Porg has been successfully applied for the detection of ancient crop fields and dwelling sites (Macphail, *et al.*, 2000; Macphail and Linderholm, 2004b; Linderholm, sd.), as well as the characterisation of dark earth (Macphail and Linderholm, 2004a).

- A third group of analyses measures the concentration of plant-available and plant-non-available phosphorus. When applying this type of analysis the significant seasonal variations should be taken into consideration (Wood, 1977; Leinweber, *et al.*, 1999). Proudfoot (1976) reviews the problems that have arisen when dealing with this kind of analysis. It appears that when different extracting agents are used, the results can vary significantly. This is only partly due to the amounts originally dissolved by the acids. The main course seems to be the chemical precipitation and adsorption during the extracting process. So here again, it is very important to describe carefully the method applied, if any comparisons are to be made. Nevertheless this type of analysis has been successfully tested on a potential burial mound (Solecki, 1951).

- Another group of analyses applies the so-called fractionation method. This implies the determination of different types (fractions) of phosphorus. This technique, developed by Chang and Jackson, and later modified by several researchers (Holliday and Gartner, 2007), is especially useful to have an idea of the concentration of respectively the iron-, aluminium- and calciumphosphate present in the soil. One such modified method is the one developed by Eidt (1985). This method distinguishes three types of inorganic phosphorus, i.e. non-occluded aluminium and iron phosphates (fraction I), occluded aluminium and iron phosphates (fraction II) and calcium phosphates (fraction III), and finally the total concentration of phosphorus (Lillios, 1992; Miller and Gleason, 1994; Smyth, *et al.*, 1995). It is based on the differential solubility of the major phosphate forms (Woods, 1977). Although this method is advantageous as it gives us more insight in the importance of the different fractions, there is in fact still much uncertainty concerning the exact nature of the analysed fractions (Proudfoot, 1976; Holliday and Gartner, 2007). Moreover the method is rather time consuming and expensive to carry out. Nevertheless the results obtained by Eidt (see Lillios, 1992) and Lillios (1992) on the study of past agricultural activities are, although often debated, rather interesting.

This amalgam of existing phosphorus methods makes the comparison between different sites very difficult if not impossible. Each method will reveal data that might be different for different types of soil, while the application of certain methods will even give different results depending on the sampling period.

The problem of the huge diversity in laboratory (or field) techniques has been tackled by many authors (see for example: Proudfoot, 1976; Sjöberg, 1976; Macphail, *et al.*, 2000; Goldberg and Macphail, 2006: 344-350; Holliday and Gartner, 2007).

Their main conclusion seems to be that the data should be accompanied by a text that clearly explains the applied method and its specificity (e.g. what is precisely analysed). As Wells, *et al.* (2000) remark the non-indication of the applied method is one of the major problems making it difficult or even impossible to draw comparisons between different sites or even different regions.

## Interpretation of the data

The results of the phosphorus analyses are stored in a database and mostly presented graphically.

In some cases, some basic statistics are applied on the acquired phosphorus data. Kyhlberg (1993) recommends grouping of the data based on the median values and the quartile deviation of the data, and not according to predefined categories. In cases of chronologically simple structures mean value and standard deviation can be used (Kyhlberg, 1993).

Cavanagh, *et al.* (1988a; 1988b) propose the statistical technique of point change analysis to define site/structure boundaries.

Parnell, *et al.* (2002) calculated the background control concentration based on the average of the lowest concentrations measured. However, this background control concentration is not based on data from off-site samples. Entwistle, et al. (2000), however, calculated the background control concentration based on the mean of the off-site samples.

## Experimental data

Several authors argue that to be able to interpret the phosphorus data, reference material is needed (e.g. Macphail, *et al.*, 2000; Fernández, *et al.*, 2002; Middleton, 2004; Linderholm, 2007). This necessity becomes very obvious when we look at the phosphorus study Dietz (1956) made on an Indian habitation site. The interpretation of the result was very limited and uncertain because of a lack of experimental data to explain the differences in phosphorus content.

Linderholm (2007) mentions traditional managed cultural soils, or sites covering sufficient long periods to show changes when compared with the background soil, but not heavily turbated.

Several ethnoarchaeological studies have been set up to enhance the interpretation level of the phosphorus data. As such there are the experiments carried out on Maya house floors (Smyth, *et al.*, 1995; Fernández, *et al.*, 2002; Terry, *et al.*, 2004) and a Mexican house floor (Middleton and Price, 1996).

Archaeological experiments (e.g. reconstructions of ancient houses, earthworks, mounds) also provide useful data on phosphorus concentrations in the soil due to particular human activities. Examples are the experiments in Overton Down (Bell and Fowler, 1996), Butser Ancient Farm (Macphail, *et al.*, 2000), Wareham (Macphail, *et al.*, 2003), and the data from experimental arable soils in Scandinavia (Macphail, *et al.*, 2000).

Although such studies do exist and can provide useful information, some remarks should be made.

For Belgium, France and Luxembourg, there is an obvious lack of such experimental data.

Hence, we have to refer to other regions, facing the problem of extrapolation. The question should be raised if we can just use the results obtained on the experiments in for example Overton Down, Butser Ancient Farm or Wareham. The same question can be raised when using the results of ethnoarchaeological studies set up in Amazonia or data provening from the or even the investigation of medieval urban contexts in Great Britain (Macphail, *et al.*, 2000) or the data obtained by the study of past and present agriculture in Agroal (Portugal) (Lillios, 1992). Moreover, it should be kept in mind that the problem of the application of different methods makes it even more difficult to apply these data.

A specific problem is that archaeological experiments show only short-term changes in the soil, but they do not explain what can happen over hundreds or thousands of years. Moreover these experiments deliver only information on what happened during site formation and shortly after. They do not take (or to a very limited extent) into consideration any post-depositional activity. To overcome this problem, planning a set of long-term experiments on the sandy soil and the loess belt, on different types of human occupation should be performed. Part of these problems can be overcome when studying for example agricultural fields, by studying well documented traditionally managed field systems (e.g. Entwistle, *et al.*, 1998; Guttmann, *et al.*, 2006).

In conclusion, such reference data, whether they come from short or medium-term archaeological experiments, traditionally managed cultural soils, ethnoarchaeological studies, can provide useful data, as they help to understand the processes responsible for phosphorus concentrations in the soil, and their link to specific human activities.

## Comparison with other types of analytical data

In order to improve the quality of the data, phosphorus analysis is often combined with other analyses:

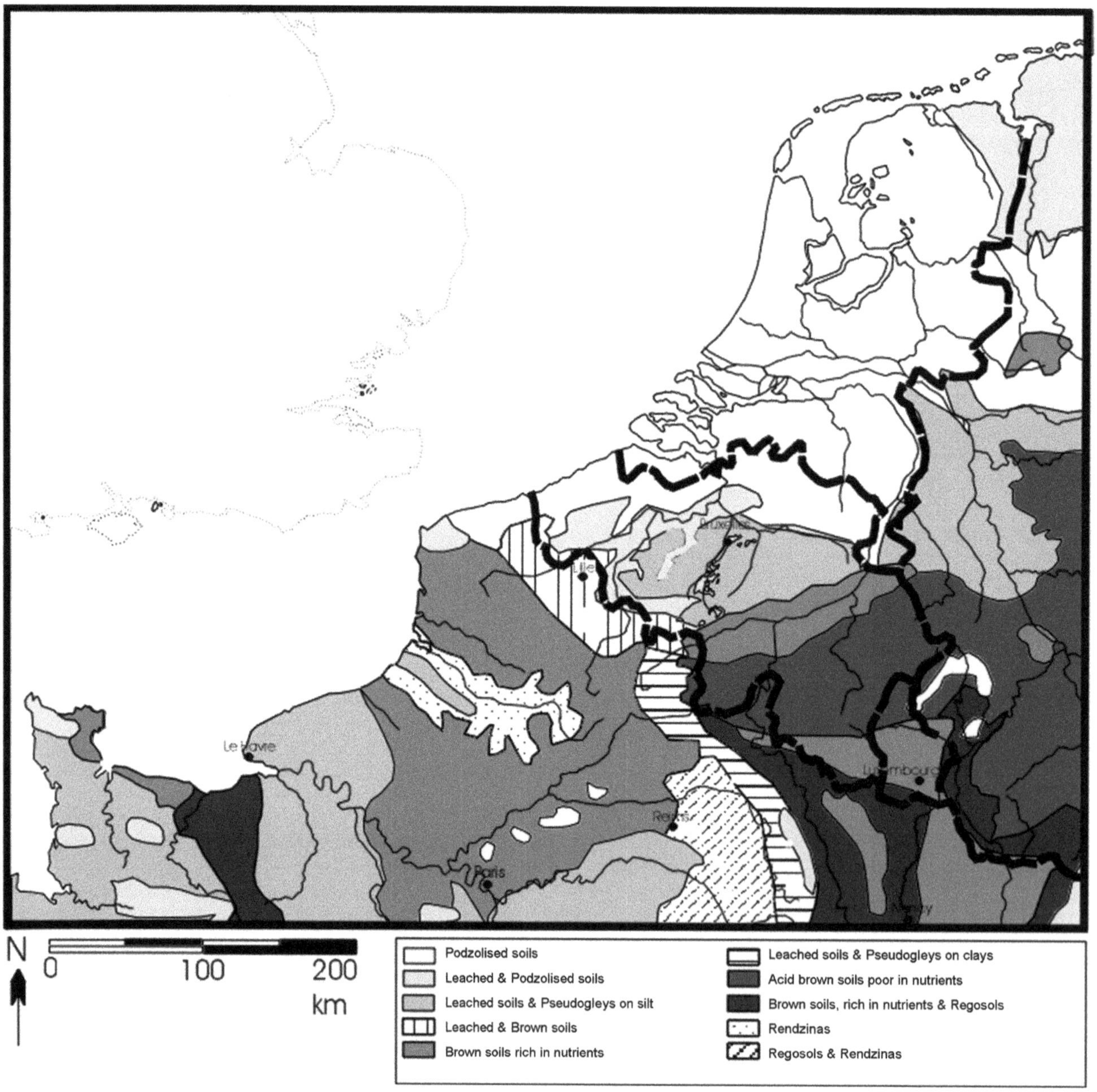


*Figure 4. soil map showing the main soil types.*

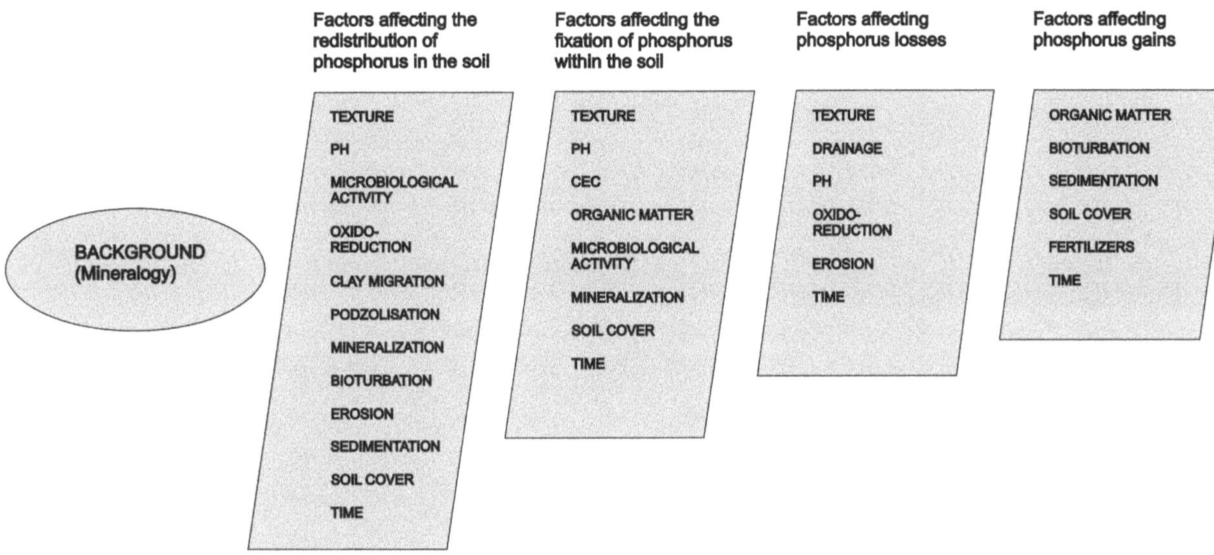


*Figure 5. checklist summarizing the important factors that can affect phosphorus concentrations in the soil.*

- chemical analyses (elemental analyses, organic carbon, nitrogen, pH, CEC, ...) (Berlin, *et al.*, 1977; Ottaway and Matthews, 1988; Lillios, 1992; Linderholm and Lundberg, 1994; Middleton and Price, 1996; Entwistle, *et al.*, 1998; da Costa and Kern, 1999; Derry, *et al.*, 1999; Vizcaíno and Cañabate, 1999; Entwistle, *et al.*, 2000; Wells, *et al.*, 2000; Fernández, *et al.*, 2002; Parnell, *et al.*, 2002; Alexandrovskaya and Panova, 2003; Macphail, 2003b; Middleton, 2004; Terry, *et al.*, 2004; Wells, 2004; Sampietro and Vattuone, 2005)
- physical analyses (granulometry) (Davidson, 1973; Foss, 1992; Lillios, 1992; Crowther, 1997; Simpson, 1997; Macphail; 2003a)
- micromorphological studies (Simpson, 1997; Macphail, *et al.*, 2000 ; Macphail and Cruise, 2001 ; Lima, *et al.*, 2002)
- magnetic susceptibility measurements (Macphail, *et al.*, 2000; Smith, *et al.*, 2001; Macphail, 2003; Macphail and Linderholm, 2004a)
- geophysical mapping (Sarris, *et al.*, 2004; Persson, 2005)

Ideally the data are also to be compared with historical documents, maps showing the distribution of the recovered archaeological materials (Smyth, *et al.*, 1995; da Costa and Kern, 1999; Parnell, *et al.*, 2002), as well as with a spatial distribution of the botanical (e.g. pollen (Cruise and Macphail, 2000), phytoliths (Sullivan and Kealhofer, 2004) and zoological remains (Sampietro and Vattuone, 2005). When all these data are available, high quality interpretations are often possible.

## Proposal for a field and laboratory methodology for phosphorus cartography on archaeological sites in Belgium, Luxembourg and Northern France

Based on the present knowledge on phosphorus behaviour and personal field expertise, a protocol has been developed for Belgium, Northern France and Luxembourg.

This region presents a temperate climate. Loess, coversand and marl compose the major types of parent material in this region. The soil map (figure 4) presents the main soil types for this region.

This research protocol includes following steps:

### *1. Field study*

Field work should at least include a careful study of the parent materials (if present), the soil development, soil moisture and redox, as well as the indication of possible perturbated areas (taphonomic study), degree of erosion, and a control on application of modern fertilizers. Figure 4 summarizes the most important factors that can affect phosphorus concentrations in the soil. As indicated, it is very important to differentiate between pre-occupational, occupational and post-occupational changes in the soil. Also the soil variability needs to be evaluated during field work. Whenever necessary, the results of the field study will be complemented by the necessary laboratory analyses, such as texture, pH, organic matter, CEC. The field work should also include the search for off-site reference profiles. These profiles should reflect the phosphorus background in the soil at the moment of occupation. It is obvious that it is easier to find appropriate reference profiles for more recent periods where historical records report on the organisation of the landscape, than for older occupations, where there is also a higher risk for post-occupational input of phosphorus.

This field study will allow to estimate the potential of phosphorus analyses for the studied area. It will also indicate areas that are bioturbated, eroded or biased in another way, and that should therefore be avoided.

### *2. Sampling*

Sampling for phosphorus cartography can only be conducted after a thorough field study, and should rely on the results of such a study. As said, it is of no use to sample biased areas.

#### Regional surveys

For regional surveys a grid system is recommended. For facility reasons the Cartesian system, where the cross-points of perpendicular lines are sampled, is preferred. However, this grid system is not to be followed too rigorous as for example bioturbated and eroded areas will be avoided.

No strict rules for sampling distances can be given, because the distance between two samples depends amongst others on the specific aim of the survey, the available time and budget and the soil properties. As a general rule, smaller sampling distances can provide more detailed results, but are more expensive and much more time consuming.

It is also important to carefully sample several reference profiles to obtain a better understanding of the phosphorus background and the soil variability.

#### Intra-site analysis

When sampling a site or a particular archaeological structure a grid system is highly recommended. Again, this system should not be followed to rigorously and disturbed areas should be avoided.

When dealing with house structures where the plan is more or less known, it may be advantageous to sample along the axe of the house as Fechner, *et al.* (2003) did when investigating long houses in Luxembourg and Belgium.

It is however crucial not to limit the sampling to the actual structure, but also to take samples outside the structure/ site (off-site reference samples). If possible some reference profiles should be sampled too.

The distance between the samples should not exceed a 1x1 m grid. Again, smaller distances may provide more detailed information, but are more expensive and time consuming.

When sampling vertical profiles, it is important to sample each recognised horizon. When necessary, additional samples can be taken within these soil horizons.

It might be wise to perform the phosphorus sampling of the whole area at once, because available phosphorus concentrations in the soil can vary throughout the year (Woods, 1977; Leinweber, *et al.*, 1999). If sampling total phosphorus or inorganic phosphorus as we recommend this point is not relevant.

Finally it should be remembered that samples should be taken with care, and contamination avoided (Devos, 2001). Therefore, profile sampling is preferred over auger sampling, because the latter method makes it more difficult to detect disturbed areas (Zimmerman, 1998).

The sampled surface should be cleaned before sampling. Samples are then taken close to the ancient surface. The reference profiles are also cleaned before sampling.

### *3. Laboratory analyses*

For our research protocol we limited ourselves, according to the situation, to two methods of phosphorus analyses. When dealing with large scale surveys a rapid test method (Eidt, 1973) is applied. Based on the results of this method, more precise analyses are carried out.

For small scale research, e.g. the investigation of one particular archaeological structure we apply the method developed by Mikkelsen (Mikkelsen and Langohr, 1996), which is a method with similarities to the one developed at Umeå (Sweden) (Macphail, *et al.*, 2000; Macphail and Linderholm, 2004b; Linderholm, 2007). This method analyses both the Ptot and the Pi content of the sample (whereby the Porg content is calculated as the difference between both) and allows the calculation of the Ptot/Pi ratio. A good correlation was found between the two methods (Jari Mikkelsen and Johan Linderholm, unpublished data).

### *4. Interpretation of the data*

Detection and interpretation of patterns

The results of the phosphorus analyses are stored in a database and preferentially presented graphically.

Some basic statistics are applied on the acquired phosphorus data. Grouping of the data is performed according to the mean and/or median values of the data, and not according to predefined categories.

First the normal distribution of the phosphorus concentrations of the samples taken outside the structure/ site is checked.

Studying one specific structure, a 'mean control background value' is calculated. This value is obtained by taking the mean of all the samples taken outside the structure (e.g. the reference samples). Then the standard deviation (s) of these data is calculated. All the data from samples within the structure that exceed the value of the [mean + s] are possibly enriched in phosphorus. All data that are lower than the [mean – s] are possible impoverished in phosphorus.

By mapping these results a first idea of the interior space organisation of the structure will be obtained. The same principle can be adapted to the entire site, but care should be taken that when dealing with different parent materials or horizons, only data of the same parent material or horizon are used to calculate the 'mean background control values' of that particular parent material or horizon.

*5. **Presentation of the phosphorus maps***

Basically, results can be presented in simple Excel-files. However, to present results more nicely and users-friendly, several computer programs can be applied (GIS-systems, Surfer,...).

*6. **Confrontation of the phosphorus data with other relevant data and synthesis***

As demonstrated in the literature review, the quality of the results depends often on the integration of other data.

It is therefore recommended to 'back up' the phosphorus data with:

- chemical analyses (elemental analyses, organic carbon, nitrogen, pH, CEC, ...)
- physical analyses (granulometry)
- micromorphological studies
- magnetic susceptibility measurements
- geophysical mapping

Such analyses can complement the data collected during field work and help to understand the soil processes affecting phosphorus concentrations in the soil. They can also contribute to identify particular human activities.

Ideally the data are also to be compared with historical documents, maps showing the distribution of the recovered archaeological materials, as well as with a spatial distribution of the botanical (e.g. seeds, phytoliths) and zoological remains.

## Conclusions and future recommendations

The critical literature review demonstrates the importance of carefully interpreting the phosphorus data. It is therefore essential that all parameters and variables are considered that might influence the data; already during field prospection this should be considered. Only when the impact of the variables and parameters has correctly been evaluated, the potential of phosphorus cartography (vertical and/or horizontal) can truly be estimated, and an optimal sampling strategy can be developed. Cross-examination of the data with other information such as chemical data, magnetic susceptibility, pollen, micromorphology, etc. is recommended in order to avoid mis- or overinterpretation of observed patterns.

Whenever the results of the phosphorus-mapping are compared successfully with archaeological results, such a robust data set can help to detect in a rather detailed way the ancient activity areas.

Finally, in order to confirm the kind of activities that took place in the different recognised areas, experimental data can be of great help to understand the processes of phosphorus accumulation and their relation with specific human activities. We therefore recommend the further development of these experiments.

## Acknowledgements

The authors would like to express their gratitude for the financial support granted by the Brussels Capital Region. In particular Robert Baes for his interesting ideas and comments on an early draft of this paper. Prof. Dr. Roger Langohr, University of Ghent should be acknowledged for his valuable comments and guidance. Finally our referee Dr. Geertrui Louwagie's valuable comments and suggestions are sincerely appreciated.

## References

Alexandrovskaya, E. and Panova, T., 2003. History of the soil, cultural layer, and people in medieval Moscow, *Revista Mexicana de Ciencias Geológicas*, 20, 3: 289-294.

Arrhenius, O. 1931. Markanalysen I arkeologins tjänst, *Geologiska Föreningens Förhandlingar*, 53.

Ayala, G., Canti, M., Heathcote, J., Sidell, J. and Usai, R., 2004. *Geoarchaeology. Using earth sciences to understand the archaeological record*, Swindon.

Baes, R. and Fechner, K., 2003. Etude géopédologique du site archéologique d'Altwies - "Op dem Boesch" (Grand-Duché de Luxembourg). Etat de la question, *Bulletin de la Société Préhistorique Luxembourgeoise*, 23-24: 163-179.

Baes, R. and Fechner, K., in preparation. Two Early Neolithic villages with cultivating surfaces in the Moselle valley: a pedological study. In: *Archaeology, soil- and life sciences applied to houses, gardens and courtyards. Proceedings of the Session 'From microprobe to spatial analysis – Enclosed and buried surfaces as key sources in Archaeology and Pedology' organised at the European Association of Archaelogists, 12th Annual Meeting, Krakow-Poland. 19th to 24th September 2006*. (=British Archaeological Reports, International Series).

Bakkevig, S., 1980. Phosphate Analysis in Archaeology – Problems and Recent Progress, *Norwegian Archaeological Review*, 13, 2: 73-100.

Bell, M. and Fowler, P.-J. (eds.), 1996. *The Experimental Earthwork Project 1960-1992*, York. (= Council for British Archaeology, Research Report, 100).

Berlin, G.L., Ambler, J.R., Hevly, R.H. and Schaber, G.G., 1977. Identification of a Sinagua agricultural field by aerial thermography, soil chemistry, pollen/plant analysis, and archaeology, *American Antiquity*, 42, 4: 588-600.

Berlin, G.L., Ambler, J.R., Hevly, R.H. and Schaber, G.G., 1977. Identification of a Sinagua agricultural field by aerial thermography, soil chemistry, pollen/plant analysis, and archaeology, Errata, *American Antiquity*, 43, 2: 318.

Bjelajac, V., Luby, E.M. and Ray, R., 1996. A Validation

Test of a Field-Based Phosphate Analysis Technique, *Journal of Archaeological Science*, 23: 243-248.

Black, C.A., 1968. *Soil-Plant Relationships*, 2nd ed., New York.

Boer, G. and Exaltus, R.P., 1994. *Invloed van verzuring en verdroging op grondsporen; deel II: Analyse van de chemische en micromorfologische samenstelling van grondsporen op klei- en zandgrond*, Amersfoort. (= ROB Interne Rapporten, 22).

Bolender, D. and Steinberg, J., 1998. *The Mosfell Archaeological Project, Survey Report: Phosphate testing in Mosfellssveit, September 1998*. (http://64.233.183.104/search?q=cache:ZTMxlJGSgMMJ:tyr.ioa.ucla.edu/MAP/1998%2520Report.pdf+mosfellssveit&hl=nl&gl=be&ct=clnk&cd=4).

Brady, N.C. and Weil, R.R., 1999. *The nature and properties of soils. Upper Saddle River*, Prentice Hall.

Cavanagh, W.G., Buck, C.E. and Litton, C.D., 1988a. The interpretation of noisy data from archaeological field survey: phosphate analysis, *Environmental Geochemistry and Health*, 10, 3/4: 92-95.

Cavanagh, W.G., Hirst, S. and Litton, C.D., 1988b. Soil Phosphate, Site Boundaries, and Change Point Analysis, *Journal of Field Archaeology*, 15: 67-83.

Collective, in press. *Le site rubané d'Ormeignies « Le Pilori » (Hainaut). Etudes et Documents*. (= Série Fouilles).

Conway, J.S., 1983. An investigation of soil-phosphorus distribution within occupation deposits from a Romano-British hut group, *Journal of Archaeological Science*, 10: 117-128.

Cross, A.F. and Schlezinger, W.H., 1995. A literature review and evaluation of the Hedley fractionation: Applications to the biogeochemical cycle of soil phosphorus in natural ecosystems, *Geoderma*, 64: 197-214.

Crowther, J., 1997. Soil Phosphate Surveys: Critical Approaches to Sampling, Analysis and Interpretation, *Archaeological Prospection*, 4: 93-102.

Cruise, G. and Macphail, R., 2000. Microstratigraphical Signatures of Experimental Rural Occupation Deposits and Archaeological Sites. In: Roskams, S. (ed.), *Interpreting Stratigraphy. Site evaluation, recording procedures and stratigraphic analysis. Papers presented to the Interpreting Stratigraphy Conferences 1993-1997*, Oxford: 183-191. (= British Archaeological Reports International Series, 910).

da Costa, M.L. and Kern D.C., 1999. Geochemical signatures of tropical soils with archaeological black earth in the Amazon, Brasil, *Journal of Geochemical Exploration*, 66: 369-385.

Davidson, D.A. 1973, Particle size and phosphate analysis, evidence for the evolution of a tell, *Archaeometry*, 15: 143-152.

Delgado, A. and Torrent, J., 2000. Phosphorus Forms and Desorption Patterns in Heavily Fertilized Calcareous and Limed Acid Soils, *Soil Science Society of America Journal*, 64: 2031-2037.

Dercon, G., Davidson, D.A., Simpson, I.A., Dalsgaard, K. and Spek, T., 2003. The nature and formation of Anthrosols in NW Europe: a comparison between 3 countries. In: Boschian, G. (ed.), *Second International Conference on Soils and Archaeology, Pisa, 12th-15th May, 2003, Extended Abstracts*, Pisa: 22-24.

Derry, A.M., Kevan, P.G. and Rowley S.D.M., 1999. Soil Nutrients and Vegetation Characteristics of a Dorset/Thule Site in the Canadian Arctic, *Artic*, 52, 2: 204-213.

Devos, Y., 2001. *Proposition d'un manuel pour l'échantillonnage à l'intérieur et autour des sites archéologiques*, Université Libre de Bruxelles, unpublished report.

Devos, Y., 2002a. *Une perspective diachronique de l'environnement à Bruxelles et les nouvelles méthodes pour valoriser le patrimoine archéologique. Etude archéopédologique du site de la rue d'Une Personne. Evaluation de la pollution ancienne*, Université Libre de Bruxelles, unpublished report.

Devos, Y., 2002b. *Une perspective diachronique de l'environnement à Bruxelles et les nouvelles méthodes pour valoriser le patrimoine archéologique: Synthèse de l'étude archéopédologique du site de la rue du Midi*, Université Libre de Bruxelles, unpublished report.

Devos, Y., 2003. *Une perspective diachronique de l'environnement à Bruxelles et les nouvelles méthodes pour valoriser le patrimoine archéologique: étude archéopédologique du site de l'Hôtel d'Hoogstraeten-Lalaing et le site de la rue Villa Hermosa*, Université Libre de Bruxelles, unpublished report.

Devos, Y., Fechner, K., Vrydaghs, L., Deligne, F. and Degraeve, A., 2007. Contribution of archaeopedology to the palaeoenvironmental reconstruction of (pre-) urban sites at Brussels (Belgium). The example of the Treurenberg site. In: Boschian, G. (ed.), *Proceedings of the Second International Conference on Soils and Archaeology, Pisa, May 12-15, 2003*, Pisa: 145-151. (= Atti della S.T.S.N. - Memorie serie A, anno 2007 - volume CXII).

Devos, Y., Vrydaghs, L., Fechner, K., Laurent C., Degraeve, A. and Modrie, S., this volume. *Buried Anthropic Soils in the Centre of Brussels (Belgium): Looking for Fields in a (Proto-) urban Context.*

Dietz, E.F. 1956, Phosphorus accumulation in soil of an Indian habitation site, *American Antiquity*, 22, 4: 405-409.

Djodjic, F., Börling, K. and Bergström, L., 2004. Phosphorus Leaching in Relation to Soil Type and Soil Phosphorus Content, *Journal of Environmental Quality*, 33: 678-684.

Dunning, N.P., 1993. Ancient maya Anthrosols: Soil Phosphate testing and Land Use. In : Foss, J.E., Timpson, M.E. and Morris, M.W. (eds.), *Proceedings of the First International Conference on Pedo-archaeology, February 16-20, 1992, Knoxville*: 203-210. (= University of Tennessee special publication 93-03).

Dupuoey, J.L., Dambrinne, E., Laffite, J.D. and Moares, C., 2002. Irreversible impact of past land use on forest soils and biodiversity, *Ecology*, 83, 11: 2978-2984.

Edwards, K.J., 1983. Phosphate analysis of soils associated with the Old Kinord field and settlement system, Muir

of Dinnet, Aberdeenshire, *Proceedings of the Society of Antiquaries of Scotland*, 113: 620-627.
Eidt, R.C., 1973. A rapid chemical field test for archaeological site surveying, *American Antiquity*, 38, 2: 206-210.
Eidt, R.C., 1985. *Advances in abandonment settlement analysis: Application to prehistoric anthrosols in Columbia, South America*, Milwaukee.
Engelmark, R. and Linderholm, J., 1996. Prehistoric land management and cultivation : a soil chemical study, *6th Nordic conference on the Application of Scientific Methods in Archaeology, Esjberg 1993*: 315-322. (= PACT).
Entwistle, J.A., Abrahams, P.W. and Dodgshon, R.A., 1998. Multi-Element Analysis of Soils from Scottish Historical Sites, Interpreting Land-Use History Through the Physical and Geochemical Analysis of Soil, *Journal of Archaeological Science*, 25: 53-68.
Entwistle, J.A., Abrahams, P.W. and Dodgshon, R.A., 2000. The Geoarchaeological Significance and spatial variability of a range of physical and chemical soil properties from a former habitation site, Isle of Skye, *Journal of Archaeological Science*, 27: 287-303.
Farswan, Y.S. and Nautiyal, V., 1997. Investigation of Phosphorus Enrichment in the Burial Soil of kumaun, Mid-Central Himalaya, India, *Journal of Archaeological Science*, 24: 251-258.
Fechner, K., 2005. Etude des sols noirs des fonds de vallées de Bussy-Lettrée. In : Lagatie, C. and Vanmoerkerke J. (coord.), *Europort Vatry : les pistes de l'archéologie. Quand la plaine n'était pas déserte…*, Langres: 85-88.
Fechner, K., in collaboration with Bocquillon, H., Bruant, J ., De Lil, A., Devos, Y., Fercoq du Leslay, G., Loicq, S., Grisard, J., Mikkelsen, J.H., Mondolini, A. and Riquier, V., this volume. Case studies of phosphorus mapping in Late La Tène to Roman enclosures and ponds between Rhine and Seine.
Fechner, K., Baes, R., Langohr, R., Louwagie, G., Mikkelsen, J., Slachmuylder, J.-L. and Devos, Y., 2003. *Pre-medieval plough horizons in High and Middle Belgium, Luxembourg and Northern France. A status quaestionis*, presentation at the Second international conference on soil and archaeology, Pisa, 12th-15th May, 2003. In: Boschian, G. (ed.), *Second International Conference on Soils and Archaeology, Pisa, 12th-15th May, 2003, Extended Abstracts*, Pisa: 29.
Fechner, K. and Devos, Y., 2006. *Etude pédologique et cartographie des phosphates sur le site de Lieusaint « ZAC de la Pyramide – quartier F » (Île de France). Rapport final d'analyse et d'interprétation*, INRAP Paris, Unpublished report.
Fechner, K. and Slachmuylder, J.L., 2009. Les « sols noirs » en relation avec l'homme sur le tracé champenois du TGV-Est. Compréhension pédologique et malacologique, approche statistique et essai d'interprétation archéologique. In : Vanmoerkerke, J. (ed .), *Le bassin de la Vesle du Bronze final au Moyen Age à travers les fouilles du TGV-est*, Chalons-en-Champagne: 335-361.
Fechner, K., Mikkelsen, J.H., Baes, R., Devos, Y., Hus, J. and Vrydaghs, L., 2007. Les cartographies de phosphates et de phytolithes dans quatre maisons néolithiques du Nord de la France, de Belgique et du Luxembourg, *Archaeologia Mosellana*, 7.
Fechner, K., de Lil, A., Clavel, V., Teheux, E., in press. Cartographies de batiments néolithiques du Nord de la France, de Belgique et du Luxembourg. *Revue archéologique de Picardie.*
Fechner, K., Devos, Y., Mikkelsen, J.H., Baes, R., Hus J., De Lil, A., in preparation. Phosphate mapping and magnetic susceptibility in Neolithic to Medieval (long)houses and courtyards in Belgium, Luxemburg and Northern France. In: *Archaeology, soil- and life sciences applied to houses, gardens and courtyards. Proceedings of the Session 'From microprobe to spatial analysis – Enclosed and buried surfaces as key sources in Archaeology and Pedology' organised at the European Association of Archaeologists, 12th Annual Meeting, Krakow-Poland. 19th to 24th September 2006.* (=British Archaeological Reports, International Series).
Fernández, F.G., Terry, R.E., Inomata, T. and Eberl, M., 2002. An Ethnoarchaeological Study of Chemical Residues in the Floors and Soils of Q'eqchi' Maya Houses at Las Pozas, Guatemala, *Geoarchaeology*, 17, 6: 487-519.
Foss, J.E., 1992. Appendix: The Soils of Hadrian's Villa, *American Journal of Archaeology*, 96: 595-597.
Foss, J.E., Lewis, R.J., Timpson, M.E. and Lee S.Y., 1995. Appendix: Soils of Thuburbo Maius, *American Journal of Archaeology*, 99: 573-576.
Gardiner, M.J. and Walsh, T., 1966. Comparison of soil materials buried since Neolithic times with those of the present day, *Proceedings Royal Irish Academy*, 65: 29-35.
Gobat, J.-M., Aragno, M. and Matthey, W., 1998. *Le Sol vivant. Bases de pédologie; Biologie des sols*, Lausanne. (= Collection gérer l'environnement, 14).
Goldberg , P. and Macphail, R.I., 2006. *Practical and Theoretical Geoarchaeology*, Malden, Oxford, Carlton.
Guttmann, E.B., Simpson, I.A. and Davidson, D.A., 2005. Manuring Practices in Antiquity: A Review of the Evidence. In: Smith, D.N., Brickley, M.B. and Smith, W. (eds.), *Fertile Ground. Papers in Honour of Susan Limbrey*, Oxford: 68-76. (= Symposia of the Association for Environmental Archaeology, 22).
Guttmann, E.B., Simpson, I.A., Donald A. Davidson, D.A. and Dockrill, S.J., 2006. The Management of Arable Land from Prehistory to the Present: Case Studies from the Northern Isles of Scotland, *Geoarchaeology*, 21, 1: 61-92.
Hassan, F.A., 1981. Rapid Quantitative Determination of Phosphate in Archaeological Sediments, *Journal of Field Archaeology*, 8: 384-387.
Heron, C., 2001. Geochemical Prospecting. In: Brothwell, D.R. and Pollard, A.M. (eds.), *Handbook of Archaeological Sciences*, Chichester: 565-573.
Holliday, V.T. and Gartner, W.G., 2007. Methods of soil P analysis in archaeology, *Journal of Archaeological Science*, 34: 301-333.

Jarvis, I., 1992. Sedimentology, geochemistry and origin of phosphatic chalks: the Upper Cretaceous deposits of NW Europe, *Sedimentology*, 39: 55-97.

Jarvis, I., 2006. The Santonian-Campanian phosphatic chalks of England and France, *Proceedings of the Geologists' Association*, 117: 219–237.

Kyhlberg, O., 1993. Explicatory Aspects of Phosphate Spatial Distributions. In: Arwidsson, G., Hansson, A.-M., Holmquist Olausson, L., Johansson, B.M., Klockhofff, M., Liden, K. and Nordstrom, H.-A. (eds.), *Sources and Ressources, Studies in Honour of Birgit Arrhenius*, Rixensart: 247-264. (= PACT, 38).

Langohr, R., Mikkelsen, J. H. and Vanwesenbeeck, V., 2004. Twee Romeinse potstallen. In: *Verloren voorwerpen. Archeologisch onderzoek op het HST-traject in de provincie Antwerpen*, Antwerpen: 207-209.

Laurelut, C., Louwagie, G., 2002. Défrichements et mises en culture d'un secteur marginal à l'âge du fer : la côte oxfordienne entre Poix-Terron et Villers-le-Tourneur sur le tracé de l'A34 (Ardennes), *Mémoire de la Société Archéologique Champenoise,* 16/ 1: 2-14.

Leinweber, P., Meissner, R., Eckhardt, K.-U. and Seeger, J., 1999. Management effects on forms of phosphorus in soil and leaching losses, *European Journal of Soil Science*, 50: 413-424.

Leonardi, G., Miglavacca, M. and Nardi, S., 1999. Soil Phosphorus Analysis as an Integrative Tool for Recognizing Buried Ancient Ploughsoils, *Journal of Archaeological Science*, 26: 343-352.

Letkeman, L.P., Tiessen, H. and Campbell, C.A., 1996. Phosphorus transformations and redistribution during pedogenesis of western Canadian soils, *Geoderma*, 71: 201-218.

Lillios, K.T., 1992. Phosphate fractionation of soils at Agroal, Portugal, *American Antiquity*, 57, 3: 495-506.

Lima, H.N., Schaefer, C.E.R., Mello, J.W.V., Gilkes, R.J. and Ker, J.C., 2002. Pedogenesis and pre-Colombian land use of "Terra Preta Anthrosols" ("Indian black earth") of Western Amazonia, *Geoderma*, 110: 1-17.

Lindsay, W.L., Vlek, P.L.G., Chien, S.H., 1989. Phosphate minerals. In: *Minerals in Soil Environments*, Madison: 1089-1130. (= Soil Science Society of America, Book Series, 1).

Linderholm, J., sd. *The Umeå chemical method.* (http://www.umu.se/archaeology/forskning/index_eng.html)

Linderholm, J., 2003. Soil chemical surveying: a path to a different understanding of Prehistoric sites and societies in Northern Sweden. In: Boschian, G. (ed.), *Second International Conference on Soils and Archaeology, Pisa, 12th-15th May, 2003, Extended Abstracts*, Pisa: 114-117.

Linderholm, J., 2007. Soil Chemical Surveying: A Path to a Deeper Understanding of Prehistoric Sites and Societies in Sweden, *Geoarchaeology*, 22, 4: 417-438.

Linderholm, J. and Lundberg, E., 1994. Chemical Characterization of Various Archaeological Soil Samples using Main and Trace Elements determined by Inductively Coupled Plasma Atomic Emission Spectrometry, *Journal of Archaeological Science*, 21: 303-314.

Lippi, R.D., 1988. Palaeotopography and Phosphate Analysis of a Buried Jungle Site in Ecuador, *Journal of Field Archaeology*, 15: 85-97.

Louwagie, G. and Langohr, R., 1999. *A34. Section Poix-Terron/Villers-le-Tourneur (France). Rapport archéopédologique.* Unpublished report (Universiteit Gent).

Louwagie, G., Noens, G. and Devos, Y., 2005. *Onderzoek van het bodemmilieu in functie van het fysisch-chemisch kwantificeren van de effecten van grondgebruik en beheer op archeologische bodemsporen in Vlaanderen, Eindrapport, Oktober 2005.* (http://www.vlm.be/PDF/beheerovereenkomsten/060303+studie+arch+bodemsporen/Eindrapport.pdf ).

Macphail, R.I., 2003a. Soil microstratigraphy: a micromorphological and chemical approach. In: Cowan, C. (ed.), *Urban development in north-west Roman Southwark. Excavation 1974-90*, London: 89-105. (= MoLAS Monograph, 16).

Macphail, R.I., 2003b. Attaining robust interpretations of archaeological "soils": examples from rural and urban contexts. In: Boschian, G. (ed.), *Second International Conference on Soils and Archaeology, Pisa, 12th-15th May, 2003, Extended Abstracts*, Pisa: 60-63.

Macphail, R.I. and Cruise, J., 2001. The Soil Micromorphologist as team Player. A multianalytical Approach to the Study of European Microstratigraphy. In: Goldberg, P., Holliday, V.T. and Ferring, C.R. (eds.), *Earth Sciences and Archaeology*, New York: 241-267.

Macphail, R.I., Cruise, G., Engelmark, R. and Linderholm, J., 2000. Integrating Soil Micromorphology and Rapid Chemical Survey Methods: new developments in reconstructing past rural settlement and landscape organisation. In: Roskams, S. (ed.), *Interpreting Stratigraphy, Site evaluation, recording procedures and stratigraphic analysis*, Oxford: 71-80. (= British Archaeological Reports International Series, 910).

Macphail, R.I., Crowther, J., Acott, T.G., Bell, M.G. and Cruise, J.M., 2003. The Experimental Earthwork at Wareham, Dorset After 33 Years: Changes to the Buried LFH and Ah Horizons, *Journal of Archaeological Science*, 30: 77-93.

Macphail, R.I. and Linderholm, J., 2004a. "Dark earth": recent studies of "Dark earth" and "dark-earth-like" microstratigraphy in England, UK. In: Verslype, L. and Brulet R. (eds.), *Terres noires, Dark earth, Actes de la table-ronde internationale tenue à Louvain-la-Neuve, les 09 et 10 novembre 2001*, Louvain-la-Neuve : 35-42. (= Collection d'archéologie Joseph Mertens, XIV).

Macphail, R.I. and Linderholm, J., 2004b. Neolithic land use in south-east England: a brief review of the soil evidence. In: Cotton, J. and Field D. (eds.), *Towards a New Stone Age: aspects of the Neolithic in south-east England*, York: 29-37. (= Council for British Archaeology Research Report, 137).

Mattingley, G.E.G. and Williams, R.J.B., 1962. A note on

the chemical analysis of a soil buried since Roman times, *Journal of Soil Science*, 13: 254-258.

Middleton, W.D., 2004. Identifying chemical activity residues on prehistoric house floors: a methodology and rationale for multi-elemental characterization of a mild extract of anthropogenic sediments, *Archaeometry*, 46, 1 : 47-65.

Middleton, W.D. and Price, T.D., 1996. Identification of Activity Areas by Multi-element Characterization of Sediments from Modern and Archaeological House Floors Using Inductively Coupled Plasma-atomic Emission Spectroscopy, *Journal of Archaeological Science*, 23: 673-687.

Mikkelsen, J.H. and Langohr, R., 1996. A pedological caracterization of the Aubechies soil, a well preserved soil sequence dated to the earliest neolithic agriculture in Belgium. In: *Proceedings of the 13th international congress of prehistoric and protohistoric sciences, Vol. 3*, Forli: 143-150.

Mikkelsen, J.H., Langohr, R., Boas, N.A. and Macphail, R.I., 2006. Land Use and Environmental Degradation in Bronze Age Settlements, Eastern Jutland, Denmark. In: Engelmark, R. and Linderholm, J. (ed.), *Proceedings from the VIII Nordic conference on the applications of Scientific Methods in Archaeology Umeå 2001*, Umeå: 81-92. (= Archaeology and Environment, 21).

Mikkelsen, J.H., Langohr, R., Macphail, R., Vanwesenbeeck, V. and Verbeek, C., 2003. *Roman Potstal byres, a case study from the coversand belt of northern Belgium*, Poster presented at the Second International Conference on Soils and Archaeology, Pisa, 12th-15th May, 2003.

Mikkelsen, J. H., Langohr, R. and Vanwesenbeeck, V., 2002. *Bodemkundig onderzoek op de archeologische site Ekeren-Het Laar, Provincie Antwerpen, België. Project: Archeologische onderzoek van het HSL-traject Antwerpen- Nederlandse grens, opdrachtgever: Provinciebestuur van Antwerpen*, Gent. (= Archeo-Bodemkundig rapport nr. 1, Laboratorium van Bodemkunde, Universiteit Gent).

Milek K., 1999. Settlement in thin section: the application of sediment micromorphology to the interpretation of Norse farmsteads in Iceland. In: Fabech, C. and Ringtved, J. (eds.), *Settlement and Landscape. Proceedings of a Conference in Århus, Denmark, May 4-7, 1998*. Moesgård: 497-498.

Miller, N.F. and Gleason, K.L., 1994. Fertilizer in Identification and Analysis of Cultivated Soil. In: Miller, N.F. and Gleason, K.L. (eds.), *The archaeology of garden and field*, Philadelphia: 25-43.

Ottaway, J.H., 1984. Persistence of organic phosphates in buried soils, *Nature*, 307: 257-259.

Ottaway, J.H. and Matthews, M.R., 1988. Trace element analysis of soil samples from a stratified archaeological site, *Environmental Geochemistry and Health*, 10, 3/4: 105-112.

Parnell, J.J., Terry, R.E. and Nelson, Z., 2002. Soil Chemical Analysis Applied as an interpretive Tool for Ancient Human Activities in Piedras Negras, Guatemala, *Journal of Archaeological Science*, 29: 379-404.

Persson, K., 2005. *Integrated Geophysical-Geochemical Methods For Archaeological Prospecting, TRITA-LWR LIC 2027*. Licentiate Thesis in Land and Water Resources Engineering. KTH Architecture and the Built Environment, Stockholm. (http://www.lwr.kth.se/Publikationer/PDF_Files/LWR_LIC_2027.pdf)

Prøsch-Danielsen, L. and Simonsen, A., 1988. Principal Components Analysis of Pollen, Charcoal and Soil Phosphate Data as a Tool in Prehistoric Land-use investigation at Forsandmoen, southwest Norway, *Norwegian Archaeological Review*, 21, 2 : 85-102.

Proudfoot, B., 1976. The analysis and interpretation of soil phosphorus in archaeological contexts. In: Davidson, D.A. and Shackley, M.L. (eds.), *Geoarchaeology, Earth Science and the Past*, London: 93-113.

Provan, D.M.J., 1971. Soil Phosphate Analysis as a Tool in Archaeology, *Norwegian Archaeological Review*, 4: 37-50.

Quine, T.A., 1995. Soil analysis and archaeological site formation studies. Barham, A.J. and Macphail R.I. (eds.), *Archaeological Sediments and Soils. Analysis, Interpretation and Management*, London: 77-98.

Salas, A.M., Elliott, E.T., Westfall, D.G., Cole, C.V. and Six, J., 2003. The Role of Particulate Organic Matter in Phosphorus Cycling, *Soil Science Society of America Journal*, 67: 181-189.

Sampietro, M.M. and Vattuone, M.A., 2005. Reconstruction of Activity Areas at a Formative Household in Northwest Argentina, *Geoarchaeology*, 20, 4: 337-354.

Sandor, J.A., Gersper, P.L. and Hawley, J.W., 1990. Prehistoric agricultural terraces and soils in the Mimbres area, New Mexico, *World Archaeology*, 22, 1: 70-83.

Sarris, A., Galaty, M.L., Yerkes, R.W., Parkinson, W.A., Gyucha, A., Billingsley, D.M., Tate, R., 2004. Geophysical prospection and soil chemistry at the Early Copper Age settlement of Vésztö-Bikeri, Southeastern Hungary, *Journal of Archaeological Science*, 31: 927-939.

Scalenghe, R., Edwards, A.C., Ajmone Marsan, F. and Barberis, E., 2002. The effect of reducing conditions on the solubility of phosphorus in a diverse range of European agricultural soils, *European Journal of Soil Science*, 53, 3: 439-447.

Schlezinger, D.R. and Howes, B.L., 2000. Organic Phosphorus and Elemental Ratios as Indicators of prehistoric Human Occupation, *Journal of Archaeological Science*, 27: 479-492.

Semmel, A., 1995. Bodenkundliche Hinweise auf Ackernutzung und intensive Bodenerosion um 8000 B.P. im Rhein-Main-Gebiet, *Archäologisches Korrespondenzblatt*, 25: 157-63.

Simpson, I.A., 1997. Relict Properties of Anthropogenic Deep Top Soils as Indicators of Infield Management in Marwick, West Mainland, Orkney, *Journal of Archaeological Science*, 24: 365-380.

Simpson, I.A., Dockrill, S.J., Bull, I.D., Evershed, R.P., 1998. Early Anthropogenic Soil Formation at Tofts Ness, Sanday, Orkney, *Journal of Archaeological Science*, 25: 729-746.

Sjöberg, A., 1976. Phosphate Analysis of Anthropic Soils, *Journal of Field Archaeology*, 3: 447-454.

Smith, H., Marshall, P. and Pearson, M.P., 2001. Reconstructing house activity areas. In: Albarella, U. (ed.), *Environmental Archaeology: Meaning and Purpose*, Dordrecht, Boston and London: 249-270. (= Environmental Science and technology Library, 17).

Smyth, M.P., Dore, C.D. and Dunning, N.P., 1995. Interpreting Prehistoric Settlement Patterns: Lessons from the Maya Center of Sayil, Yucatan, *Journal of Field Archaeology*, 22: 321-347.

Solecki, R.S., 1951. Notes on soil analysis and archaeology, *American Antiquity*, 13: 254-256.

Spek, T., Groenman-van Waateringe W., Kooistra, M. and Bakker, L., 2003. Formation and land-use history of Celtic fields in north-west Europe – an interdisciplinary case study at Zeijen, The Netherlands, *European Journal of Archaeology*, 6, 2: 141-173.

Staüble, H. and Lüning, J., 1999. Phosphatanalysen in Bandkeramischen Häusern, *Archäologisches Korrespondenzblatt*, 2: 169-187.

Stevenson, F.J., 1986. *Cycles of Soil: carbon, nitrogen, phosphorus, sulphur, micronutrients*, New York, Chichester, Brisbane, Toronto, Singapore.

Sullivan, K.A. and Kealhofer, L., 2004. Identifying activity areas in archaeological soils from a colonial Virginia house lot using phytolith analysis and soil chemistry, *Journal of Archaeological Science*, 31, 12: 1659-1673.

Terry, R.E., Hardin, P.J., Houston, S.D., Nelson, S.D., Jackson, M.W., Carr, J. and Parnell, J., 2000. Quantitative Phosphorus Measurement: A Field Test Procedure for Archaeological Site Analysis at Piedras Negras, Guatemala, *Geoarchaeology*, 15, 2: 151-166.

Terry, R.E., Fernández, F.G., Parnell J.J. and Inomata, T., 2004. The story in the floors: chemical signatures of ancient and modern Maya activities at Aguateca, Guatemala, *Journal of Archaeological Science*, 31: 1237-1250.

Tyler, G. & Olsson, T. 2001. Concentrations of 60 elements in the soil solution as related to the soil acidity, *European Journal of Soil Science*, 52: 151-165.

Van Vliet, B., Fagnart, J.P., Langohr, R., Munaut, A.V., 1992. Importance de la succession des phases écologiques anciennes et actuelles dans la différenciation des sols lessivés de la couverture loessique d'Europe occidentale : argumentation stratigraphique et archéologique, *Science du Sol*, 30: 75-93.

Van Vliet-Lanoë, B., 1990. The genesis and age of the argillic horizon in Weichselian loess of northwestern Europe, *Quaternary International*, 5: 49-56.

Vissac, C., 2005. Study of a historical garden soil at the Grand-Pressigny site (Indre-et-Loire, France); evidence of landscape management, *Journal of Cultural Heritage*, 6: 61-67.

Vizcaíno, A.S. and Cañabate, M.L., 1999. Identification of Activity Areas by Soil Phoshorus and Organic Matter Analysis in Two Rooms of the Iberian Sanctuary "Cerra El Pajarillo", *Geoarchaeology*, 14, 1: 47-62.

Wells, E.C., 2004. Investigating activity patterns in prehispanic plazas: weak acid-extraction ICP-AES analysis of anthrosols at classic period El Coyote, Northwestern Honduras, *Archaeometry*, 46, 1: 67-84.

Wells, E.C., Terry, R.E., Parnell, J.J., Hardin, P.J., Jackson, M.W. and Houston, S.D., 2000. Chemical Analyses of Ancient Anthrosols in Residential Areas at Piedras Negras, Guatemala, *Journal of Archaeological Science*, 27: 449-462.

White, E.M., 1978. Cautionary note on soil phosphate data interpretation for archaeology, *American Antiquity*, 43, 3: 507-508.

White, R.E., 1997. *Principles and Practice of Soil Science. The Soil as a Natural Resource*, Oxford, London, Edingburgh, Malden and Carlton Victoria.

Woods, W.I., 1977. The quantitative analysis of soil phosphate, *American Antiquity*, 42, 2 : 248-252.

Zimmerman, W.H., 1998. La cartographie de phosphates au service de l'archéologie de l'habitat (trad. Anne Nissen-Jaubert). In: Dabas, M., Deletang, H., Ferdière, A., Jung, C. and Zimmerman, W.H. (eds.), *La prospection*, Paris : 207-216. (= Collections archéologiques).

# CHAPTER 3
# CASE STUDIES OF PHOSPHORUS MAPPING IN LATE LA TÈNE TO ROMAN ENCLOSURES AND PONDS BETWEEN RHINE AND SEINE

*Kai Fechner (1, formerly: 2/3)*
*E-mail: Kai.Fechner@inrap.fr*
*in collaboration with Bocquillon, H. (1), Bruant, J. (1), Caparros, T. (1), Clavel, V. (1), Créteur, Y. (1), Devos, Y. (3), Fechner G., Fercoq du Leslay, G. (4), Ghigny A.-C. (5), Grisard J. (1), de Lil, A. (2, 6), Loicq S. (2, 6), Mahé-Hourlier, N. (1), Mikkelsen, J.H. (2, 7), Mondolini, A. (1), Riquier V. (1)*

(1) INRAP, Institut national de Recherches Archéologiques préventives
(2) RooTS, Research Team in Archaeo- and Palaeo-Sciences
(3) Centre de Recherches Archéologiques, Université Libre de Bruxelles
(4) Centre archéologique départemental de la Somme, Ribemont-sur-Ancre
(5) Service Public wallon, Service de l'Archéologie
(6) Recherches et Prospections archéologiques en Wallonie, asbl
(7) Research Institute for Nature and Forest, Geraardsbergen.

## 1. Introduction

The understanding of the spatial organisation and the identification of activity areas within enclosures, areas of land separated from its surroundings by ditches, earthworks, walls or fences, is often problematic (e.g. Wieland, 1995). The relation with the surrounding area is also often poorly understood.

One way to study these enclosures is the application of phosphorus analyses that reflect the amount of, mostly, anthropic enrichment by elements as bones and excrements (see Devos *et al.*, this volume). A phosphorus concentration map is obtained by the analysis of soil samples sampled at regular distances by laying out a lateral grid over the excavation (fig. 1).

In this paper we will present several case-studies of late La Tène and Roman enclosures excavated between Rhine and Seine (fig. 2-3) and subjected to phosphorus mapping during the last 10 years. In the concerned areas, the systematic intervention of pedology on archaeological sites allowed a reasoned selection of the sites that were best adapted to this method and to the involved archaeological questions and investments of time and finances. Based on a long tradition of phosphorus-mapping in archaeology (see Devos *et al.*, this volume), the choice to generalize this approach was made with the archaeologists and the decision-makers, mostly in large-scale projects of rescue excavations.

Special attention will be given to the relation of the enclosures with ponds and the immediate surroundings, as the presence of a pond can in some become an essential and central element for the functional interpretation of an occupation.

Finally, the article also aims at outlining the potential contribution of phosphorus-mapping to future sites.

## 2. Methods

The detailed field and laboratory methodology of phosphorus mapping that has been applied on the studied sites is presented in Devos *et al.* (this volume).

Phosphorus analyses have been realized in the laboratory of Soil Science of University of Ghent. Both the Ptot (total) and the Pi (inorganic) content of the sample are measured with the method using sulphuric acid ($H_2SO_4$, see Mikkelsen, 1997), in order to avoid the liberation of phosphorus inside the silicate structures of the mineral fraction, liberated by stronger acids like hydrogen fluoride (HF). The latter method has only been used for the samples of the site of Ribemont-sur-Ancre and the reference profile of Nangis as they are treated by the INRA soil laboratory of Arras. This is usually the case when a phosphorus analysis is limited to few samples, rather than giving a priority to phosphorus mapping: the concerned samples are then treated in combination with the other types of analyses that are proposed in Arras. On the type of soils encountered on the two concerned sites (decarbonated loess), there is no risk of contamination by a geological source of phosphorus, unlike for instance on calcareous sub-soils (see e.g. samples on limestone of Dampierre-le-Chateau (Marne, France), treated comparatively in both laboratories: Fechner et al., in preparation).

In some cases, mentioned in the respective chapters of the case-studies, rapid phosphorus tests according to Eidt (1985) have been used to evaluate the potential of phosphorus mapping on a larger number of samples before total or selective phosphorus analyses. For the sites and samples that were not subjected to laboratory analyses, some less precise information has in some cases been gained from these tests.

Other punctual chemical analyses, used according to field questioning (presence of chemical environments related to specific activities or soil environments,...) and especially in some reference profiles, involve the quantitative determination of the pH, organic carbon, iron, calcium carbonates and bases as calcium, magnesium, potassium and sodium. The analyses have been done in the INRA soil laboratory of Arras (France) and the methods have been chosen according to the soil context and the purpose (Baize, 1988).

It should be noted that the term "past pollution", "accumulation" or "enhanced phosphorus content" are used here in the sense of phosphorus enrichments that are related to the past occupations, in opposition to the term "contamination" which is related to post-depositional or recent phenomena.

## 3. Case-studies

### *3.1. The Late Iron Age enclosure of Buchères (Aube, France, fig. 4-5)*

The Late La Tène enclosure of Buchères « Parc logistique de l'Aube » (archaeologists: Vincent Riquier and Julien

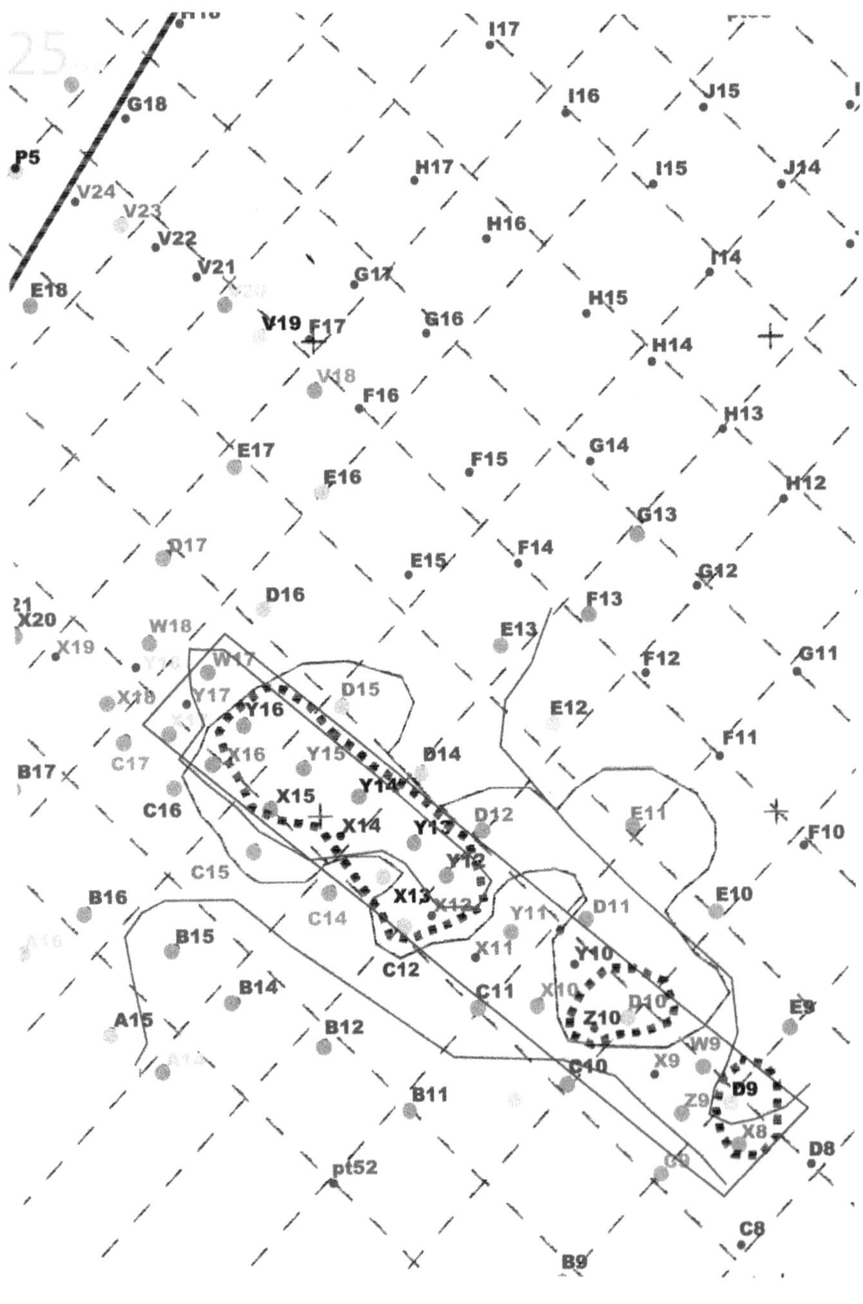


*Figure 1. Example of a grid of samples in an enclosure. Part of the enclosure of Saint-Brice-sous-Forêt, with a denser sampling in a rectangular building; rounded and stippled lines represent the half-way interval between samples that are richer and poor in phosphorus, respectively for the phosphorus tests (thick stippled line) and for the phosphorus analyses (thin continuous line).*

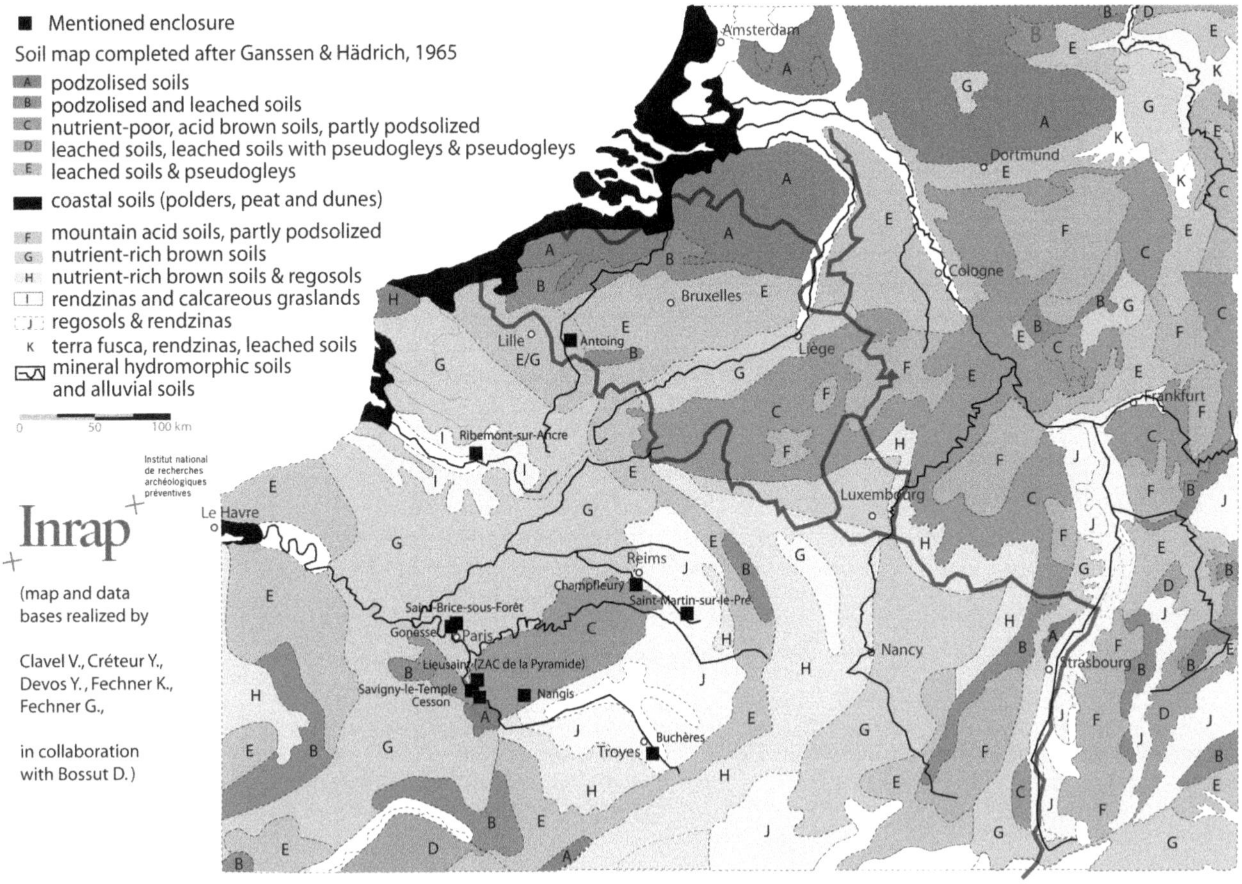


*Figure 2. Distribution of the studied sites on the simplified soil map (after Ganssen and Hädrich, 1965).*

Grisard, INRAP) has been partly excavated and sampled for phosphorus mapping. It is situated on a flat plateau, on calcareous loess ("Cca"-horizon) whose 50 top centimeters are decarbonated with the formation of thin clay eluvial (clay export or "E") and illuvial (clay import or "Bt") horizons. It appears that these decarbonated horizons have never been formed in some lower topographical contexts of the site, a situation also found on the nearby site of Marnay-sur-Seine (Fechner *et al.*, in preparation). These and other differences in soils have been taken into consideration for the interpretation of phosphorus cartographies in these sites.

The field study identified some soil bioturbation (humiferous pockets in the top of the "Cca"-horizon) and thus potential homogenization of the phosphorus content. The high mean ratio of total phosphorus over inorganic phosphorus indicates the input of more recent organic phosphorus. This will influence the final amounts measured in the laboratory compared to sites without such post-depositional modifications. As this input appears to be rather constant in the analyses of Buchères, it does not influence the intra-site comparison of the results.

In order to better read the limits of the archaeological structures (dark humus on light-coloured calcareous loess), the decarbonated horizons were partly scraped away during the excavation. However phosphorus tests suggest that no phosphorus was present in the underlying very calcareous loess. All the phosphorus has been concentrated in the ancient surface soil and the shallow "E" and "Bt"-horizons, as phosphorus will immediately be strongly fixed at the contact of the highest calcium rich soil horizons instead of moving further down in the natural soil (see experimental stable at Butser Iron Age Farm: Macphail and Goldberg, 1995: 4-8; Macphail, 2000: 66; Macphail *et al.*, 2000: 76; see numerous archaeological examples on calcareous loess and substrata: Fechner *et al.*, 2003; Fechner *et al.*, in preparation). Therefore, after systematic phosphorus tests, quantitative analyses were limited to the samples taken in the areas where the decarbonated and clay enriched top horizons were still preserved. Nevertheless, only a very slight to negligible enrichment of the inside of the enclosure can be evidenced, reminding of a slightly stronger anthropisation without strongly polluting activities (fig. 5).

## Comparisons and interpretation

For the good understanding of the phosphate map of the enclosure of Buchères, it is useful to benefit from data from another site situated on the same type of soils and in a comparable geographical and climatic setting: the Roman enclosure of Saint-Martin-sur-le-Pré is the only such site that has been subjected to phosphorus analyses, even they were very limited (archaeologist: Michel Kasprzyk; Fechner, 2009). This enclosure has not been studied in detail, as its shallow surface horizons ("Bt" horizon on

| SITE | Range of total phosphorus | Mean all samples | Mean ratio P total/ P inorganic | Total number of samples |
|---|---|---|---|---|
| La Tène III: | | | | |
| 1. Buchères 2006 (F)** | Ca. 600-1350 ppm | 861 | 1,74 | 35 |
| La Tène III-Early Roman : | | | | |
| 2.Ribemont-sur-Ancre (F)*** | Ca. 1000-4000 ppm (1) | 1637 | - | 50 |
| La Tène III-Roman: | | | | |
| 3. Champfleury (F)* | Ca. 400-2750 ppm | 1106 | 1,42 | 184 |
| 4. Savigny-le-Temple (F)** | Ca. 109-1018 ppm | 278 | 2,34 | 114 |
| Roman: | | | | |
| 5 . Saint-Brice-sous-Forêt (F)** | Ca. 250-1850 ppm | 504 | 1,99 | 52 |
| 6. Nangis (F)*** | Ca. 500-1700 ppm | 874 | 1,42 | 45 |
| 7. Gonesse (F)** | Ca. 600-9500 ppm | 1903 | 1,43 | 25 |
| 8. Saint-Martin-sur-le-Pré (F)** | Ca. 1100-1150 ppm | 1122 | 1,56 | 3 |
| 9. Cessons «Plaine du Moulin à Vent» (F)** | Ca. 550-3450 ppm | 995 | 1,52 | 107 |
| 10. Antoing (B)*** | Ca. 400-1400 ppm | 751 | 1,51 | 55 |
| Roman-Early Medieval: | | | | |
| 11. Lieusaint "Zac de la Pyramide" (F)** | Ca. 1000-10000 ppm | 2125 | 1,39 | 109 |

*Figure 3. Inventory of studied sites and data. Legend: *not decarbonated (except some localized pockets). **slightly decarbonated. ***deeply decarbonated. n: deeply bioturbated and rather low mean value. B: Belgium. F: France. n: very high mean value. Ppm: parts per million*
*(1) data of the laboratory of INRA in Arras, otherwise Ghent University.*

undeeply decarbonated loess) were once again scraped away to reach the top of the calcareous loess. However, the phosphorus tests might indicate a stronger past pollution of the inside compared to the outside and the very few analyses indicate a much higher amount of total phosphate than in Buchères. The bioturbation of the top of the calcareous subsoil is comparable in both sites. Moreover, the phosphorus of Saint-Martin-sur-le-Pré is not due to not due to differential infiltrations as observed on loessic sites, as the calcareous loess is never affected in depth by phosphorus (see above for process). The difference between both sites appears to be effective and due to differences in man-made phosphorus enrichments.

By this comparison, it can be reinforced that the contact between calcareous and decarbonated loess is very little enriched in Buchères, pointing to an enclosure that is very clean concerning phosphorus accumulation or at least not dedicated to a particular activity that produces such phosphorus enhancement, as the presence of bones, of excrements, of large quantities other organic matter or ash, the high density of certain animals.

### *3.3. The Late La Tène cultural enclosure of Ribemont-sur-Ancre (Somme, France, fig. 5-10)*

The Late La Tène enclosures of Ribemont-sur-Ancre is also treated in a separate article (Fercoq du Leslay *et al.*, 2009; this volume). Here, we will concentrate on the presentation of the quantitative and field pedological data related to the enclosure and the comparison of the results with the other studied enclosures.

The quadrangular and rounded enclosures (fig. 6) are situated on a plateau with decarbonated lœss (archaeologists: Jean-Louis Brunaux, Gérard Fercoq du Leslay, CNRS and Centre archéologique départemental de la Somme). In three quarters of the quadrangular enclosure, the surface and subsurface horizon of the occupation is preserved below the bone deposits and the Roman raisings of the surface, for the construction of the Roman sanctuary. The other quarter has been completely destroyed by the leveling done by the Romans in order to establish a flat surface for this construction.

The correlation of high total phosphorus with high calcium and calcium carbonate contents has permitted to detect specific sectors of the sanctuary that are probably the areas of bone manipulation or exposure, as the physical remains of bones are also present in most cases. However, there are some local areas without such bone discoveries (fig. 8). The post-depositional decomposition of these bones can be excluded because of the calcareous fill-up that covered the whole site early in Roman times; this change in chemical environment allowed to stop the alteration of bones that is common in acid to neutral loess soil. It intervened sufficiently early to avoid advanced decomposition in the period between bone deposition and protection by calcareous materials. These areas of enhanced phosphorus content without bones probably point to bone displacement by Man after their use, even if other causes for this enrichment cannot be completely excluded.

Note that the phosphorus tests show a clear south-north opposition (see fig. 9). The analyses of total phosphorus

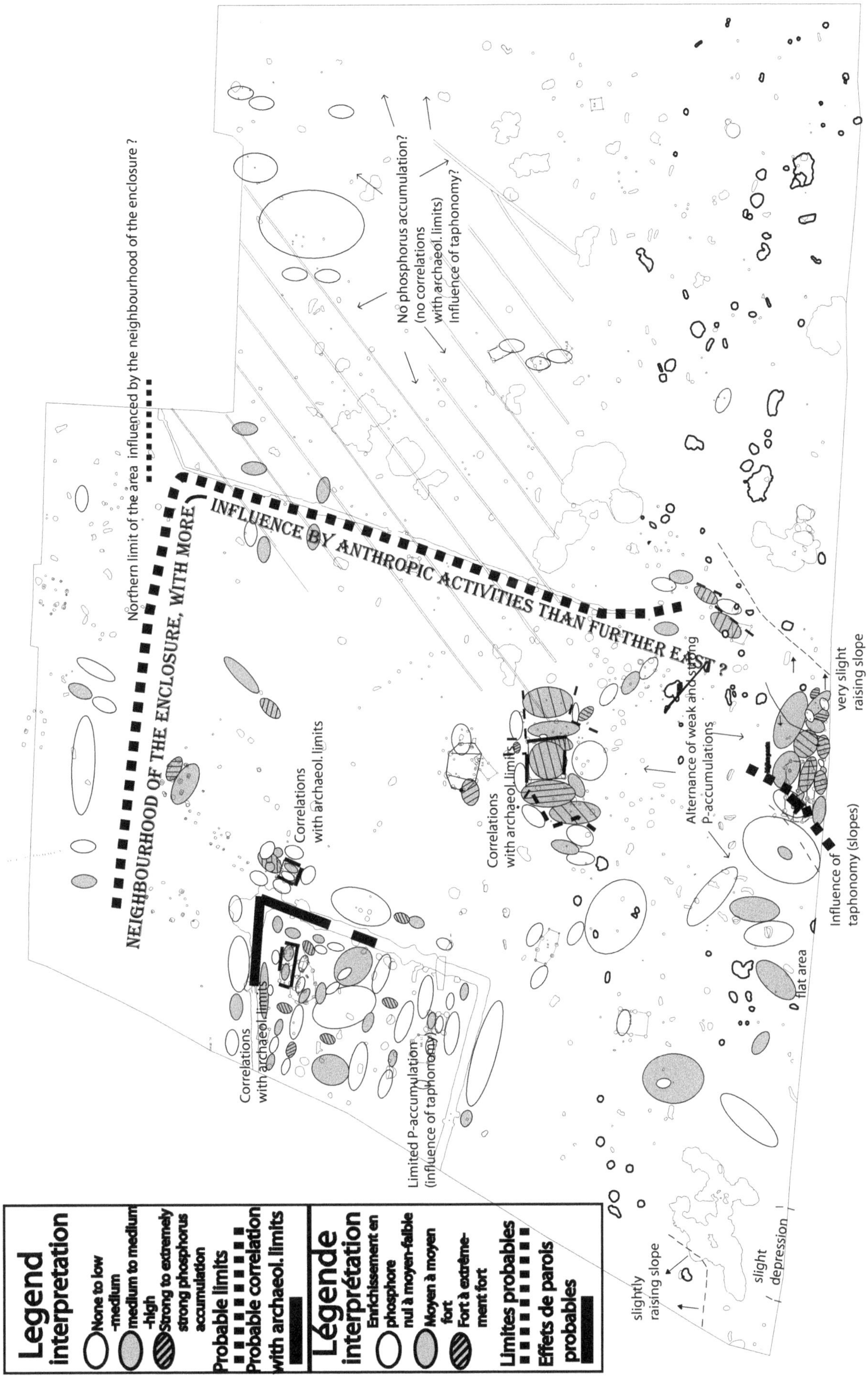


*Figure 4. Map of phosphorus analyses and tests in and around the enclosure of Buchères*

| | *Min. and max. Ptot* | *Min. and max. Porg* | *Mean* | *Mean Ptotal/ Pinorg ratio* | *Total number of samples* |
|---|---|---|---|---|---|
| Outside values | 781 – 925 | 290 - 441 | 851 | 1,77 | 6 |
| Inside values without buildings | 643 – 1336 | 180 – 559 | 861 | 1,71 | 24 |
| Inside buildings | 744 – 1100 | 199 – 473 | 871 | 1,86 | 6 |

*Figure 5. Range of the results of analyses of total P of Buchères:*

are less numerous. They present some exceptions with quite some enrichment by phosphorus in the northern part that are however a little less affected than their equivalents in the southern part. Moreover, these samples could be of Roman age.

It should be noted that some deep homogenization by the roots of bush vegetation was active locally, which may have produced a lowering of the absolute values of total phosphate. In this site, we have no information on the organic and inorganic fractions of the total phosphate to control the impact of organic sources and turbations.

## Interpretation

For the discussed reasons, the phosphorus-map points to the presence of large areas strongly affected by bones. All the areas of strong phosphorus enhancement coincide with the southern half of the enclosure, including the bone construction preserved just outside of the enclosure, at its south-east corner. The partial lack of bone remains in the area enriched in phosphorus makes it possible that some bones have been displaced by Man already in the Late Iron Age or Early Roman times. The precision given by these analyses could point to some well delimited areas of bone manipulation or exposure in the central and south-eastern parts.

The areas of bone-manipulation can be completed by the mapping of another soil phenomenon characterized by a marked local bioturbation, due to the presence of trees and/ or shrubs and mostly poor in phosphorus (fig. 10). These areas are detected by the coincidence of palynological, malacological data and field pedological observations of abnormal bioturbation (Fercoq *et al.*, 2009; Fercoq *et al.*, this volume; Fechner *et al.*, 2004). They are limited to the north-central part of the enclosure, where they would coincide with the presence of a "sacred forest" and, possibly, to the partly filled ditch in that northern side of the enclosure. The latter phenomenon is still active in the first Roman levels and can also have another interpretation: here the bioturbation could also coincide with the installation of a natural vegetation in a less frequented part of the site.

### *3.4. The Late Iron Age and Roman enclosure of Champfleury (Marne, France, fig. 11-14)*

The Late Iron Age and Roman enclosure of the sector 4 of the site of Champfleury "La fosse aux Fromages" (archaeologist: Hervé Bocquillon, INRAP) has been studied by a particularly large number of phosphorus analyses and tests. This was justified by its particular good preservation, in a large dry valley floor on limestone and limestone gravel, including local preservation of the ancient surface horizon. The enclosure includes six buildings that are discussed by Fechner *et al.* (in preparation). Some of these are Roman and it is probable that the enclosure has been used until that period. It cannot be assured that the enclosure as it has appeared is complete or if it has a second part along its southern side, in an area that was not accessible to excavation. This supposed south border of the enclosure is marked only by the remains of a discontinuous earth and stone levee, while the northern sides have one or several, more or less continuous ditches.

The table of results (fig. 12) shows that the mean of the ratios of total over inorganic phosphorus is closer to 1 than to 2, indicating the limited influence of organic phosphorus. This suggests that the analytical data are scarcely influenced by later bioturbation and deep homogenization. The enclosure is installed on a large and flat dry valley floor, characterized by gravel, marl and limestone, with local preservation of the humiferous palaeo-surface horizon. These lateral changes explain that only the map of *inorganic* phosphorus was well correlated with archaeological limits of the enclosure, confirming the dating of the phosphorus enrichments.

Organic part of phosphorus tends indeed to be influenced by the presence and absence of the preliminary humiferous surface. Statistical treatment of the data (see Devos *et al.*, this volume, adapted from Entwhistle *et al.*, 2000) confirmed the significant character of the contrast between inside and outside the enclosure. Some of the individual buildings inside the enclosure show significantly higher or lower *total* phosphate-levels (allowing to propose particular uses: Fechner *et al.*, in preparation) compared to the surrounding area within the enclosure, but all of them are richer in phosphorus than the areas around the enclosure.

## Interpretation

The first result is a significant and very marked past pollution of the inside of the enclosure as opposed to its outside. This enrichment sometimes reaches 3000 ppm which is outstanding for a sample in the natural soil. It is

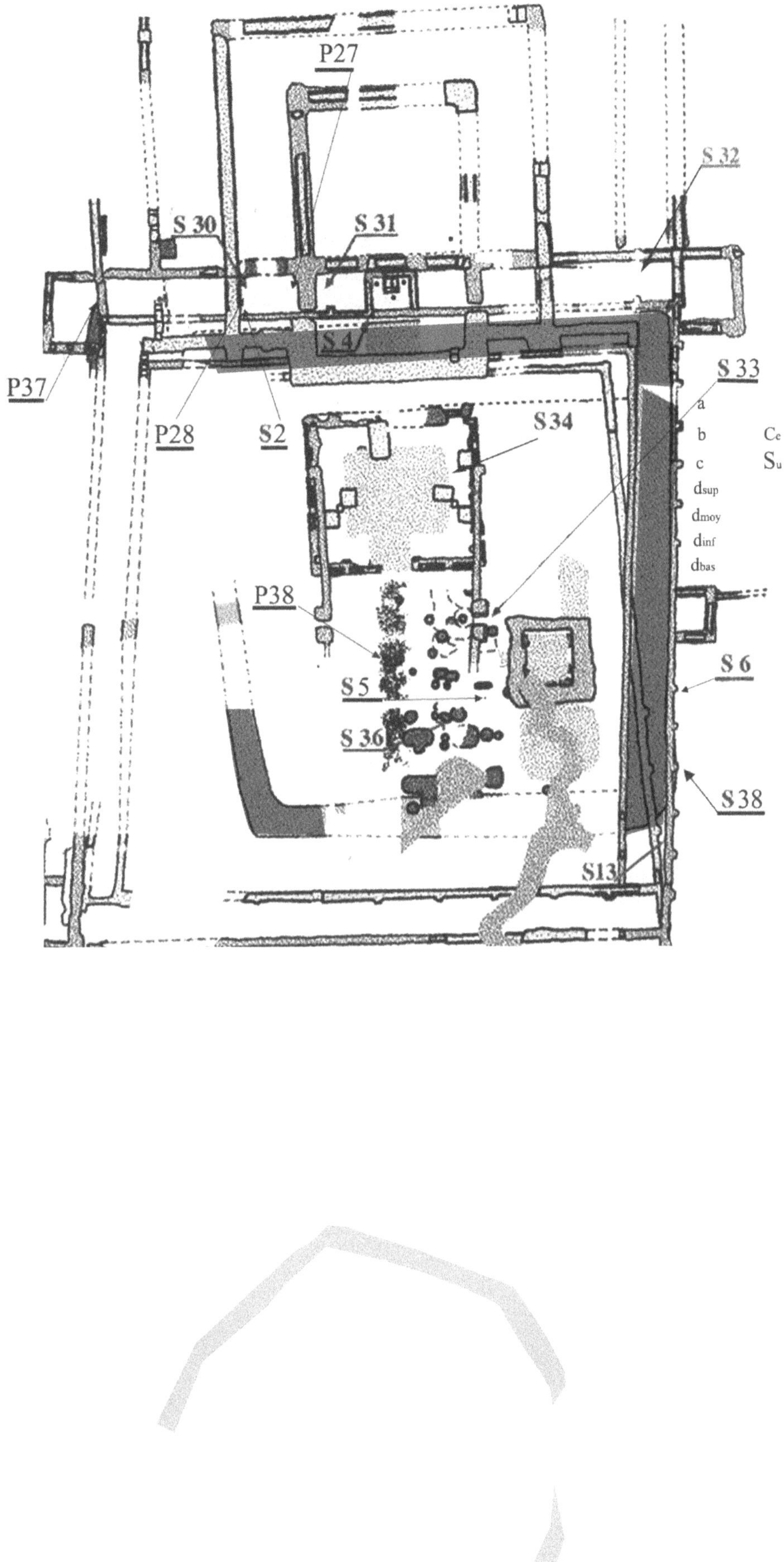


*Figure 6. Archaeological plan of the enclosures of Ribemont-sur-Ancre: in dark grey, the quadrangular enclosure is covered by the lighter traces of the Roman period; the rounded enclosure is indicated in light grey in the lower part of the figure (see also Fercoq et al., this volume).*

part be due to the exceptional preservation of the ancient surface in this sector. As mentioned above, on such a calcareous soil, only the top centimeters of the calcareous subsoil are enriched by the man-induced enrichments that intervened at its surface (see experimental stable at Butser Iron Age Farm: Macphail and Goldberg, 1995: 4-8; Macphail, 2000: 66; Macphail *et al.*, 2000: 76): a new mineral called apatite is rapidly formed when phosphorus and calcium meet, which makes it impossible for the phosphorus to penetrate any deeper in the soil.

A second result is the following: the map of inorganic phosphate might indicate that the enclosure as it has appeared is complete along its southern side, as no phosphorus-rich samples appear anymore here. The clearness of this difference and the parallelism with the other three sides make it improbable that this area would be part of another part of the enclosure, but without enhanced phosphorus content.

Finally, it should be mentioned that a localized enrichment of inorganic phosphate is observed in front of one of the clearest interruptions of the enclosure ditch, at the north-eastern border.

The lack of coincidence with bone remains points in the direction of an organic source of phosphorus, as both the systematic displacement of bones by Man and their natural decomposition, on these very calcareous soils, are excluded. It points to an activity related to high amounts of organic phosphorus, possibly of fecal origin. A map of magnetic susceptibility would have been another independent argument to establish a link between organic matter and phosphorus (Fechner et al., in preparation: Dampierre-le-Château), but has only been initiated (see Fechner et al, in preparation).

This might indicate an enclosure dedicated to animal husbandry, even if other arguments as parasitology or archaeological indications are lacking so far, even if we cannot completely exclude other activities that produce an enrichment by organic phosphorus. Finally, it has to be stressed that high amounts can also be due to the duration of enhanced phosphorus contenting activities (for instance: Fechner and Devos, 2006). Taking all this into account, the Late Iron Age to Roman enclosure of Champfleury remains a case dominated by strong and/or prolonged polluting activities compared to other contemporary cases on calcareous subsoils.

### *3.5. The Late La Tène to Roman enclosures of Savigny-le-Temple (Seine-et-Marne, France, fig. 15-18)*

The Late La Tène to Roman enclosures of Savigny-le-Temple "ZAC du Mont Blanc" (archaeologist: Jean Bruant, INRAP) are situated on a plateau and top of slopes with bioturbated clayey sands. The studied surface is very large and includes some areas with some more erosion that is responsible for some lower phosphorus levels; however, the analyses of reference samples and their correlation with the horizons of a well preserved reference profile (fig. 16) show a deep homogenization of the content of phosphorus. This homogenization gives us the possibility to compare the data over most of the studied surface including little eroded parts on slight slopes (except were erosion reaches more than 65 cm).

Indeed, the mean of the ratios of total over inorganic phosphorus (fig. 15: more than 2) shows very strong influence of organic phosphorus, which is in close correlation with the field observations of strong bioturbation and the deep homogenization of the reference profile. The amount of organic phosphate is much higher than elsewhere as shown by the very high Ptot/Pinorg ratios (total phosphorus compared to inorganic phosphorus, see the comparison of means in figure 3). The significant consequences are the lowering of both the absolute values of the analyses compared to other sites, of the intensities of the phosphorus tests and of the contrasts between the richer and the poorer areas of the mapping.

The two main enclosures (1 and 2) are joined by one of their borders and have buildings in their internal space. The others enclosures are more or less isolated but connected by a large ditch system. It has been shown that a progressive extension is highly probable (Jean Bruant, pers. comm.).

| | *Min. and max. Ptot* | *Min. and max. Porg* | *Mean* | *Mean Ptotal/ Pinorg ratio* | *Total number of samples* |
|---|---|---|---|---|---|
| Outside values except outer bone construction | 1050-1750 | - | 1312 | - | 17 |
| Inside values | 1200-1900 (3700[1]) | - | 1886 | - | 18 |
| Outer bone construction | 1100-3500 | - | 1780 | - | 7 |
| Ditch filling | 1350-2000 | - | 1666 | - | 5 |
| Reference profile | 1100-1650 | - | 1239 (1320[2]) | - | 6 |

(1) for profile P18, probably Early Roman
(2) without taking into account the buried "A1" surface horizons (poorer in phosphorus than the "Bt" –horizon).

*Figure 7. Range of the results of analyses of total P in and around the enclosure of Ribemont-sur-Ancre (data of the laboratory of INRA in Arras)*

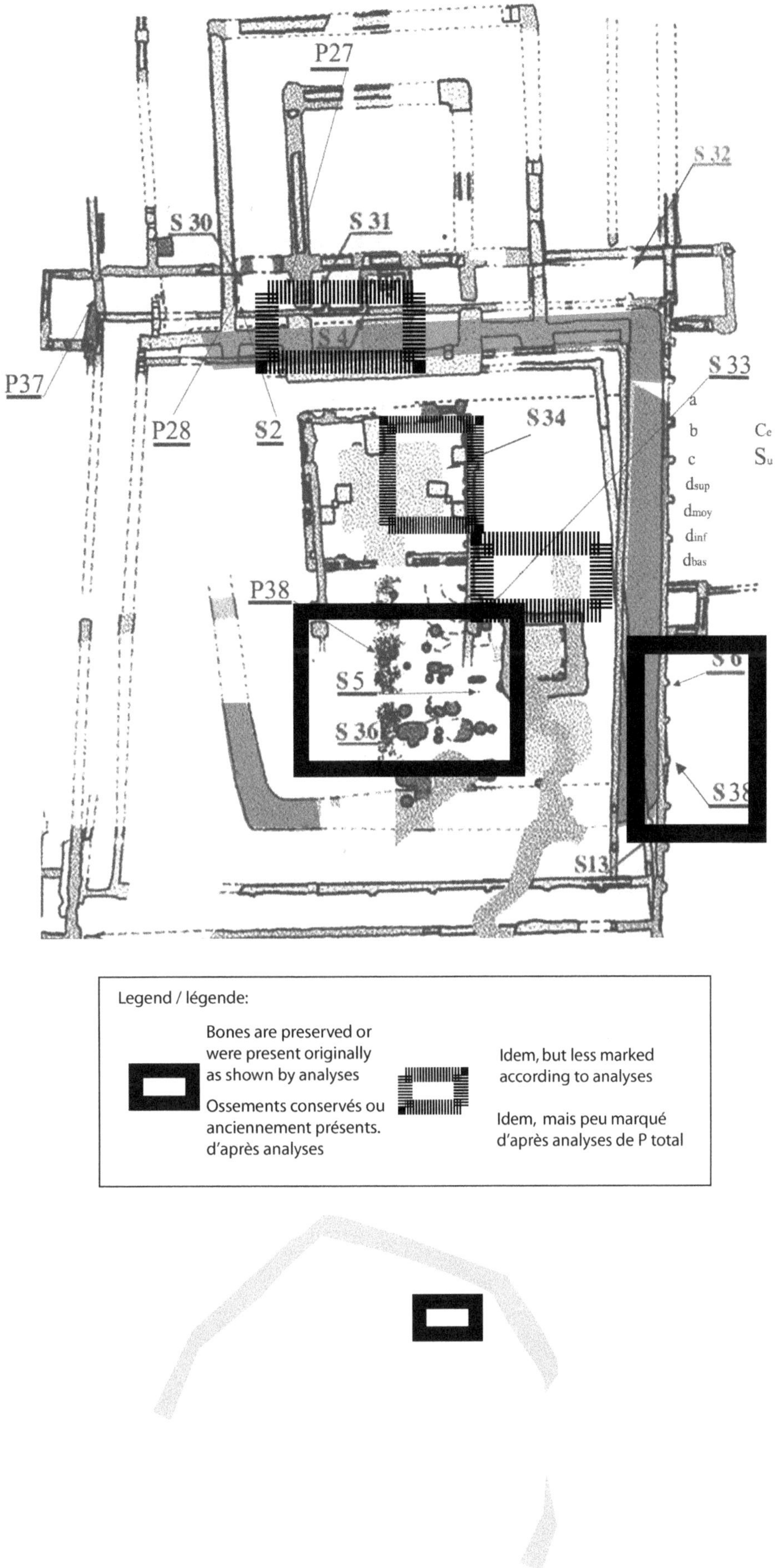


*Figure 8. Interpretation of the phosphate mapping of Ribemont-sur-Ancre. In the pre-Roman surface horizon of the enclosure, preserved below the Roman upfilling, the very high phosphorus accumulations are associated to high calcium contents.*

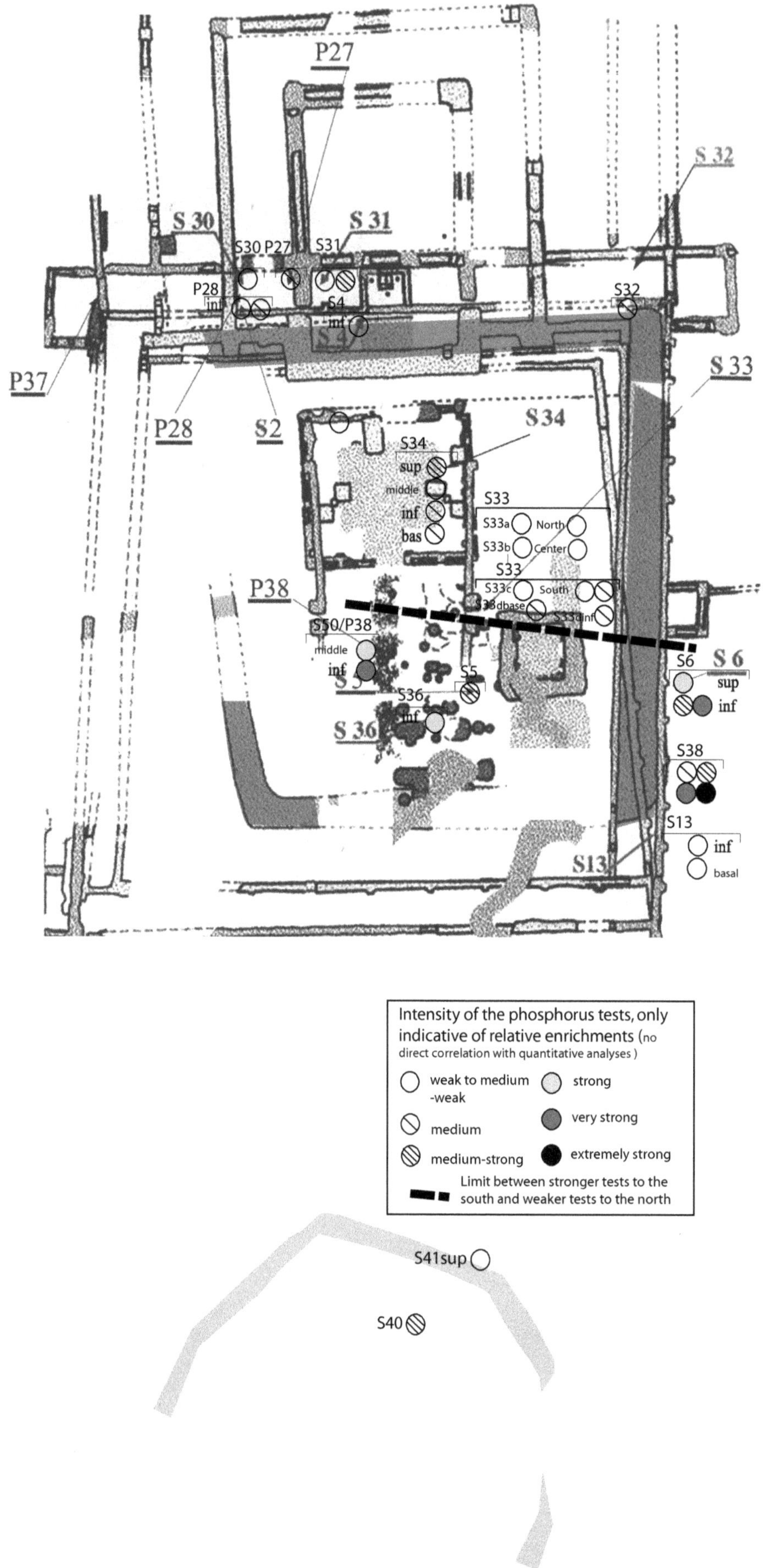


*Figure 9. Phosphorus-tests of Ribemont-sur-Ancre. Indicated by the filled dots, the higher phosphorus-tests are clearly limited to the southern half of the quadrangular enclosure and the annex bone structure outside its south-east border*

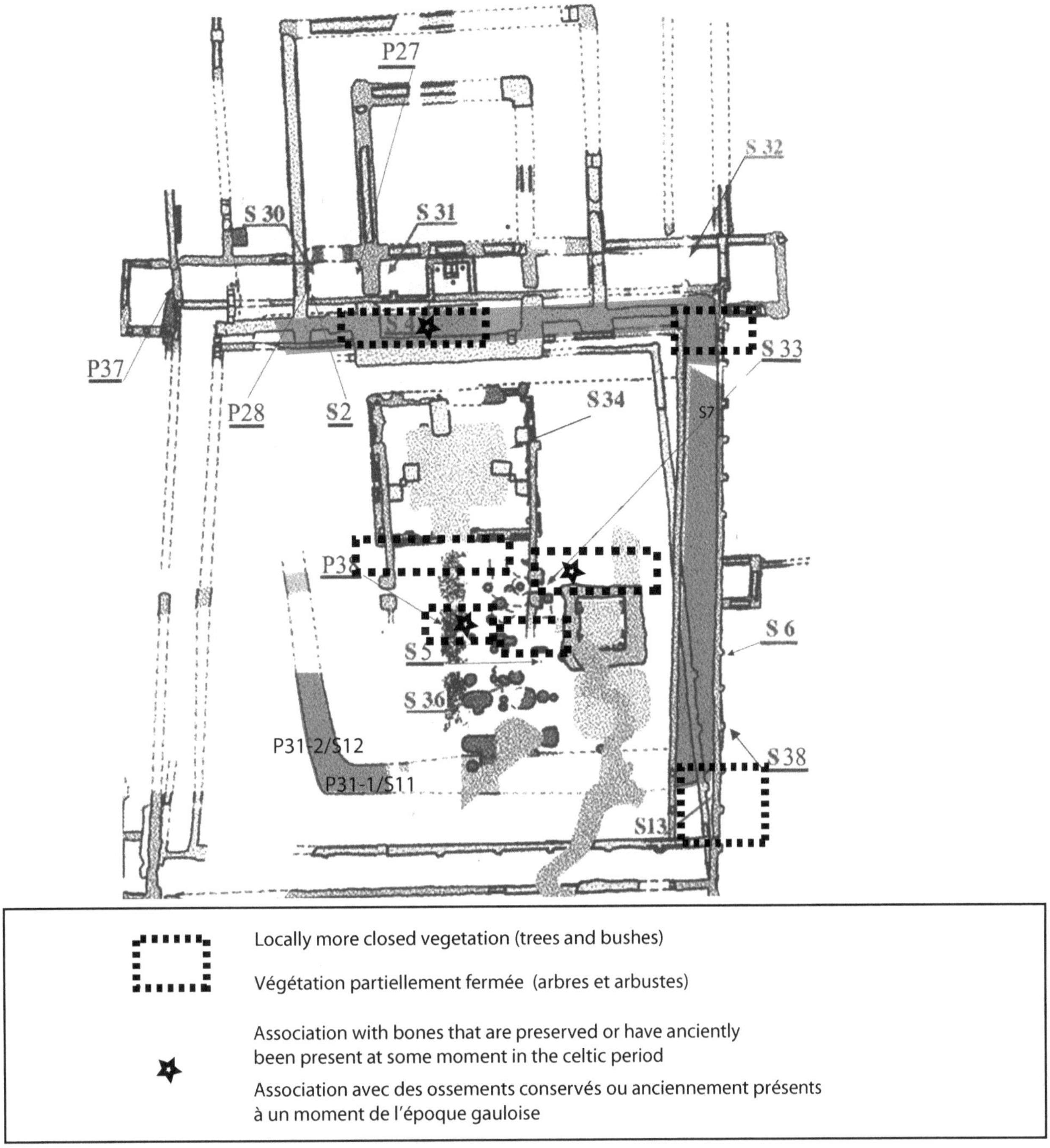


*Figure 10. Confrontation of field pedology, soil analyses, pollen and malacological data in the palaeosurfaces. The mapping of all the indicators allowed to map sectors of specific activity, including former bone deposits and the supposed Iron Age and Roman "sacred forest".*

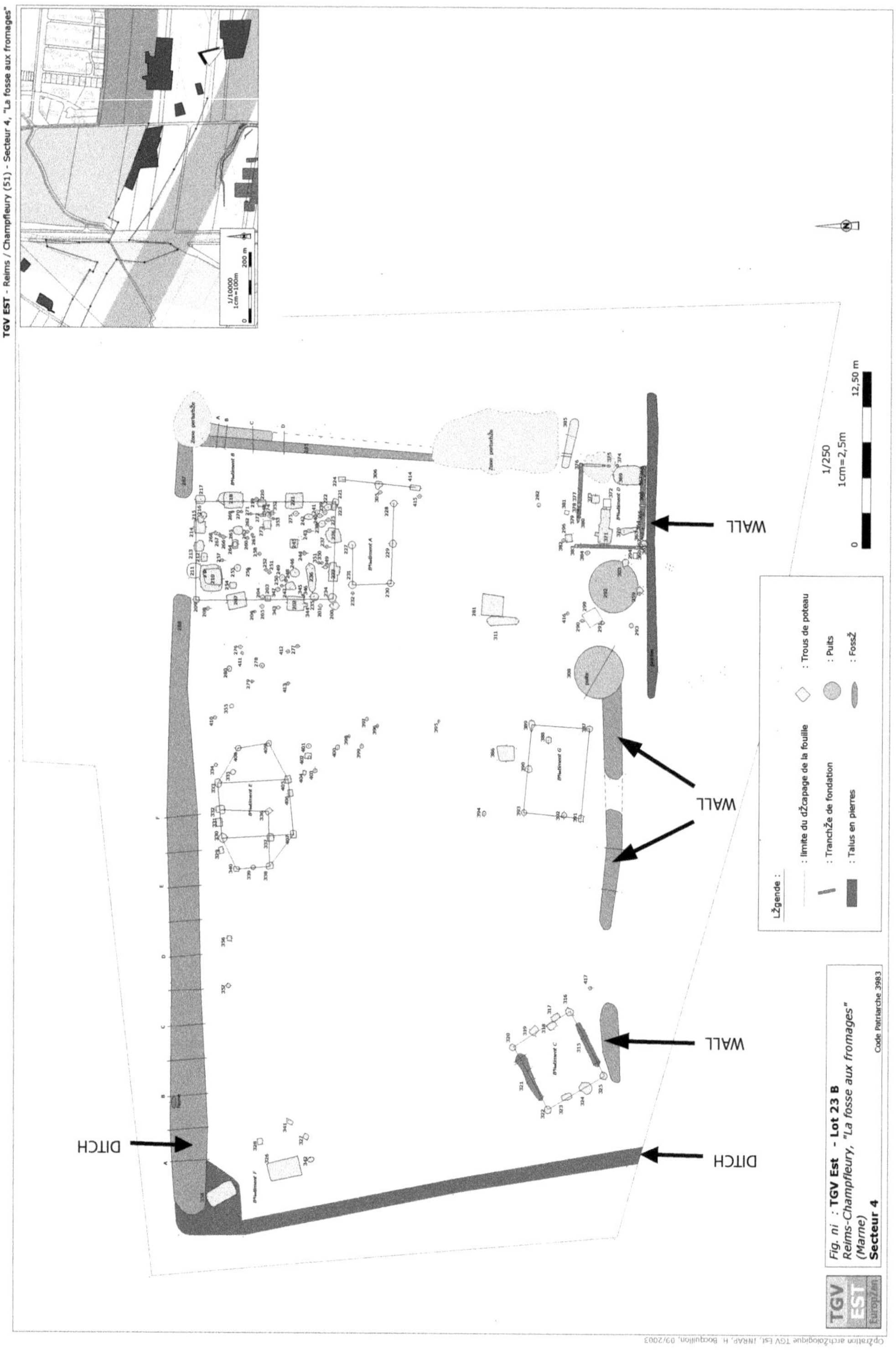


*Figure 11 Archaeological plan of the enclosure of Champfleury*

| | ***Min. and max. Ptot*** | ***Min. and max. Porg*** | ***Mean*** | ***Mean Ptotal/ Pinorg ratio*** | ***Total number of samples*** |
|---|---|---|---|---|---|
| Outside values | 442 – 1710 (2721[1]) | 9 – 740 | 865 | 1,39 | 59 |
| Inside values except buildings | 548 – 1994 | 13 – 1242 | 1270 | 1,42 | 95 |
| Inside buildings | 567 – 1863 | 7 - 596 | 1039 | 1,36 | 25 |

([1]) In the interruption of the outside ditch

*Figure 12. Results of analyses of total P of Champfleury, analytical results of the reference profile*

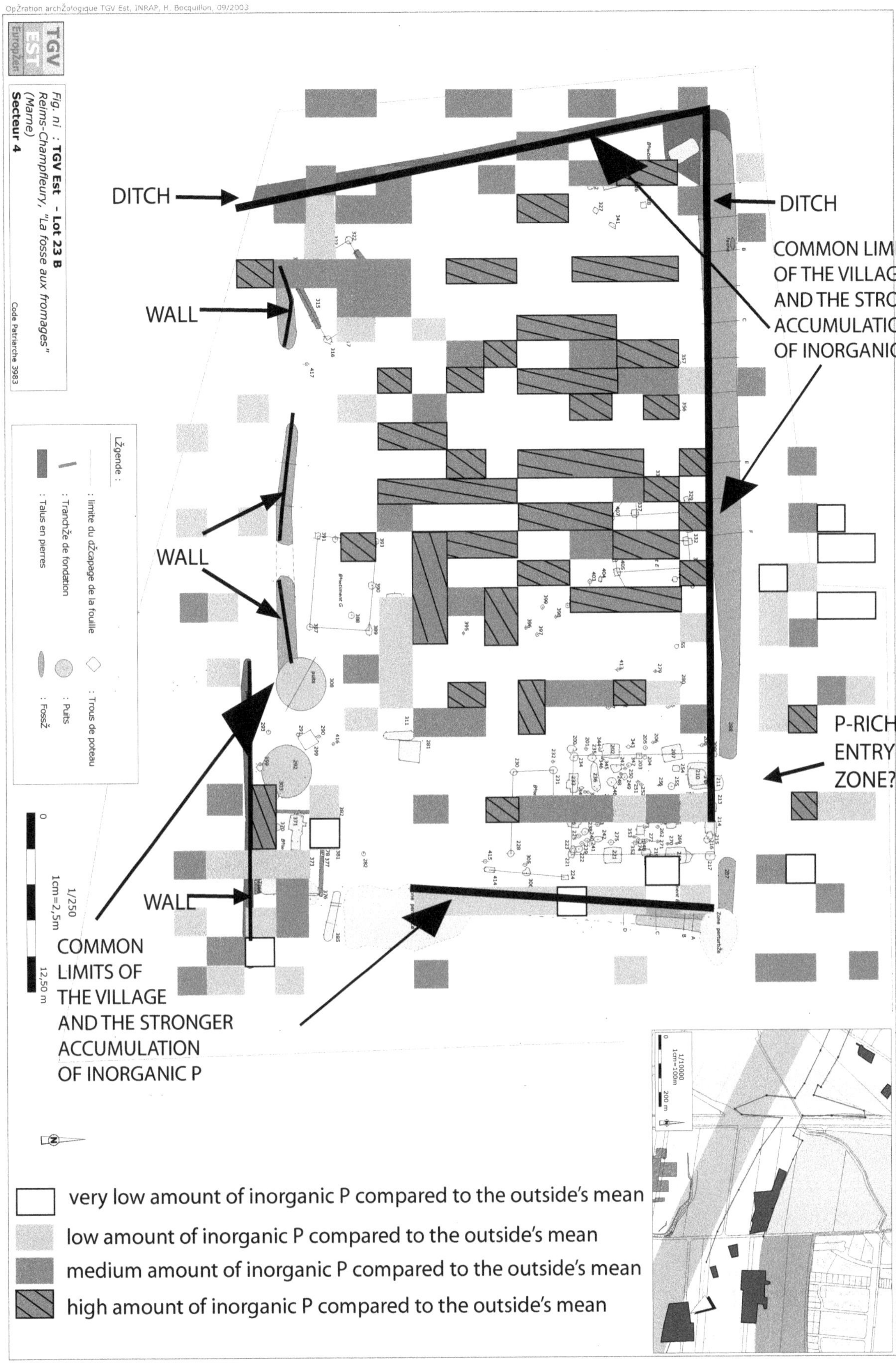


*Figure 13. Representation of the statistical treatment that takes into account the relative amounts of inorganic phosphate compared to the mean of the values outside the enclosure. It shows the contrasts with the outside of the enclosure.*

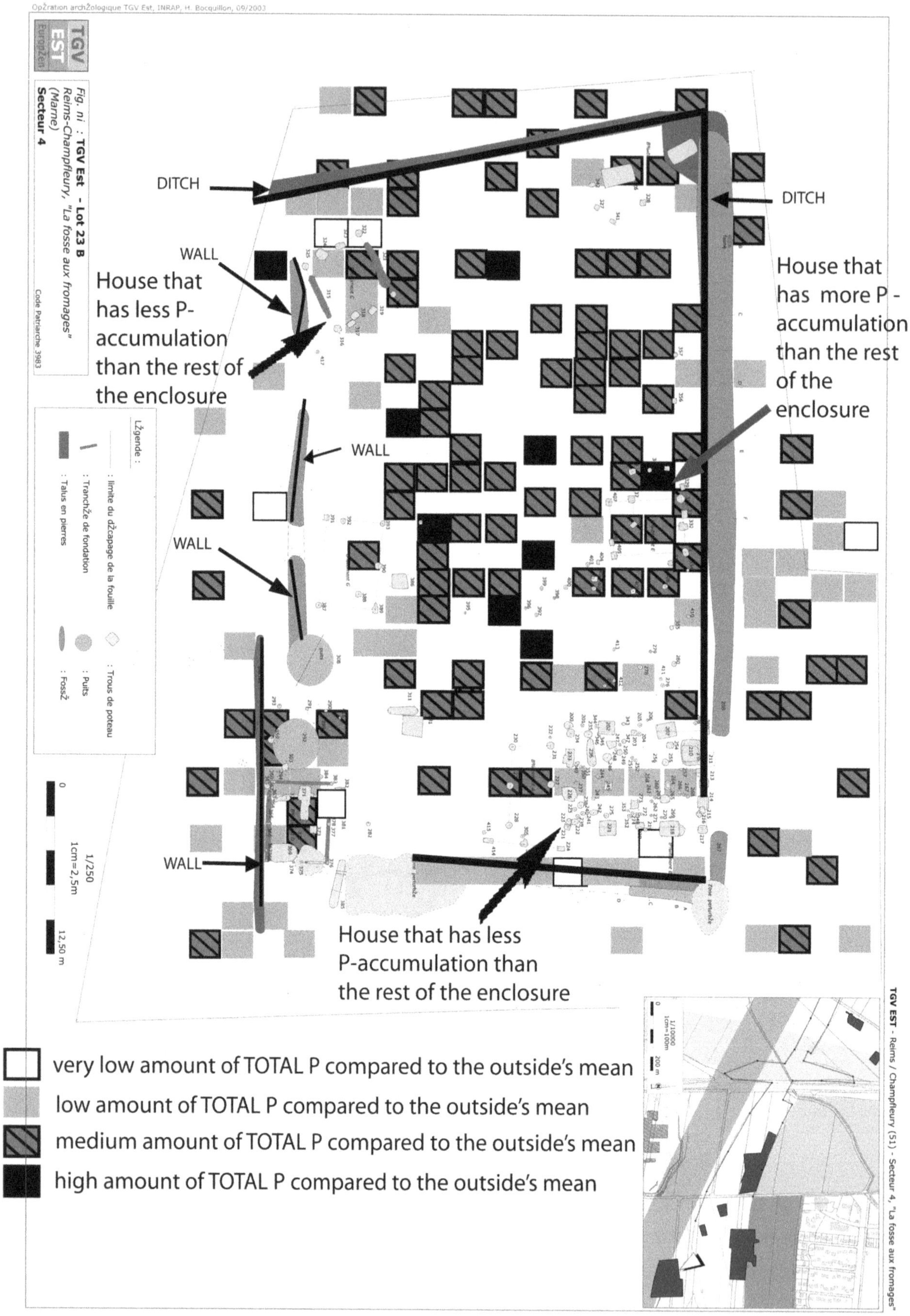


*Figure 14. Representation of the statistical treatment that takes into account the relative amounts of total phosphate compared to the mean of the values outside the enclosure. It confirms the contrasts with the outside of the enclosure and differentiates between more or less polluted houses inside the enclosure.*

The phosphorus tests show rather low contrasts, only local features as parts of the houses and of the enclosures appear to be more affected by phosphorus, as well as, locally, inside the Roman road or pathway that crosses the site (fig. 18).

The main enhancement is found locally, outside both enclosures 1 and 2, including in particular the entry zone of enclosure 1 (fig. 17). The main trend that is to be noted is the lack of phosphorus enhancement inside enclosures 1 and 2 that contrasts with the outside and local phenomena in the enclosures. The local phenomena are difficult to date as they are not correlated with any archaeological limit; they are possibly posterior or anterior spots.

Note that the general trend of enclosure 3 and its surroundings goes towards a slightly stronger phosphorus enrichment but is difficult to date because of the lack of clear correlation with archaeological limits.

The lack of coincidence with bone remains (which are not decomposed naturally on this site) points in the direction of an organic source of phosphorus, even though a map of magnetic susceptibility would have been a stronger argument (see Fechner *et al.*, in preparation).

It has to be insisted that the results of phosphorus mapping can hardly be correlated with one or another of these sub-phases (Jean Bruant, pers. comm.). Indeed, the phosphorus enhancement can be formed in relation with later reutilizations of the same surface with identical delimitation, rather than during the main phase of use. For instance, the former ditches of the enclosure might be filled and those limits permuted into simple hedges or parcel limits. In the present case, the enclosure dug in the Late Iron Age could have been enriched in phosphorus until or even only until one of the Roman periods.,

There are no archaeological indications that this enhancement would due to processes that are even later, namely posterior to the Roman occupation. Indeed, the association of the phosphorus enhancement with different sides of enclosures 1 and 2, is different from the persistence of an isolated limit as a pathway or a parcel limit which is more probable over such a long period and without clear Medieval archaeological materials.

#### Interpretation

The absence of phosphorus in the enclosures 1 and 2 indicates that no prolonged or intense past polluting activities have been associated to the inside surface, neither in Late Iron age nor in Roman (or later) times. One or the other would have marked the soil until now, as there is no soil factor here that permits the elimination or export of phosphorus (see also Devos *et al.*, in this volume). The very limited and localized enrichment in phosphorus of some of the buildings tends to confirm this.

The Late Iron-Age to Roman site of Savigny-le-Temple might be a case of an enclosure where activities associated with phosphorus enrichment where avoided and left outside the enclosed area, as opposed to other contemporary cases and this in spite of the fact that it has been used over long periods. The ditches and/or hedges could have been used to maintain cows or other mammals outside the enclosures 1 and 2. Any past polluting activities as those involving the manipulation of excrements, of large amounts of other types of organic matter, of bones (e.g. Fechner and Kleiner, 1996; Gillet, in press) have also been avoided in those two enclosures.

### *3.6. The Roman enclosure and the giant pits or "ponds" of Saint-Brice-sous-Forêt (Val d'Oise, France)*

#### 3.6.1. The enclosure of Saint-Brice-sous-Forêt (Val d'Oise, France, fig. 19-22)

The Roman enclosure of Saint-Brice-sous-Forêt "Chapelle Saint-Nicolas" (archaeologist: Vanessa Rouppert, INRAP) and its building are treated in a separate article (Rouppert *et al.*, this volume), and some pedological aspects specific to the building are found in Fechner *et al.* (in preparation). The specificity of the present text is the presentation of the quantitative and field pedological data related to the enclosure and the comparison of the results with the other studied enclosures.

The enclosure is situated on a plateau and top of slope, with soils that are developed on loess that are only decarbonated to a limited depth.

As in the site of Savigny-le-Temple (see above), the analyses of reference samples and their correlation with the horizons of a well preserved reference profile (fig. 21) show a deep homogenization of the content of phosphorus. This is also shown by the high mean of the ratios of total over inorganic phosphorus (very strong influence of organic phosphorus), which is in close correlation with the field observations of strong bioturbation and the deep homogenization of the reference profile.

This results also here in the possibility to compare the data over most of the studied surface (except were erosion reaches more than 50 cm). Another consequence is the strong lowering of both the absolute values of the analyses compared to other sites, of the intensities of the phosphorus tests and of the contrasts between the richer and the poorer areas of the mapping (see comparative table in figure 3).

The amount of organic phosphorus is still higher than elsewhere in some sectors between and outside the ditches and it is even extreme for the data coming from outside the enclosure. This might indicate a zone with initial stronger influence by organic phosphates or with recent bioturbation (as for instance the presence of a large number of deep roots), is thus difficult to interprete from the archaeological point of view.

The lack of coincidence with bone remains (which are no or little decomposed naturally on this site) points in the

| | *Min. and max. Ptot* | *Min. and max. Porg* | *Ptotal/ Pinorg ratios* | *Mean Ptotal/ Pinorg ratios (total number of samples)* |
|---|---|---|---|---|
| Outside values including entry | *154*– 349 | *102*– 208 | 230 (2) | 2,63 (n=5) |
| Entry | 332 | 178 | - | 2,16 (n=1) |
| Roman road | 333 | 210 | - | 2,71 (n=1) |
| Inside values of enclosures 1+2 (exc. buildings) | 156 - 637 | 46 – 160 | 266 (encl. 1+2) | 2,0 (n=15) |
| Inside buildings (enclos. 1+2) | 109 - 593 | 29 - 283 | 255 (encl. 1+2) | 2,35(n=70) |
| Reference profile (1) | 184 – 349 | 114 – 208 | 264 | 2,72 (n=3) |

Note that the outside values are scarce and limited to atypical situations. (1): without present-day plowing horizon. (2) based on samples of the gallery, the base of the profile and the entry

*Figure 15. Range of the results of analyses of total P of Savigny-le-Temple, analytical results of the reference profile*

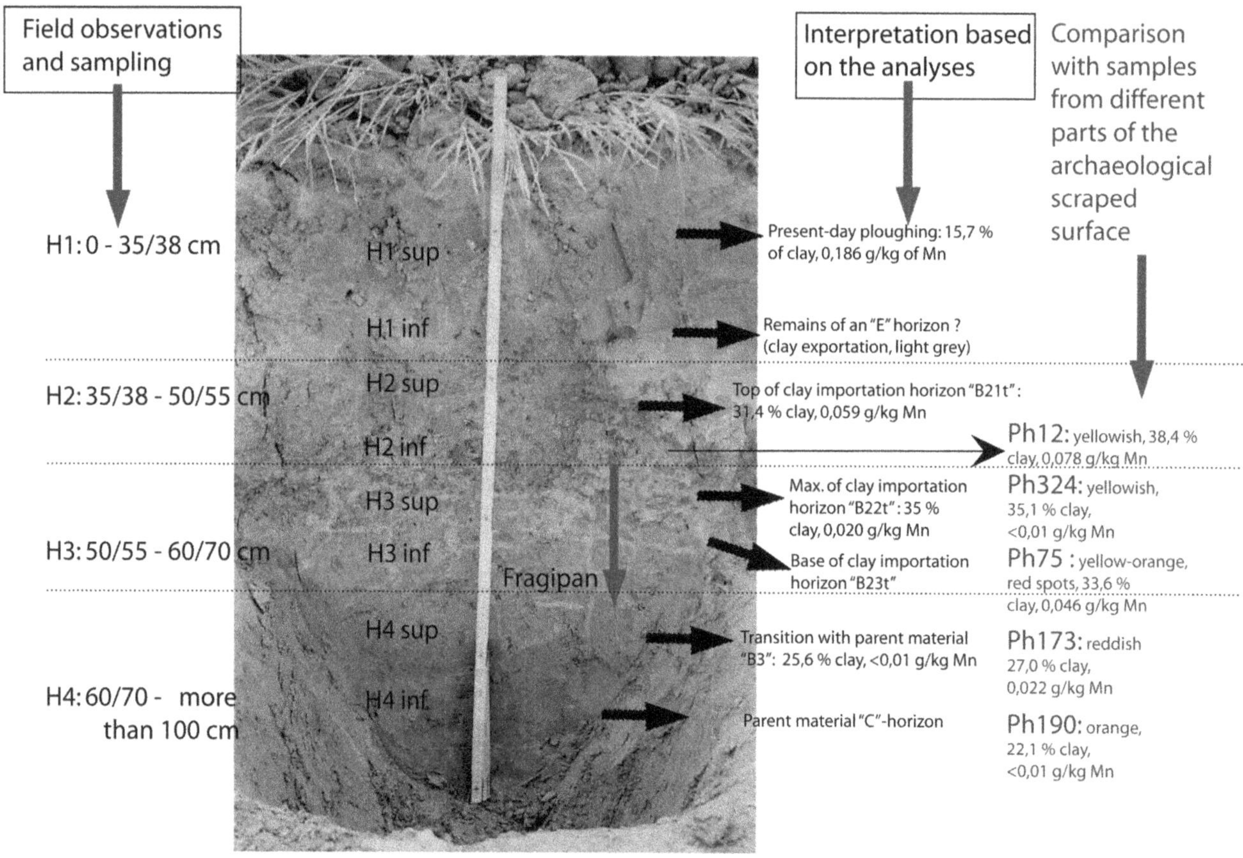


*Figure 16. Reference profile of Savigny-le-Temple that illustrates the differential erosion of the scraped surface on the site and the deep homogenization of the P content. The latter that allows to achieve good results (with low amounts of total P) in most parts.*

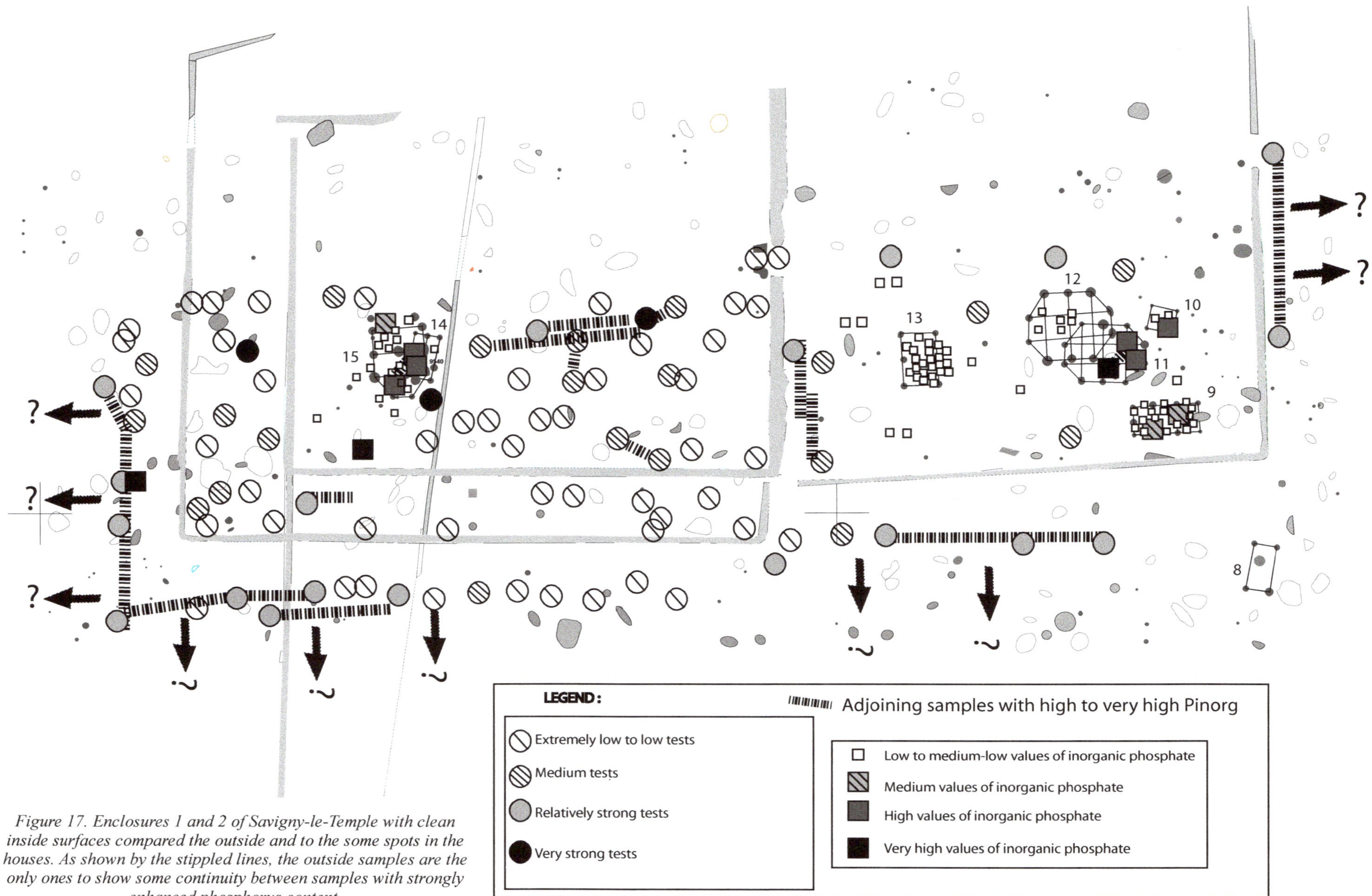


*Figure 17. Enclosures 1 and 2 of Savigny-le-Temple with clean inside surfaces compared the outside and to the some spots in the houses. As shown by the stippled lines, the outside samples are the only ones to show some continuity between samples with strongly enhanced phosphorus content.*

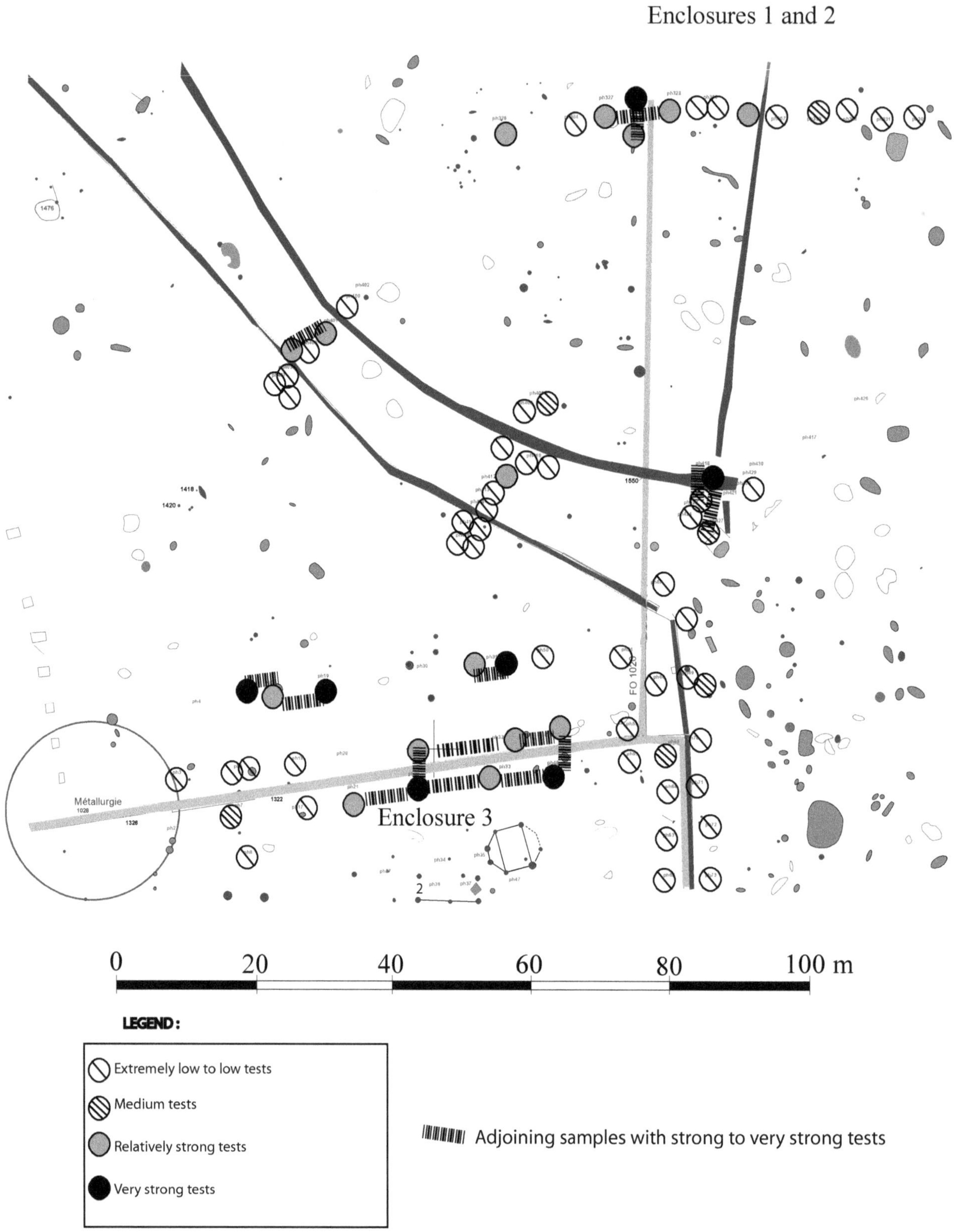


*Figure 18. Enclosure 3 of Savigny-le-Temple, without clear associations of the enhanced phosphorus content (equal tests on both sides of all limits), except for the Roman pathway and its entry.*

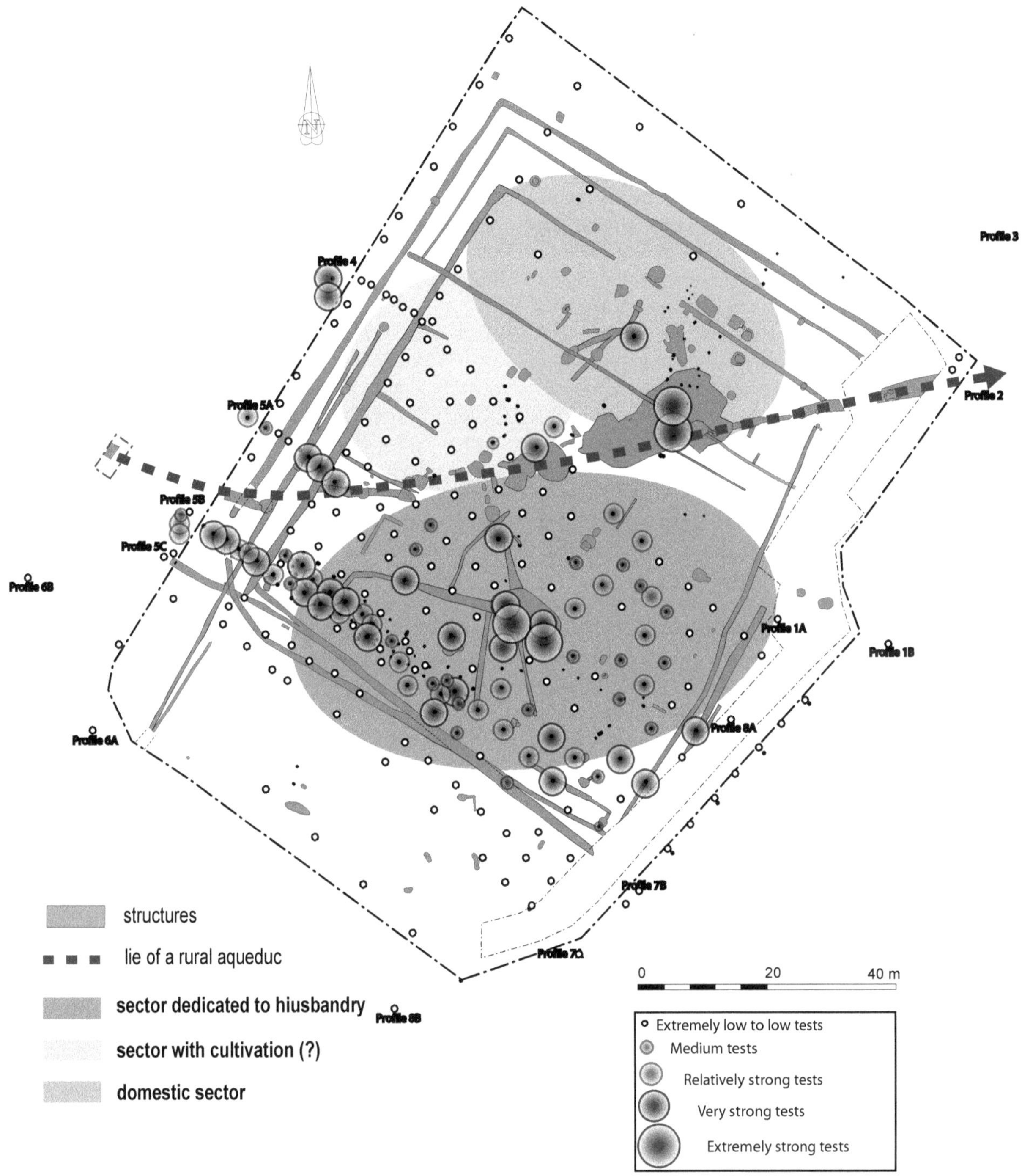


*Figure 19. Map of phosphorus tests of Saint-Brice-sous-Forêt (see also analyses in Rouppert et al., this volume: fig. 4)*

direction of an organic source of phosphorus, even though a map of magnetic susceptibility would have been a stronger argument.

Some tests for parasites in the postholes of the longhouse were negative unfortunately, due to preservation conditions.

### 3.6.2. The large pits or "ponds" of Saint-Brice-sous-Forêt (Val d'Oise, France, fig. 20; 23) and their role in the enclosure

In the enclosure of Saint-Brice-sous-Forêt "Chapelle Saint-Nicolas", two parts are distinguished by the archaeologists on the basis of both archaeological and phosphorus data (Rouppert et al., this volume). Each of these parts has produced one large pit whose stratigraphy has been studied in the field, in thin sections and by laboratory analyses. The question was whether they could be assimilated to ponds or other types of large pits. It was also asked whether the interpretations of the large pits might have a relation with the ones of the enclosure and might confirm the latter's bi-polarisation. Their initial interpretation as ponds was abandoned in both cases because of the lack of any direct and indirect indication of stagnating water. Both large pits have comparably rather high phosphorus contents, but in different stratigraphical positions: the pit 1117 that is associated to the house and connected to its supposed stable has this enrichment in the subsoil, below a much cleaner, rather recent (3rd century) fill (fig. 23). The enhanced phosphorus content is thus related to the activities that intervened before this fill up and even before the last cleaning of the original fill(s). A relation with the husbandry activities is proposed and acceptable, even if the total amount and optical signal are not spectacular. This is not due to the post-depositional homogenization of the soils in this case: in the deep pits a rather low ratio between total and inorganic phosphate is observed, unlike in the higher natural horizons used to map the enrichment in phosphorus of the enclosure. It is probably due to the cleaning of the base and the fact that the analyzed sample is well below the original pit base, in the natural soil (rather than to the limited duration of use of the pit). The fact that the analyzed layer, below the pit, did not suffer from waterlogging excludes concentrations of phosphorus as known in some septic pits, wells, ponds, etc. (see below).

The second large pit (n°1196) presents this enhanced phosphorus content in the dark fill itself, where thin sections indicate the presence of bone fragments in what seems to be a colluvial infilling originating from gardening activities (fig. 22). The enrichment is thus due to an activity that took place elsewhere, and that is compatible with the interpretation as domestic area of that part of the enclosure. The pit itself originates from quarrying and/or scraping sediment to fill back the deep trench of the supposed aqueduct that crossing the depression. Again, in this pit the ratio between total and inorganic phosphate is rather low and does not favor any lowering of the values by deep homogenization (as the one affecting the scraped surface).

#### Interpretation of the enclosure and its structures

The Roman enclosure or ditched village of Saint-Brice-sous-Forêt presents levels of phosphorus contamination that are higher than around, but also vary inside a longhouse, between the different large pits and inside the living surface of the village (Rouppert *et al.*, this volume). The southern part of the enclosure has clear associated accumulations, including below a large pit (n°1117) and in the longhouse contrast with the less enriched areas in the north. The latter include another large pit (n°1196) that was probably an extraction area for the construction of an aqueduct, possibly converted into a gardening area with soil enrichment (?) including bones in a later stage. An organic source is probable and might point to a large distribution related to excrements.

As these particular accumulations of phosphorus coincided with some architectural and chronological particularities, a stable, a pit for collecting the dung (fig. 23) and an area related to husbandry (fig. 22) are reconstructed in the eastern part of the surface of the enclosure. It is essentially the confrontation arguments that cross phosphorus and archaeological data that has permitted to go so far in the interpretation. This confrontation is essential to go beyond the mere observation of a synchronous area of enhanced phosphorus content and to argument a more precise interpretation, as has been done in some German and Dutch examples of Bronze to Middle Age houses (Behre, 1998; Hausbau in Norddeutschland, 2004; Zimmermann, 2004a; Zimmermann, 2004b).

### *3.7. The site of Cessons «Plaine du Moulin à Vent » (Seine-et-Marne, France, fig. 24-26)*

In Cessons «Plaine du Moulin à Vent», Zone Nord (archaeologist: Alexandra Mondolini, INRAP), another Roman enclosure on decarbonated loess with development of a Luvisol (natural soil with a clear clay-exportation over a clear clay-enriched horizon) has yielded comparable inner enrichments. The site is situated on a plateau and the top of a slope. The values are comparable to those of Saint-Brice-sous-Forêt and the contrast with the outside is equally evident. Again, one building has a spectacular impact in terms of phosphorus on its immediate borders, even if its inside has not been analyzed, only tested (see Fechner *et al.*, in preparation) and especially one small part

| | *Min. and max. Ptot* | *Min. and max. Porg* | *Mean* | *Mean Ptot/PE* | *Total number of samples* |
|---|---|---|---|---|---|
| Outside values | 257– 453 | 199 - 256 | 293 without ditch and deep profiles | 2,77 | n=5 |
| Inside values except building | 268 - 836 | 45 - 242 | 418 | 2,16 | n=22 |
| Inside building | 272 - 856 | 143 – 277 | 464 | 1,85 | n=22 |
| Reference profile | 318 – 914 | 208 – 256 | 385 without deep profile | 2,33 | n=2 |
| Pit n° 1196 | 1834 | 281 | - | 1,18 | 1 |
| Pit n° 1117 | 1134 - 1222 | 184 -226 | 1178 | 1,21 | 2 |

Legend: (1) In reference profile P5A. (2) In the base of the reference profile P4.

*Figure 20. Range of the results of analyses of total P in Saint-Brice-sous-Forêt, analytical results of the reference profile and in the large pits.*

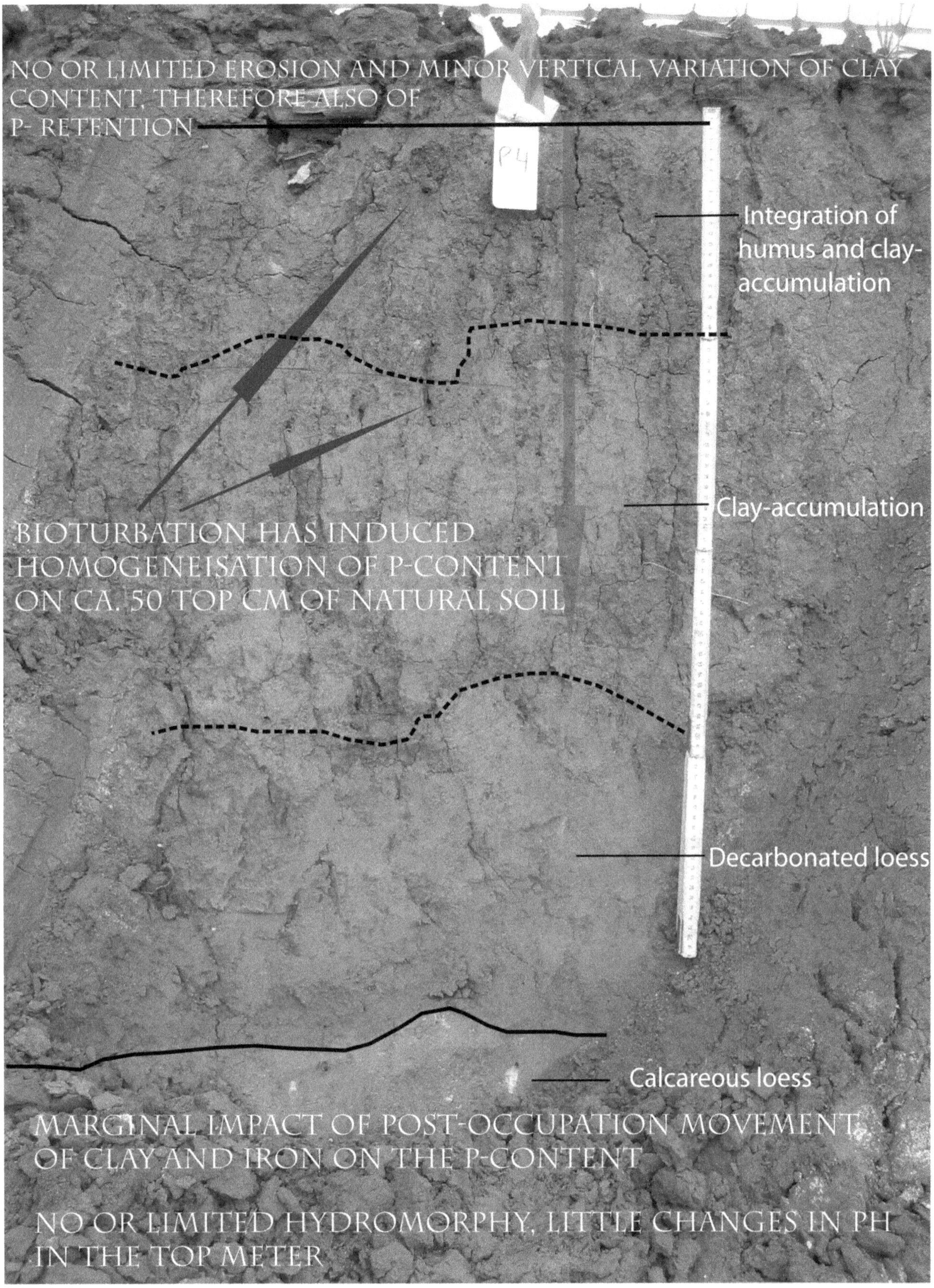


*Figure 21. Reference soil profile, illustrating some of the parameters that have been checked in Saint-Brice-sous-Forêt*

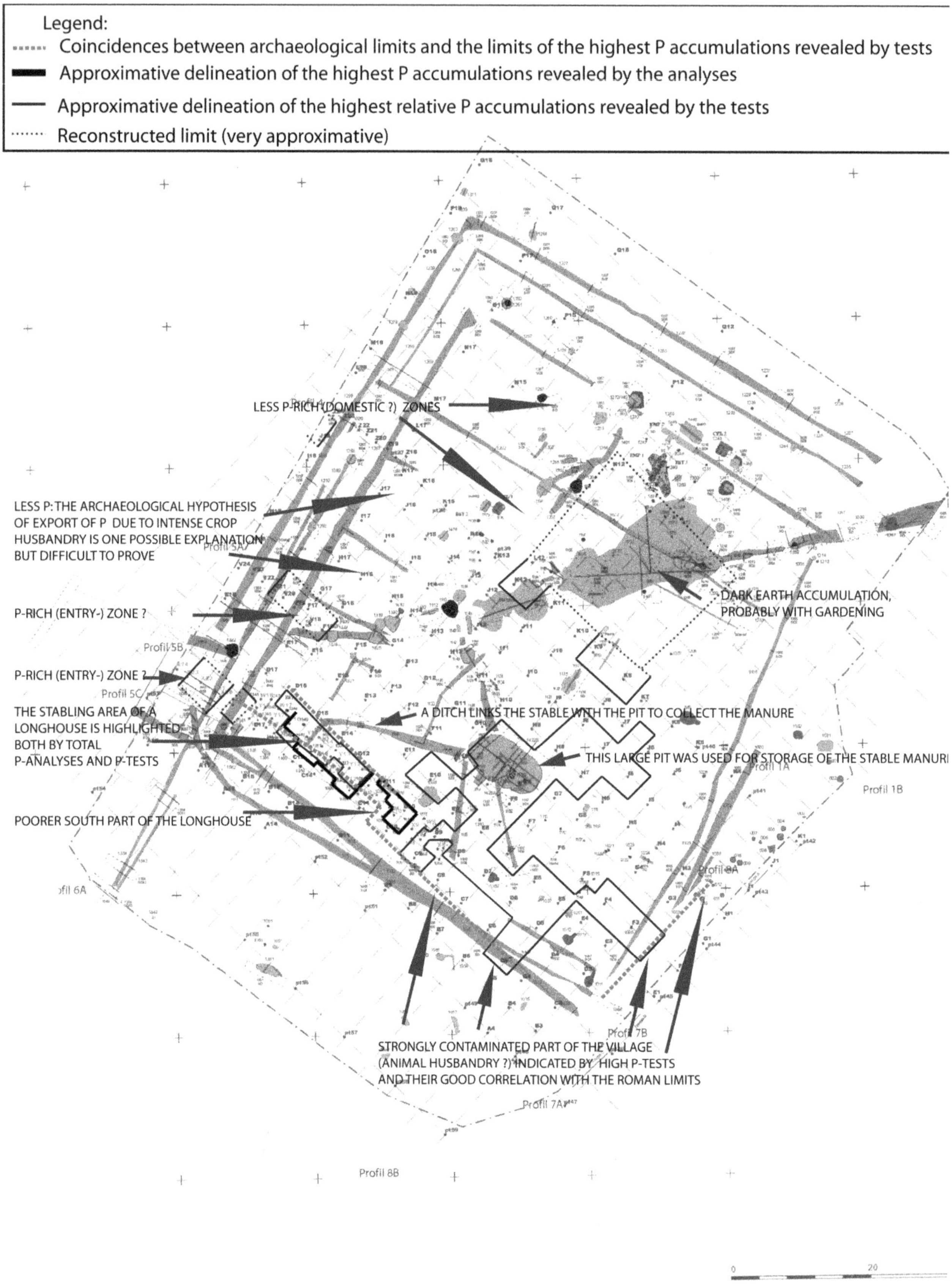


*Figure 22. General interpretation of the phosphorus-map of the Roman enclosure of Saint-Brice-sous-Forêt (Val d'Oise, France): a stable house, a pit for collecting dung and a slightly younger zone with gardening can be reconstructed (acknowledgements : Vanessa Rouppert and Jean-Yves Dufour, INRAP).*

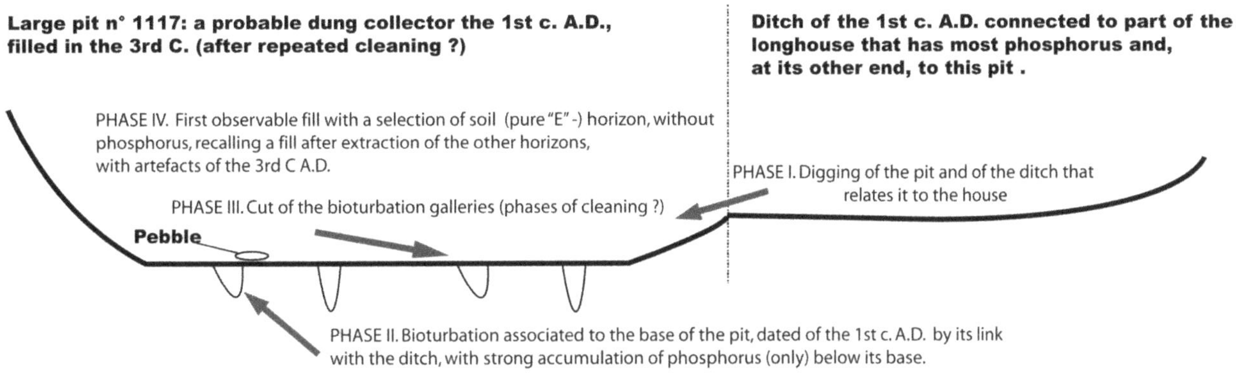


*Figure 23. Schematic representation of the stratigraphic events of the large pit n° 1117 of Saint-Brice-sous-Forêt*

shows a very strongly enhanced phosphorus content. By comparison with the borders of the building of Saint-Brice-sous-Forêt (Rouppert *et al.*, this volume; Fechner *et al.*, in preparation), this part of the building might be related to an activity related to husbandry. Here, unfortunately, no other arguments as magnetic susceptibility, parasites, architecture, associated structures are available and there is only indirect indication in favor of an organic source of that phosphorus enrichment (lack of associated bone accumulations). However, the amount of bioturbation is much less pronounced than in the sites of Savigny-le-Temple and Saint-Brice-sous-Forêt, as shown by the fact that the organic phosphate is only a little higher than elsewhere (see comparison of means of Ptot/Pinorg ratios in figure 3). This means also, that unlike in these sites, there is no deep introduction of phosphorus in the natural soil. This explains that the impact of erosion on the slight slope which shows less phosphorus enrichments. It can be noted that in Cessons «Plaine du Moulin à Vent», the Ptot/Pinorg ratios are low (less than 1,5) for the samples inside the enclosure, and that the lowest ratios concern the richest samples, situated around the larger building.

On the other side, between and outside the ditches, the mean of these ratios is clearly higher, which seems to point to some particular influence of organic phosphorus, probably to deeper homogenization by bioturbation (see above: sites of Savigny-le-Temple and Saint-Brice-sous-Forêt). As along parts of the enclosure of Saint-Brice-sous-Forêt (see above), the deeper homogenization will have lowered the absolute values of phosphorus, possibly resulting in a stronger contrast with the values inside the enclosure today than originally. This particular phenomenon could indicate a specific soil environment, as for instance the presence of a large number of deep roots (e. g. bioturbation due to trees or bushes as in Ribemont-sur-Ancre, discussed above). The fact that both Cessons and Saint-Brice-sous-Forêt show that the impact of organic phosphorus has been stronger outside the enclosure might point to specific past activities (deep rooting related to pasture land?) in relation with the enclosure rather than to later bioturbation processes. Future case-studies might confirm what has to remain a working hypothesis for the moment.

### *3.8. The site of Gonesse (Val d'Oise, France, fig. 27)*

The site of Gonesse "ZAC Entrée sud de Gonesse" (archaeologist: Thierry Caparros, INRAP) is situated in the same area as Saint-Brice-sous-Forêt, on comparable shallow loess deposits. It coincides with a top slope that also presents a shallow decarbonatation. The phosphorus tests have presented a very sharp and well associated contrast between the inside and the outside samples, limited to a east-west transect in the top, little eroded, part of the enclosure. This forms a line of samples that is on a similar topographical curve. The laboratory data explain this clear contrast: the values are extremely high (up to 9500 pm), reminding only of the pond of Lieusaint "Zac de la Pyramide" (see below and Fechner and Devos, 2006) or of the base of a septic pit in Brussels "Sainte Catherine" (Fechner, 1995). The exceptional degree of phosphorus enhancement is strongly argumented by the fact that we are outside any area of hydromorphy that allows the concentration of oxides, among others of phosphorus, unlike in these two other cases of exceptional enrichment by phosphorus. In spite of the limited amount of analyses, the investigated northern part of the enclosure of Gonesse has to be considered as a sector with very polluting activities, possibly related to bone manipulation or manufacture (see above: pollutions in Ribemont-sur-Ancre, in Blicquy in mentioned by Fechner and Kleiner, 1997; Gillet, in press and in Orp-le-Grand mentioned by Fechner *et al*, 2006). While we have not encountered elsewhere such enrichments related to animal breeding or long-term presence of dung, the questions may be raised whether the concentration and repetition of such activities could produce accumulations that recall those due to altered bone.

### *3.9. The Roman enclosure and pond of Nangis (Seine-et-Marne, France, fig. 28-29)*

The comparison site of Nangis "ZAC des Roches" (archaeologist: Nadine Mahé-Hourlier, INRAP) is a Roman enclosure on a plateau and a slight top slope with decarbonated loess affected by a well preserved Luvisol (natural soil with a clear clay-exportation over a clear clay-enriched horizon). Here, the pond in one of the enclosure's corners (fig. 28) showed no phosphorus accumulations

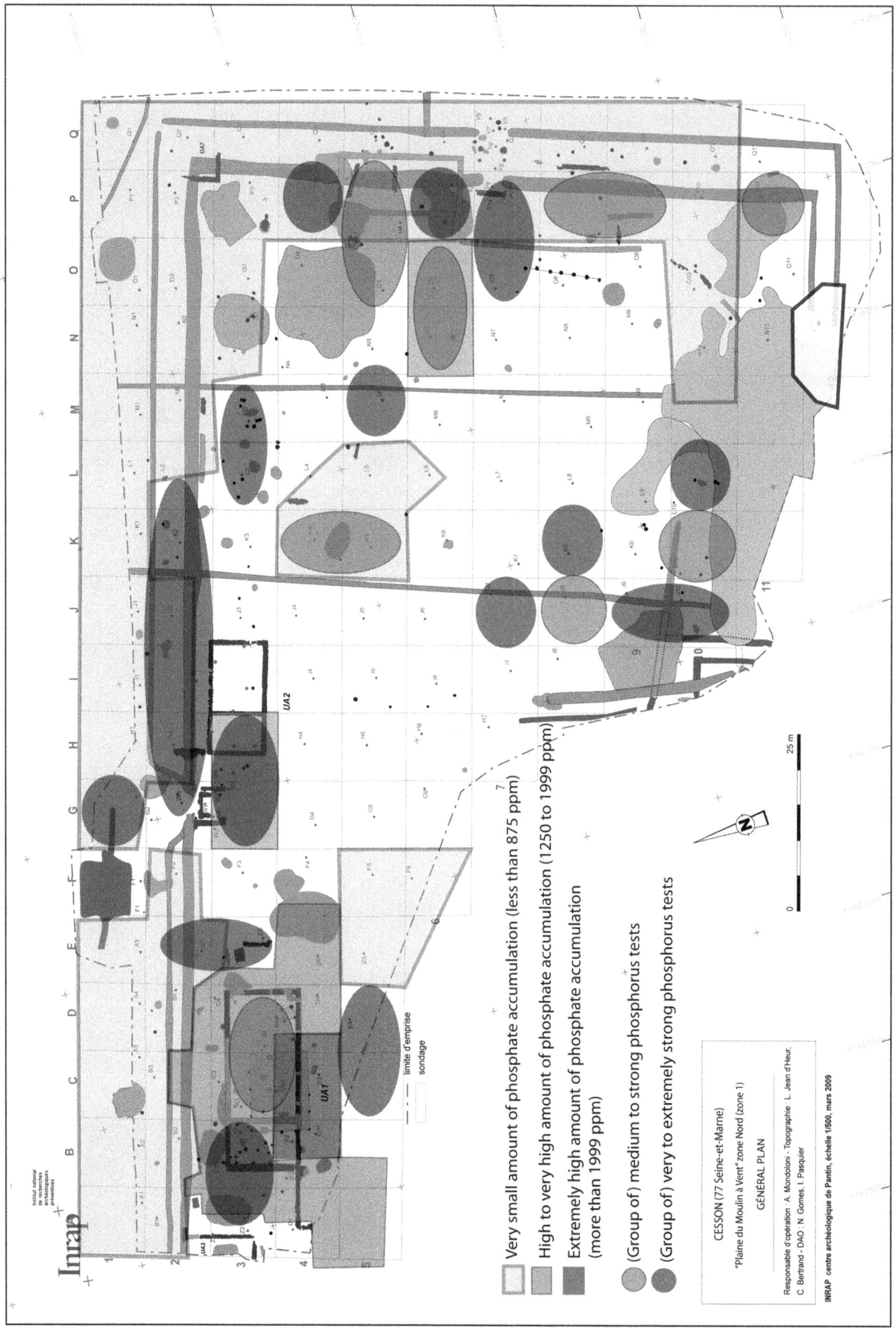


*Figure 24. Map of phosphate analyses of the Roman enclosure of Cessons «Plaine du Moulin à Vent» and confrontation with the highest phosphorus tests*

| | *Min. and max. Ptot* | *Min. and max. Porg* | *Mean* | *Mean Ptotal/ Pinorg ratio* | *Total number of samples* |
|---|---|---|---|---|---|
| Outside values | 566 – 879 | 186 – 406 | 746 | 1,74 | 15 |
| Inside values | 693 – 1473 (3444[1]) | 151 – 493 | 1133 | 1,41 | 57 |
| Values close to the main building | 1760 - 3444 | 223 – 377 | 2217 | 1,16 | 6 |
| Between the ditches | - | - | 762 | 1,85 | 15 |
| Entries in the ditch | 947 | 238 | - | 1,33 | 1 |
| Reference profile | 980 – 1168 | 239 - 253 | 1074 (2) | 1,3 (2) | 2 (2) |

(1) Highest value around the large building. (2) without the present surface horizon.

*Figure 25. Range of the results of analyses of total P of the Roman enclosure of Cessons «Plaine du Moulin à Vent», analytical results of the reference profile*

in the field, although iron accumulations and crusts were omnipresent on the scraped surface around that pond and are usually associated to phosphorus accumulations when there is any. The tests show no relative accumulation of phosphorus in the surface of the enclosure as compared to its outside (fig. 28-29). On the eastern border and at its corners, some stronger tests (in black) might even indicate a relative enrichment outside compared to inside the enclosure. The phosphate analyses show a systematic but limited enrichment of the enclosure's surface compared to its outside (fig 28: grey squares). It is very limited in amounts, reminding of a slightly stronger anthropisation without strongly polluting activities. There is no perturbation or deep homogenization to explain these low amounts, neither in the field nor according to the mean of the ratios between total and inorganic phosphate.

The large pond was completely surrounded by numerous thick and hard crusts of iron and manganese, without any visible traces of phosphorus. In the slightly more clayey base level of the pond, the total amounts of phosphate are comparable to the ones of the two large pits of Saint-Brice-sous-Forêt (see above) and below Iron Age silos of many sites on deeply decarbonated loess (Fechner *et al.*, 2006). For the pond of Nangis, there is thus rather little phosphorus, taking into account that a pond is a place of particular accumulations of oxides as iron, manganese and phosphorus by water logging. We can conclude that the pond confirms the interpretation as an enclosure without activities that involve *in situ* animal breeding or manipulation of large amounts of excrements or bones, all indicated by the absence of significant enhancement of the phosphorus content.

### *3.10. The site of Lieusaint "Zac de la Pyramide" (Seine-et-Marne, France, fig. 30-31)*

The settlement of Lieusaint "Zac de la Pyramide" (archaeologist: Alexandra Mondolini, INRAP) is treated here in order to stress the importance of the study of ponds and of parasites for the understanding of past pollution in settlements and enclosures (Fechner and Devos, 2006). The studied part of the site is only a minor part of a larger enclosure. It is surrounded by a large number of ditches, some of form the border of the enclosure. It is situated on a lower slope with a silty material with a Luvisol (natural soil with a clear clay-exportation over a clear clay-enriched horizon).

The presence of a pond can in some become an essential and central element for the functional interpretation of an occupation. In the case of Lieusaint "Zac de la Pyramide", both the pond and the adjacent ditches and pits have specific parasites and a long-term accumulation of abnormal amounts of total phosphate (more than 10.000 ppm in the main ditches and in the pond), that are visible to the eye. The extreme values can clearly be associated to the concentration of phosphorus by water logging. In the drier structures, enrichment with phosphorus never coincides with the presence of bones, although the latter are well preserved where present. This points to an organic source of phosphorus, possibly related to husbandry. The enrichment of successive layers of the largest ditch, separated by a much cleaner fill, indicates that both the Roman and Early medieval phases have known this accumulation of phosphorus.

Note that the reference profile is the only one that shows clear indications of strong influence of organic phosphate (high mean of ratios between total and inorganic phosphates), thus a lowering of the values by a strong homogenization due to bioturbation. It is noteworthy that, as in Saint-Brice-sous-Forêt, the structures have no such indicators. This is probably due to the better burying of the fillings as compared to the natural soil.

### *3.11. The Roman funeral enclosure of Antoing (Hainaut, Belgium fig. 32-33)*

In Antoing "Enclos funéraire" (archaeologist: Anne-Cécile Ghigny, Ministère de la Region wallonne), a Roman enclosed cemetery on deeply decarbonated loess with a Luvisol (natural soil with a clear clay-exportation over a clear clay-enriched horizon) was studied by the same

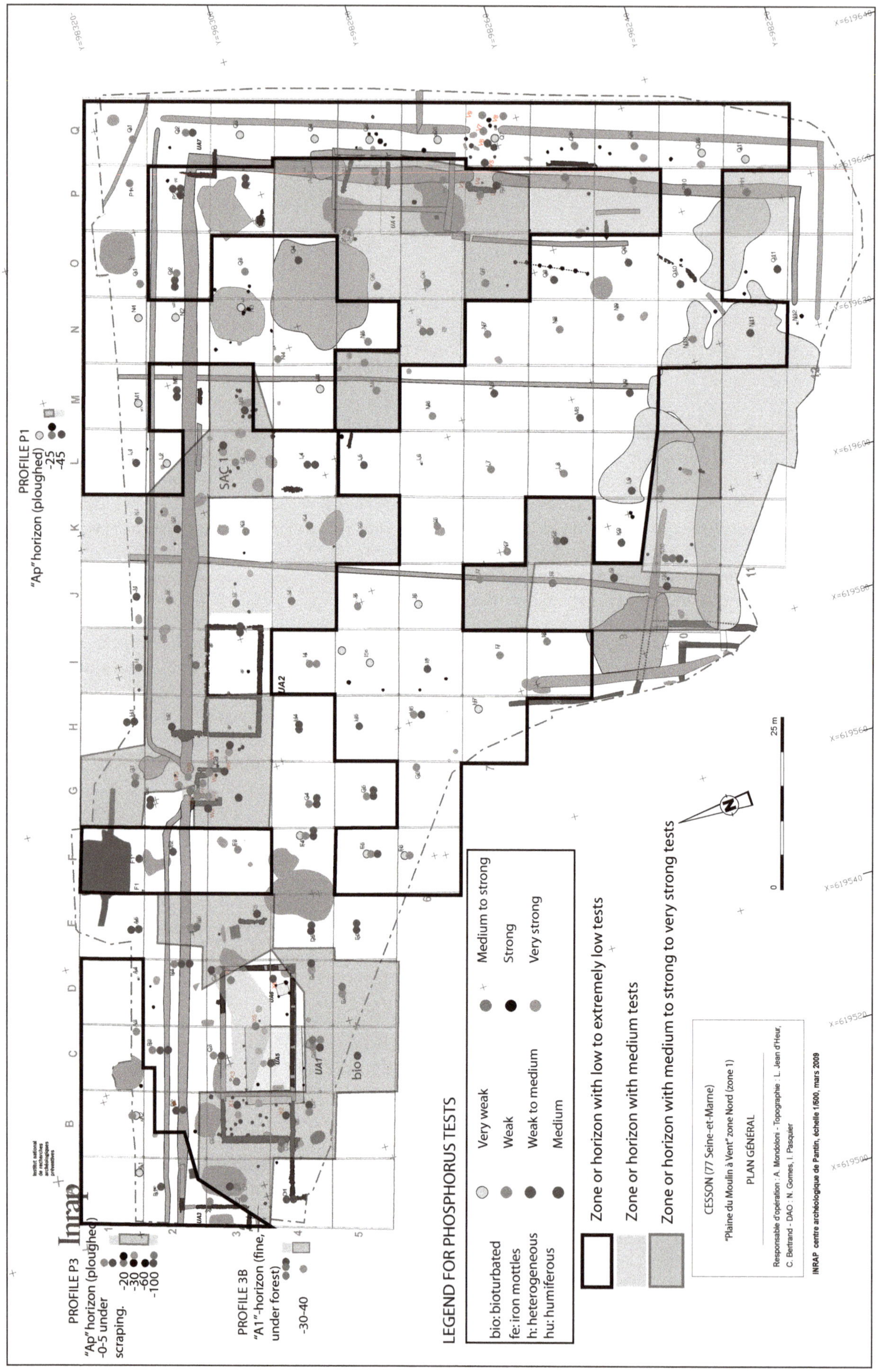


*Figure 26. Map of phosphorus tests of the Roman enclosure of Cessons «Plaine du Moulin à Vent»*

| | ***Min. and max. Ptot*** | ***Min. and max. Porg*** | ***Mean (ppm)*** | | |
|---|---|---|---|---|---|
| Outside values | 611-1073 | 225-395 | 775 | 1,76 | n=11 |
| Inside values | 994-9422 | 32-390 | 2731 | 1,19 | n=15 |

*Figure 27. Range of the results of analyses of total P in the enclosure of Gonesse*

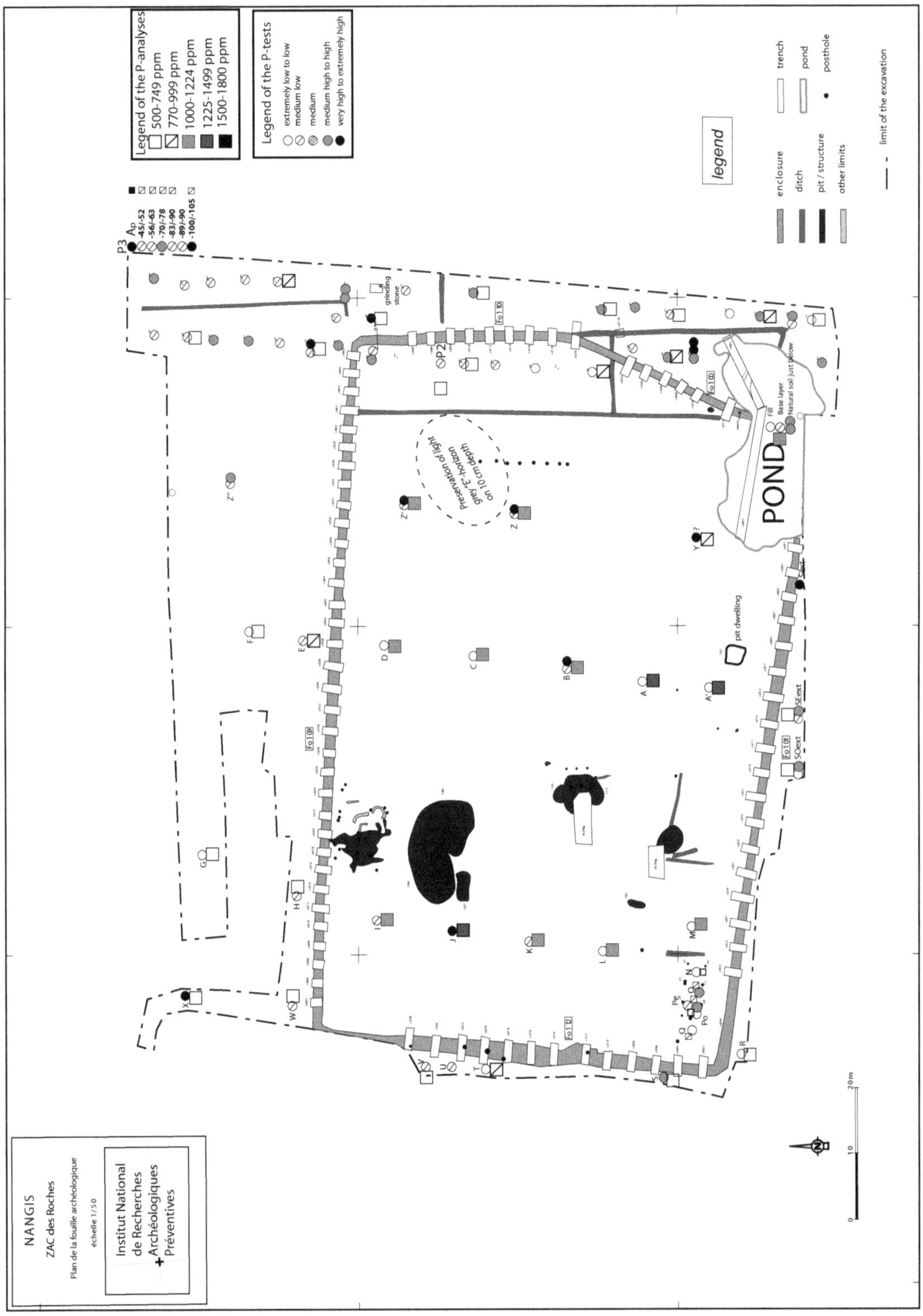


*Figure 28. Map of phosphate analyses of the Roman enclosure of Nangis*

| | *Min. and max. Ptot* | *Min. and max. Porg* | *Mean* | *Mean Ptotal/ Pinorg ratio* | *Total number of samples* |
|---|---|---|---|---|---|
| Outside values | 610 - 695 | 181 - 216 | 680 | 1,43 | n=21 |
| Inside values | 679-1499 | 97-979 | 1020 | 1,41 | n=21 |
| Pond | 1233 | 226 | - | 1,5 | N=1 |
| Reference profile | 810-900(1) | - | 848(1) | - | n=4 |

(1) Data from the laboratory of INRA in Arras

*Figure 29. Range of the results of analyses in the Roman enclosure of Nangis*

| | *Min. and max. Ptot* | *Min. and max. Porg* | *Mean* | *Mean Ptotal/ Pinorg ratio* | *Total number of samples* |
|---|---|---|---|---|---|
| Pits alone | 910-6673 | 12 – 1232 | 1808 | 1,52 | 14 |
| Pond | 141 - 7562 | 49-759 | 2004 | 1,38 | 61 |
| Large ditch | 404-9722 | 51-1591 | 3168 | 1,34 | 10 |
| All ditches | 404-9722 | 51-1591 | 3016 | 1,2 | 25 |
| Reference profile | 423-1192 | 102-503 | 829 | 1,91 | 8 |

*Figure 30. Range of the results of analyses of total P of Lieusaint "Zac de la Pyramide", analytical results of the reference profile*

methods. It is situated on a flat plateau site beside a Roman tumulus. Here the phosphate mapping is not correlated with the cemetery enclosure, except for minor spots in the corners. Textural anomalies and iron analyses might confirm the original presence of a tumulus or platform around a central megalithic tomb.

The data (ratio between total and inorganic phosphate) indicate limited bioturbation and homogenization, even if, in the field, a certain density of galleries was locally observed on the scraped surface. The quantitative data are thus not lowered by this factor.

The presence of the megalithic tomb has not induced activities related to bones (clean area), which have apparently been concentrated elsewhere, perhaps in the northern part that has not been studied by us (cremation and activity area ?). The very localized accumulation of phosphorus in the corners might be due to some evacuation of remains or later uses of the ruined site.

## 4. Discussion

### *4.1. Comparing the enclosures according to soil type*

The discussion of the individual cases indicates that the best way of comparing the qualitative results of the individual sites is to compare sites on comparable soil developments (fig. 34):

On calcareous substratum with little or no decalcification, only the site of Champfleury has been treated and shows rather high values that are very nicely correlated with some archaeological limits of the enclosure (and some of its buildings).

On calcareous loess with a few decimeters of decarbonation and a shallow Luvisol (see above), Buchères is rather poor compared to Saint-Martin-sur-le-Pré, but there is some homogenization that may have contributed to some lowering of the values.

On deeply bioturbated and decarbonated loess or silty materials with a Luvisol, Saint-Brice-sous-Forêt is clearly richer than Savigny-le-Temple, but the homogenization is responsible for low values (except or to a lesser extent in the deep pits). The contrast with the outside is important to note in Saint-Brice-sous-Forêt and in Cessons, as opposed to Savigny-le-Temple where there even a slight opposite tendency (enrichment outside the enclosure, along it). The long-term protection against the effects of phosphorus accumulations has thus concerned the outer side of the first two sites and the inner part of the enclosures 1 and 2 of the latter site. This involves a different, possibly even an opposed, functioning in relation with past polluting activities.

The enclosure of Cessons «Plaine du Moulin à Vent» is even slightly more affected by phosphorus than the enriched part of the enclosure of Saint-Brice-sous-Forêt, especially when taking into account the fact that there is some erosion in parts of Cessons as well as less lowering by bioturbation. A slightly longer or more intense activity related to phosphorus might thus be involved in Cessons.

On normally bioturbated and deeply decarbonated loess with a Luvisol, the enclosures of Ribemont and Gonesse have an outstanding enrichment by phosphorus. The fact that the first site is clearly correlated with dense bone deposits and we have only encountered elsewhere such enrichments related bones favors a correlation of the

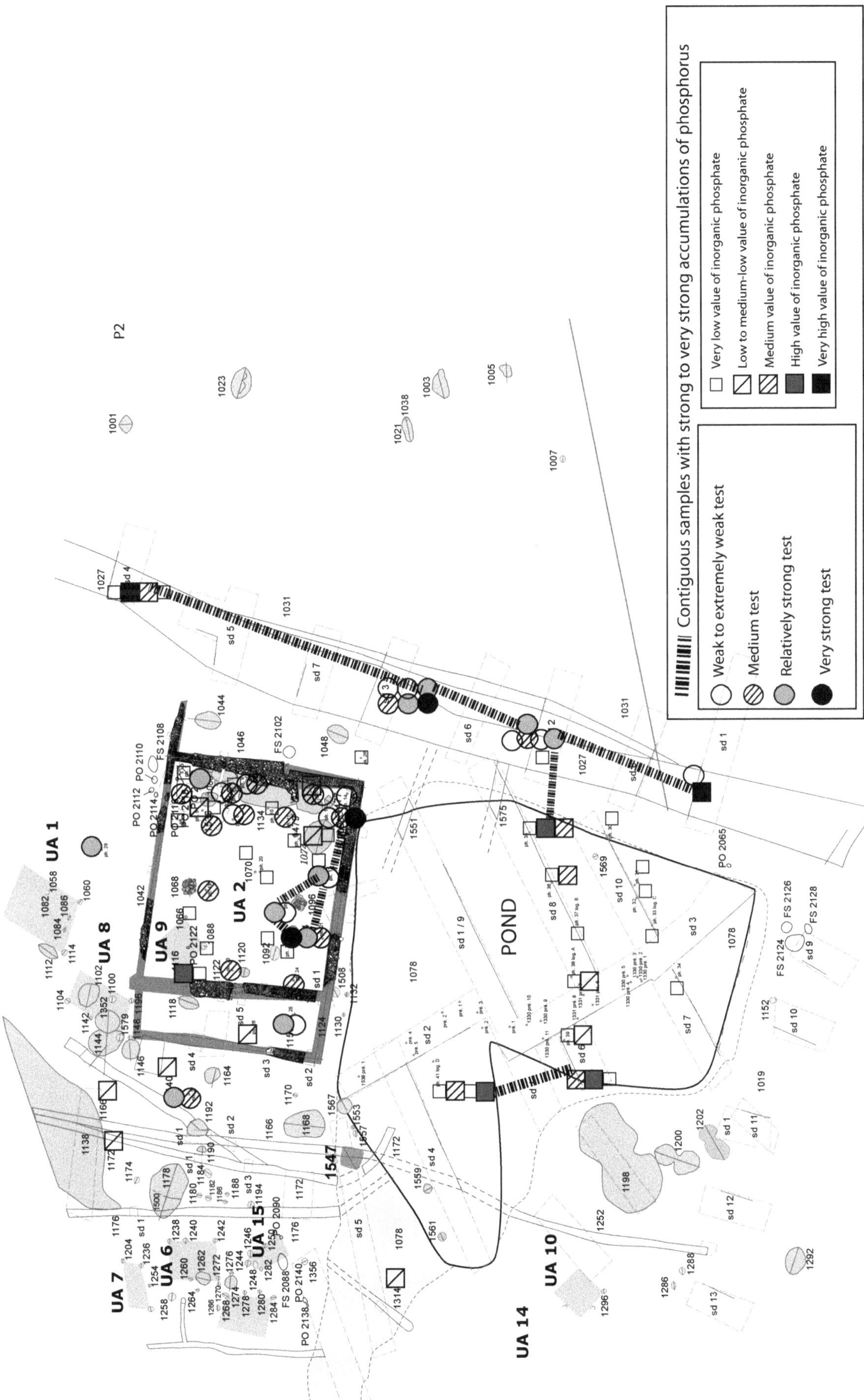


*Figure 31. Phosphorus-map of the Roman and Early Medieval enclosure of Lieusaint “Zac de la Pyramide”*

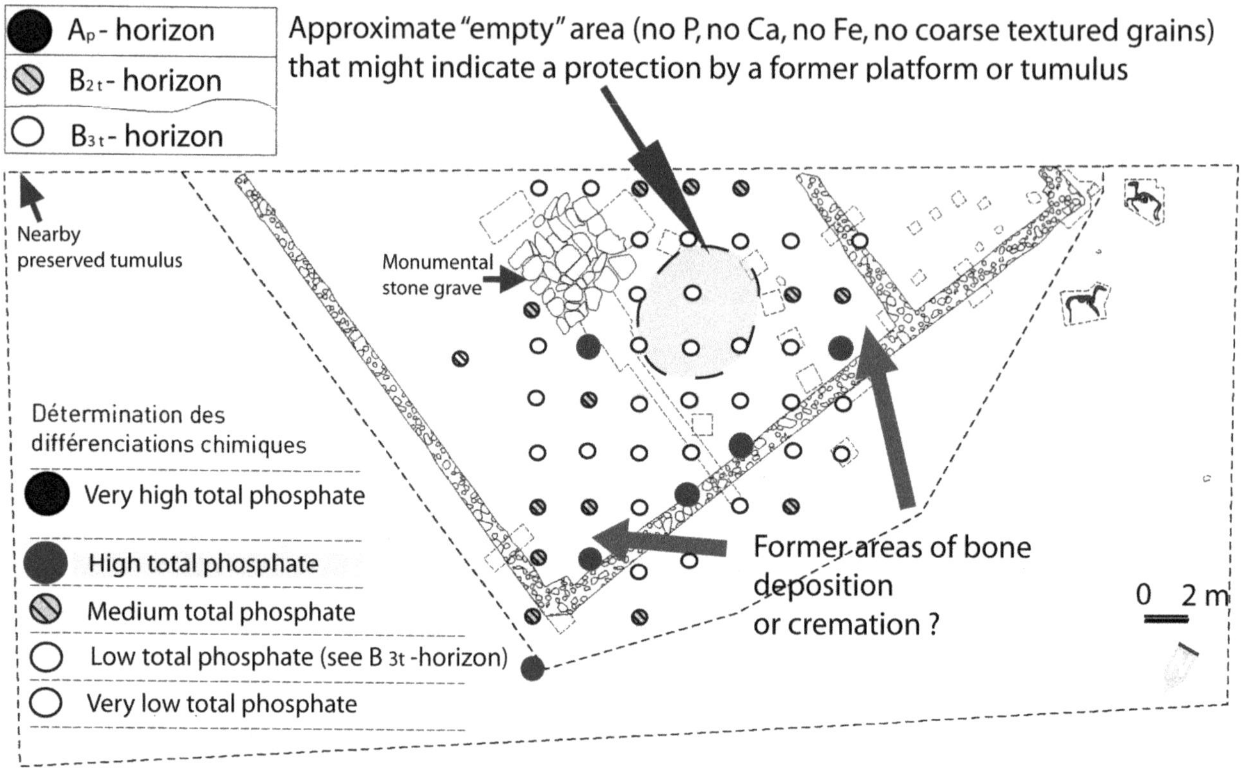


*Figure 32. Map of phosphorus-tests in the Roman funeral enclosure of Antoing. The correlation with P-mapping is less evident except for minor spots correlated with the corners of the cemetery enclosure. There might be funerary areas related to bones (cremation?) and an empty area related to the former presence of a platform or a tumulus.*

| | *Min. and max. Ptot* | *Min. and max. Porg* | *Mean* | *Mean Ptotal/ Pinorg ratio* | *Total number of samples* |
|---|---|---|---|---|---|
| Outside values | 421-863 | | 734 | 1,5 | 13 |
| Inside values | 512 – 992 | 64 – 584 | 713 | 1,5 | 35 |
| Reference profile | 421 - 1381 | 51 – 496 | 668 (1337[(1)]) | 1,13 (1,58[(1)]) | 3 (2[(1)]) |

(1) Present-day plowing layer

*Figure 33. Range of the results of analyses of total P of Antoing and analytical results of the reference profile*

second with activities related to bones. One could think of related to religious activities, as in Ribemont-sur-Ancre, or to handicaft as possibly in Blicquy. The question whether the concentration and repetition of slightly less polluting activities, as those related to organic materials could *in fine* produce accumulations that recall those due to altered bone, remains until such a case is proven.

On the same type of soil, the enclosures of Nangis and Antoing are very poor in phosphorus and have no homogenization of the soils to explain post-depostional lowering of the values, nor strong contrasts with the outside of the enclosure. They are clearly little affected by activities involving phosphorus, which allows us to situate these elsewhere. In spite of the lacking of other structures than a granary and of erosion to explain this poverty, the enclosure of Nangis would have other types of activities than animal breeding or large scale manipulation of materials related to excrements or bones. The result is even more unexpected for the enclosure of Antoing that includes a large Roman megalithic grave, is next to a tumulus and is joined to a second enclosure that is rich in incineration graves. Bones might have been expected massively on some parts of the past surface or in the soils of the studied area, which has clearly not been the case.

### *4.2. The role of ponds and large pits in enclosures*

The point of this paragraph is to discuss the possibility of differentiate between ponds related to past polluting activities and other types of ponds and large pits, contributing to the functional interpretation of their immediate environment. This applies also to enclosures in which such ponds and large pits are found, allowing the comparison with other sources of information.

| ***SITE*** | ***General range of total phosphate*** | ***Mean*** | ***Mean Ptot/ Pinorg*** | ***Number of analyses*** | ***Mean of the rich areas*** |
|---|---|---|---|---|---|
| Homogeneous loess | | | | | |
| Ribemont-sur-Ancre (F)**/*** | 1000-4000 ppm (1) | 1637 | - | 50 | 2500? |
| Antoing (B)*** | 400-1400 ppm | 751 | 1,51 | 55 | (751?) |
| Nangis (F)*** | 500-1700 ppm | 874 | 1,42 | 45 | 1020 |
| Gonesse (F)** | 600-9500 ppm | 1903 | 1,43 | 25 | 2731 |
| Cessons «Plaine du Moulin à Vent» (F)** | 550-3450 ppm | 995 | 1,52 | 107 | 1133 (inside including eroded part) 2217 (around building) |
| Very bioturbated loess | | | | | |
| Savigny-le-Temple (F)** | 109-1018 ppm | 277 | 2,34 | 110 | 256 |
| Saint-Brice-sous-Forêt (F)** | 250-1850 ppm | 504 | 1,99 | 52 | 476 |
| Calcareous loess | | | | | |
| Buchères 2006 (F)** | 600-1350 ppm | 861 | 1,74 | 35 | 871 (enclosure), 520-773 (inside houses ?) |
| Saint-Martin-sur-le–Pré (F)** | 1100-1150 ppm | 1122 | 1,56 | 3 | (1122) |
| Limestone and marl | | | | | |
| Champfleury (F)* | 400-2750 ppm | 1106 | 1,42 | 184 | 1270 (enclosure), 961 (longhouse), 1577 (rich houses), 1180 (courtyard?) |
| Dampierre-le-Château (F)* (1) | . | | | | 841-1474 |
| Large pits and ponds | | | | | |
| Saint-Brice-sous-Forêt** | | | | | 1178 (ST 1117), 1834 (ST 1196) |
| Nangis | | | | | 1233 |
| Lieusaint “Zac de la Pyramide” (F)** | 1000-10.000 ppm | 2125 | 1,39 | 109 | 3161 (large ditch), 2004 (pond ?) |

(1) Data from the laboratory of INRA in Arras

**Legend**: *Not decarbonated (except some localized pockets). **Slighly decarbonated. ***Deeply decarbonated. (1) Neolithic house on comparable soils type

*Figure 34. Comparison of the sites and data per soil type*

The question has regularly been raised whether the absence of pollution in a pond always reflects an environment without phosphorus pollution in the immediate neighborhood (A. Dietrich, pers. comm.). Could it in some cases be due to the export of phosphorus by the water-table, by a later anthropic cleaning of the layers or even be due to the limitation of the pollution to the borders of the pond, without affecting its filling ?

Here, the importance of checking, in the field and by analyses, the vertical and lateral movements of hydroxydes including phosphorus in waterlogged conditions, has been noted as it modifies the understanding of the structure and of its specific chemical and anthropic environment (see above). These movements explain the concentration and crusts of oxides around the ponds of Nangis and Lieusaint “Zac de la Pyramide” and around supposed retting pits (Fechner and Sartiaux, 2000; Ingels, in press). Another case of a pond or retting pit with strong accumulation of phosphorus has been studied in the medieval village of Vitry-sur-Orne, in a more clayey environment (David and Fechner, in preparation; Blaising and Gerard, in preparation), where both waterlogging and activities involving phosphorus have been active. In a comparison site of Villers–le-Tourneur, a large marl dissolution pocket (*dolina*) was used as a pond or drinking place in Roman times (Louwagie and Langohr, 1999): here the soil compression by drinking animals and subsequent water-saturation allowed the phosphorus to move out completely of the occupation layer and into higher and lower horizons, together with other hydroxides. In

Villers–le-Tourneur, as in Lieusaint "Zac de la Pyramide", the phosphorus was found in another place than where it had been formed and it got particularly concentrated and very apparent.

There was no such accumulation of phosphorus in and around the pond of Nangis which is confirmed by the field observations (crusts are only constituted of orange and brown oxides). Moreover, there, only 2000 ppm were reached in spite of concentration of hydroxides due to waterlogged conditions, compared to up to 10.000 ppm in the pond and large ditch of Lieusaint. For these reasons, the phosphorus enhancement in the pond of Nangis has to be considered as limited and reveals an environment without husbandry as indicted by the rest of the data on the enclosure.

The examples studied by pedology show that pond fillings reflect their immediate environment when sufficient field and laboratory data are collected in the fill and in the neighboring sediment. This encourages their future use as indicators of the type of human environment.

In conditions without water logging, on the same type of soil, lower amounts of phosphorus will mostly be found. As a consequence, when equivalent amounts of total phosphorus are found there, they are to be considered as exceptional and related to particular conditions. This is the case of the basal layer or underlying contact of grain storage pits (Fechner *et al.*, 2006) or in the two large pits of Saint-Brice-sous-Forêt. One of the latter was probably rather clean originally and has a post-depositional phosphorus accumulation due to later use with addition of bone fragments. The other large, on the contrary, was probably used in very polluting conditions, before being emptied and finally filled with a clean sediment. Note that in all these cases, no visible traces of phosphorus appear, as it has not been concentrated by water logging.

#### *4.3. Comparing the enclosures from the functional point of view*

Once the factors of soil formation that influence the quantitative results are taken into account, some functional comparisons of the different enclosures can be proposed taking into account the period, type and amount of phosphorus enrichment (fig. 35).

The Late La Tène enclosure as Buchères (southern Champaign) and Late La Tène to Roman enclosures 1 and 2 of Savigny-le-Temple (south of Paris) show the lacking of any important enhancement of the phosphorus content during their (successive) occupations as opposed to their outside borders. Although they are comparable from an archaeological point of view, these two cases are in a strong opposition with the contemporary enclosure of Champfleury, in Northern Champaign which is strongly affected by phosphorus. In particular, both the results in the enclosures and the houses show clear oppositions between the cases of Champfleury and Savigny-le-Temple. While there is a physical protection against phosphorus enrichment of the inside of the enclosure in Savigny-le-Temple (ditches, hedges against cows or other mammals, lack of past polluting activities as those with manipulation of excrements, of large amounts of other types of organic matter, of bones,...), we observe on the contrary, a an relative protection from phosphorus enrichment along the outside of the enclosure in Champfleury. Both have indications of an occupation that spans from the Late Iron Age to Roman times. The analytical values of Savigny-le-Temple are so much lower than the ones of Champfleury that an opposition is very probable in this respect too, even when taking into account very different preservation conditions.

The possible implications related to economic differences are yet to be founded on more case studies and multi-proxy approaches. In particular, the preservation of humiferous horizons in sites like Champfleury would allow tentative parasitological and palynological approaches.

In the Late La Tène sacred enclosure of Ribemont-sur-Ancre, the past manipulation of bones can be mapped and opposed to a sector related to what might be a an artificial "sacred forest", revealed by a well delimited area of discontinuous bush vegetation.

The correlation with archaeological data is essential in the Roman enclosure of Saint-Brice-sous-Forêt, where archaeological arguments coincide with pedological ones to propose a large sector of the enclosure (south-western half) related to husbandry and opposed to the other more domestic part. Here, as in Savigny-le-Temple, absolute values are strongly lowered because of a deep and strong post-depositional homogenization by bioturbation of the top horizons of the natural soil and thus of the phosphorus content. The Roman enclosure of Cessons «Plaine du Moulin à Vent» has similar amounts of phosphorus enhancement with some bioturbation; the small lateral part of the main building and its immediate surroundings are related to even stronger polluting activities, possibly related to the later use of the ruin. This site might have similar use as the husbandry sector of Saint-Brice-sous-Forêt, but the final interpretation of the source of phosphorus has yet to be confirmed by the crossing with other results.

The Roman enclosure of Gonesse shows repeated extreme values that indicate an abnormal phosphorus enrichment in spite of well drained soil conditions. The comparison with Ribemont-sur-Ancre and with other sites with bones indicates that the pollution is probably related to activities related to bones (see Gillet, in press; Fechner and Kleiner, 1996: site of Blicquy; Fechner *et al.*, 2006: Orp-le-Grand) rather than to husbandry.

In the site of Lieusaint "ZAC de la Pyramide", we also have comparable extreme values but they are not caused by bones, rather by water logging which concentrates oxides including phosphates. This is indicated by parasitology in and around a large characteristic pond and large water-

| *Periods* | *Use(s) with (strong) phosphorus accumulation* | *Use(s) without (strong) phosphorus accumulation* |
|---|---|---|
| LATE IRON AGE | - | Buchères |
| IRON AGE-EARLY ROMAN | **Ribemont-sur-Ancre** | - |
| IRON AGE-ROMAN | Champfleury | Savigny-le-Temple |
| ROMAN | Saint-Martin-sur-le-Pré (?)<br>**Gonesse**, Saint-Brice-sous-Forêt, Cessons «Plaine du Moulin à Vent» | Nangis, Antoing |
| ROMAN-EARLY MEDIEVAL | Lieusaint "Zac de la Pyramide" | |

*Figure 35. Comparison between phosphorus enhancements of the different enclosures, organized by periods (**exceptional phosphorus accumulations**).*

collecting ditches used from Roman to Early Medieval times.

On the other hand, the Roman enclosure and pond of Nangis have yielded arguments not only against past pollution of the enclosure (including the lack of post-depositional lowering of the absolute values), but also against more localized phosphorus enhancement in the main, very few, archaeological structures, as the pond. The Roman funeral enclosure of Antoing also lacks clear and large-scale enrichment by phosphorus.

In what might appear as a homogeneous archaeological feature, a contrast between three sets of Roman enclosures appears: Gonesse might show a very strong past pollution related to bones, the eastern part of the village of Saint-Brice-sous-Forêt and maybe Cessons reveal a phosphorus accumulation related to organic remains including animal dung, finally Nangis is an enclosure with none of the two types of pollution, reminding of the western part of Saint-Brice-sous-Forêt.

In some cases, the enrichment in phosphorus in front of an entry zone could be due to anterior or posterior influences, as it is rather localized. However, it appears to be a recurrent phenomenon: it is found close to the entries of the enclosures of Cessons «Plaine du Moulin à Vent» and Savigny-le-Temple, to the probable entries of the enclosure of Saint-Brice-sous-Forêt and Champfleury.

Concerning phosphorus enrichment of more limited areas as houses and yards, we refer to a future article (Fechner *et al.,* in preparation).

## 5. Synthesis

Two types of conclusions are proposed, the first concerning the results, the second the aspects of methodology related to enclosures.

Even after selection of the most compatible sites according to Devos *et al.* (this volume), the absolute data appear to be very sensible to local factors as parent material, soil development and amount of post-depositional bioturbation. The latter is often reflected by the ratio between total and inorganic phosphates. More than the direct comparison of the absolute values between sites, it is the contrast between well delimited enclosed zones and their outside that reveals convincing results. Taking this into account, we can propose another approach to the comparison of the enclosures, which is a functional one:

Organized by periods, this provides the following working hypotheses that will have to be confirmed by future works and multi-proxy approaches.

The late Iron Age to Roman enclosure of Ribemont-sur-Ancre and possibly the Roman enclosure of Gonesse would be strongly affected by bones, the first of which being a sacral enclosure. The phosphorus mapping of the first enclosure contributes to its division into a southern sector with bone manipulation, another with little or less such bones. It is dominated by other roles, possibly including a "sacred wood". For the second site, the activity related to bones comes as a surprise and has to be further thought about during the ongoing archaeological study.

Other enclosures, as the late Iron Age to Roman enclosure of Champfleury and the Roman enclosures of Saint-Brice-sous-Forêt and possibly Cessons and Saint-Martin-sur-le-Pré, might be related to husbandry. In the second of these sites, this type of activity is confirmed by other sources of information (antique texts, house plans, relation with other characteristic archeological structures) and only associated to half of the enclosure. Both here and in Cessons, there are hints in favor of a specific use of the outside of the enclosure (extensive pasture?) which have to be further investigated.

These cases are opposed to others that have been lacking any activities with strong phosphorus enhancements, as the late Iron Age enclosure of Buchères, the late Iron Age to Roman enclosures 1 and 2 of Savigny-le-Temple and the Roman enclosure of Nangis, including its large pond, as well as Antoing that is a funeral enclosure. In these cases, even on the longer term, activities involving cows or other mammals, manipulation of excrements, of large amounts of other types of organic matter, of bones have been maintained beyond the limits of the enclosure marked by the ditches and possibly hedges, when and if the site was maintained afterwards.

It is also useful to summarize some methodological aspects specific to the phosphorus mapping of enclosures (see also Devos *et al.*, this volume, for more details). The case-studies permit to allow to the following recommendations. In the future, sampling should be privileged:

1. on flat or near to flat, uneroded surface, without any present or past shallow water-table.
2. before any contamination by fresh vegetation and any animals, as birds and rabbits.
3. before scraping the surface and subsurface horizons away for a better perception of hollow structures, which is particularly dramatic on very alkaline soils and on acid soils without deep homogenization of the phosphorus content by animal and plant galleries (see also Fechner *et al.*, in preparation).
4. with the presence, at least punctually and initially, of a soil scientist in order to secure optimal field recordings and controls of the soil parameters of feasibility (Devos *et al.*, this volume).
5. using a reference profile(s) with at least a minimum of adapted physico-chemical analyses.
6. using sufficient samples in and around the mapped areas.
7. using a treatment of the data by simple statistical tests using the total, inorganic and organic fractions of the P.
8. with the awareness of the problems of comparability and reproductivity of phosphorus results between two labs or sessions of analyses.
9. with a preference for the phosphate method using sulphuric acid ($H_2SO_4$, see Mikkelsen, 1997), in order to avoid to liberate phosphorus inside the silicate structures of the mineral fraction, liberated by stronger acids like hydrogen floride (HF).
10. in combination with phytolith, pollen and parasite studies in case of preservation of large buried humiferous surfaces.

Mapping of large homogeneous surfaces can be completed by phosphorus-tests (Eidt, 1985), less precise and less easy to interpret, if the budget does not allow a complete set of analyses.

On future sites, samples should in any case be done and kept if the such favorable conditions are confirmed by a soil scientist and in order to avoid missed occasions as the enclosures of Kontich (Fechner and Langohr, 1995), Aalter (Langohr and Fechner, 1993), Enghien (Fechner, 2005) and Saint-Martin-sur-le-Pré (see above) and to multiply the possibilities to compare enclosures between each other.

## 6. Perspectives

Thanks to the strict selection of sites with compatible soil formation factors (Devos *et al.*, this volume), analytical results and phosphorus tests have yielded good results, that is, accumulations of phosphorus that have a repeated and convincing correlation with archaeological limits. This is true on very different types of soils, including very calcareous clays, limestone and loess, decarbonated loess and clays, acid sands (e.g. Roman stables in Flanders), as described in different articles (Fechner *et al.*, in preparation; Langohr *et al.*, 2004; Mikkelsen *et al.*, 2003). Soil-controlled phosphorus-mappings are a useful tool to contribute to a first proposal to oppose apparently similar enclosures and areas inside some of them. They allow to compare objectively between different sites from this point of view, and to treat large amounts of sites including the numerous rescue excavations. Rather than to propose a definitive function on the basis of our results, the presented case-studies are instruments in the hands of archaeological and historical studies that involve a larger, multi-proxy, approach. Correlations with historical, archaeological and other naturalistic data are essential in order to cross arguments in favor of the proposed hypotheses, as was possible in some isolated cases in this study (e.g. Saint-Brice-sous-Forêt) and in other studies (e.g. correlations with archaeological and palynological arguments in Behre, 1988). For the future, we also insist on the fact that more studies are needed inside each region and type of soil than the few we can compare presently.

## 7. Acknowledgements

This article would not have been possible without the helping hands of Viviane Clavel and Gisela Fechner. Our sincere acknowledgements also go to the archaeologists, the technicians and the persons responsible of the concerned sites, with special mention of Hervé Guy (INRAP Centre-Ile-de-France), Hélène Remy, Martine Soumoy and Heike Fock (Service public de Wallonie), Jan Vanmoerkerke (Direction régionale des Affaires culturelles de Champagne-Ardenne). We have also benefited from the competent advice on the oral contribution in Krakow and on this paper by Yannick Devos, Dr. Matthias Leopold, Dr. Helen Lewis and Dr. Cecilia Courbot.

## 8. References

Baize, D., 1988. *Guide des analyses courantes en pédologie*, Paris.

Behre, K.-E., 1998. Landwirtschaftliche Entwicklungslinien und die Veränderung der Kulturlandschaft in der Bronzezeit Europas. In: Hänsel, B. (Hrsg.), *Mensch und Umwelt in der Bronzezeit Europas*, Kiel: 91-109.

Blaising, J.M. and Gerard, F., in preparation. Interest of the archaeological study of « akkerbergs » for recognizing Early Medieval rural settlements in Lorraine (France, VIIIe-XXe centuries). In: Fechner, K., Devos, Y., Leopold, M. and Völkel, J. (eds.), *Archaeology, soil- and life sciences applied to houses, gardens and courtyards. Second volume of the proceedings of the session 'From microprobe to spatial analysis – Enclosed and buried surfaces as key sources in Archaeology and Pedology' organised at the European Association of Archaeologists, 12th Annual Meeting*, British Archaeological Reports. International Series.

David, C. and Fechner, K., in preparation. Man-made « dark earth » and palaeosurface horizons in France: case studies based on geophysical mapping and other soil approaches. In:

Fechner, K., Devos, Y., Leopold, M. and Völkel, J. (eds.), *Archaeology, soil- and life sciences applied to houses, gardens and courtyards. Second volume of the proceedings of the session 'From microprobe to spatial analysis – Enclosed and buried surfaces as key sources in Archaeology and Pedology' organised at the European Association of Archaeologists, 12th Annual Meeting*, British Archaeological Reports. International Series.

Devos, Y., Fechner, K. and Mikkelsen, J.H., this volume. The application of phosphorus cartography to archaeological structures: the development of a protocol. In: Fechner, K., Devos, Y., Leopold, M. and Völkel, J. (eds.), *Enclosed and buried surfaces as key sources in Archaeology and Pedology (papers of the EAA congress of Krakow)*, British Archaeological Reports. International Series.

Eidt, R.C., 1985. *Advances in abandonment settlement analysis: Application to prehistoric anthrosols in Columbia, South America*, Milwaukee.

Entwistle, J.A., Abrahams, P.W. and Dodgshon, R.A., 2000. The Geoarchaeological Significance and spatial variability of a range of physical and chemical soil properties from a former habitation site, Isle of Skye, *Journal of Archaeological Science*, 27: 287-303.

Fechner, K., 1995. Etude archéopédologique. In: Degré, S. (ed.), *Brasseries au quartier Sainte-Catherine*, Brussels : 171-185. (=Archéologie à Bruxelles, 2).

Fechner, K., 2005. Archéopédologie : utilité et résultats pour les fouilles récentes de zones d'activité économique en Wallonie. In: *Actes de la Journée d'Archéologie en Wallonie 2004; Recherches archéologiques préalables à l'aménagement des zones d'activité économique*, Liège: 147-154.

Fechner, K., Devos, Y., 2006. *Cartographie des phosphates et étude pédologique sur le site de Lieusaint, ZAC de la Pyramide – quartier F (Île de France), Rapport d'analyse et d'interprétation*, unpublished report, Institut national de Recherche en Archéologie préventive, Paris.

Fechner, K. and Langohr, R., 1995. De grachtvulling op de Alfsberg te Kontich. Bodemkundige gegevens en interpretatie. In: Annaert, R. (eds.), De *Viereckschanze* op de Alfsberg te Kontich (prov. Antwerpen): meer dan een cultusplaats, *Archeologie in Vlaanderen*, 3: 85-90.

Fechner, K. and Kleiner, F., 1997. The combination of SEM, microchemical, micromorphological and other analyses for interpreting archaeological sites in Wallonia and Brussels (Belgium). In: *Round Table III, Phosphates. Report on workshop at institute of archaeology/department of geology, university college London, Monday 24th March to Wednesday 26th March 1997*. (http://www.roots.archeo.free.fr).

Fechner, K. and Sartieaux, P., 2000. Reconstituer le paléoenvironnement à l'échelle humaine. Essai de visualisation des résultats de l'archéopédologie sur le tracé occidental du Train à Grande Vitesse en Wallonie (Belgique). In: Louwagie, G. and Langohr R. (ed.), *Thematic Day. Soils and Palaeoenvironment Reconstruction. Applications in Geo- and Archaeopedology, 10 November 1999*, Leuven: 35-56. (=Pedologie-Themata, 8).

Fechner, K., Mikkelsen, J.H., Vrydaghs, L., Baes, R. and Devos, Y., 2003. Les cartographies de phosphates et de phytolithes dans quatre maisons néolithiques du nord de la France, de Belgique et du Luxembourg (Verteilung von Phosphat- und Pflanzenresten in vier neolitischen Häusern in Nordfrankreich, Belgien und Luxemburg). In: *Pré-actes du 26ème colloque interrégional sur le Néolithique. Internéo, 8-9 novembre 2003*, Luxembourg: 80-82.

Fechner, K., Deligne, F. and Fercoq de Leslay, G. 2004. L'environnement de l'occupation celtique du site de Ribemont-sur-Ancre. Etat de la question, *Lunula. Archaeologica protohistorica*, 14: 147-159.

Fechner, K., Bosquet, D. and Blancquaert, G., 2006. Indices pédologiques liés au (re)creusement et à l'utilisation des fosses : les cas exemplaires de Remicourt (Néolithique ancien, Wallonie, Belgique) et Dourges (La Tène, Pas-de-Calais, France). In : Frère-Sautot M.-C. (eds.), *Des trous... Structures en creux pré- et protohistoriques. Actes du colloque de Dijon et Baume-les-Messieurs. 24-26 mars 2006*, Montagnac : 47-74. (=collection Préhistoires, 12).

Fechner, K., Hus, J., David, C., Devos, Y., Mikkelsen, J. H., Vrydaghs, L. and Slachmuylder J.-L., in preparation. Phosphate mapping, magnetic susceptibility and other approaches to Neolithic to Roman (long)houses. In: Fechner, K., Devos, Y., Leopold, M. and Völkel, J. (eds.), *Archaeology, soil- and life sciences applied to houses, gardens and courtyards. Second volume of the proceedings of the session 'From microprobe to spatial analysis – Enclosed and buried surfaces as key sources in Archaeology and Pedology' organised at the European Association of Archaeologists, 12th Annual Meeting*, British Archaeological Reports. International Series.

Fercoq du Leslay, G., Court-Picon, M., Loicq, S. and Fechner, K., 2009. Bilan des études environnementales à Ribemont. In: Brunaux, J.L. (eds.), *Les temples du sanctuaire gallo-romain de Ribemont-sur-Ancre,* Saint-Germain-en-Laye: 255-319.

Fercoq du Leslay, G., Court-Picon, M., Loicq, S. and Fechner, K., this volume. Environmental and archaeological studies in sanctuaries, functional and spatial interpretation : the case of Ribemont-sur-Ancre (Somme). In: Fechner, K., Devos, Y., Leopold, M. and Völkel, J. (eds.), *Enclosed and buried surfaces as key sources in Archaeology and Pedology (papers of the EAA congress of Krakow)*, British Archaeological Reports. International Series.

Ganssen, R. and Hädrich, F., 1965. *Atlas zur Bodenkunde*, Mannheim.

Gillet, E. (eds.), in press. *Les fouilles et études sur le théâtre de Blicquy. Etudes et Documents, Série Fouilles*, Namur.

Hausbau in Norddeutschland. *Archäologie in Deutschland*, 2004-2: 18-33.

Ingels, D. (eds.), in press. *Tournai/ Esplechin « Maraîche » : un habitat médiéval et moderne. Etudes et documents. Serie: fouilles*, Namur.

Langohr, R. and Fechner, K., 1993. The digging and filling of Iron Age monument ditches in northwest Belgium. The pedological approach, *Lunula. Archaeologia protohistorica*, 1: 45-50.

Langohr, R., Mikkelsen, J. H. and Vanwesenbeeck, V., 2004. Twee Romeinse potstallen. In: *Verloren voorwerpen. Archeologisch onderzoek op het HST-traject in de provincie Antwerpen*, Antwerpen: 207-209.

Louwagie, G., Langohr, R., 1999. A34. *Section Poix-Terron/Villers-le-Tourneur (France). Rapport archéopédologique*, unpublished report University Ghent.

Macphail, R. I., 2000. Soils and Microstratigraphy: a Soil Micromorphological and Micro-chemical Approach: 47-71. In: Lawson A.J., *Potterne 1982-5: Animal Husbandry in Later Prehistoric Wiltshire*, Salisbury: 47-71. (=Wessex Archaeology Report, no. 17).

Macphail, R. I. and Goldberg, J., 1995. Recent advances in micromorphological interpretataions of soils and sediments from archaeological sites. In: Barham, A.J. and Macphail, R.I. (eds.), *Archaeological Sediments and Soils. Analysis, Interpretation and Management*, London: 1-24.

Macphail, R.I., Cruise, G., Engelmark, R. and Linderholm, J., 2000. Integrating Soil Micromorphology and Rapid Chemical Survey methods: new developments in reconstructing past rural settlement and landscape organisation. In: Roskams, S. (ed.), *Interpreting Stratigraphy. Site evaluation, recording procedures and stratigraphic analysis. Papers presented to the interpreting Stratigraphy Conferences 1993-1997*, Oxford: 71-80. (=British Archaeological Reports, international Series, 910).

Mikkelsen, J.H., 1997. Laboratory method for determination of organic, inorganic and total phosphate. In: *Round Table I, Micromorphology and Phosphate. Report on workshop at institute of archaeology/department of geology, university college London, Monday 24th March to Wednesday 26th March 1997.* (http://www.roots.archeo.free.fr).

Mikkelsen, J. H., Langohr, R., Macphail, R.I. and Vanwesenbeeck, V., 2003. Roman Poststal Byres, a case study from the coversand belt of northern Belgium. In: Boschian, G. (ed.). *Second International Conference on Soils and Arcaheology. Pisa, 12th-15th May 2003. Extended abstracts*, Pisa: 118-119.

Wieland, G., 1995. Die spätkeltischen Viereckschanzen in Süddeutschland- Kultanlagen oder Rechteckhöfe? In: Haffner, A. (ed.), *Heiligtümer und Opferkulte der Kelten. Archäologie in Deutschland. Sonderheft 1995*, Stuttgart: 85-99.

Zimmermann, W. H., 2004a. Vom Pfostenloch zum Hausbefund, *Archäologie in Deutschland* 2004-2: 18-20.

Zimmermann, W. H., 2004b. Spuren aus fünf Jahrhunderten, *Archäologie in Deutschland* 2004-2: 22-25.

# Chapter 4
# Functional and Spatial Interpretation of the Sanctuary and Enclosures of Ribemont-sur-Ancre (Somme): the Role of Palaeoenvironmental Studies

*Gérard Fercoq du Leslay (1), Kai Fechner (2), Mona Court-Picon (3,4) and Sabine Loicq (4), in collaboration with Fanchon Deligne (4), André-Valentin Munaut† (5), Bénédicte Bécu (6), Roger Langohr (7), Mathieu Le Bailly (8), Stephanie Harter (8) and Françoise Bouchet (8).*

(1) Conseil général de la Somme
(2) INRAP Nord Picardie. Formerly: Research Team in Archaeo – and Palaeo-Sciences; Université libre de Bruxelles
(3) Research Unit of Palaeontology, Department of Geology and Soil Sciences, Ghent University
(4) Research Team in Archaeo – and Palaeo-Sciences
(5) Université Catholique de Louvain-la-Neuve
(6) INRAP Centre-Ile de France
(7) Department Geology and Soil Sciences, Ghent University (Geo- and Archaeopedology)
(8) Université de Reims.

## I. General introduction

### *I.1. Site location and aims of the study*

The archaeological site of Ribemont-sur-Ancre is located twenty five kilometres north-east of Amiens (fig 1). Its installations are developed on the north side of the river Ancre, about ten kilometres from its confluence with the Somme (fig 2). Archaeological remains are spread out over 1 km long area, on a slope all the way up to the edges of the present-day village ; the total site area is 70 ha. This vast set of structures was part of an agglomeration that included a theatre, the baths, many buildings, and a sanctuary on the highest part of the hill. Its occupation span ranges from 270 BC (La Tène C2) to 370 AD (late Roman period).

The northern zone of the site, which occupies a dominant position, has been excavated almost completely. It is the most sacred part, because the Roman temple is located there; it is also the oldest part and has its origin in a set of enclosures of the third century BC (Brunaux et al., 1999: 213; Brunaux (ed.) et al., 2009: 35-36).

The religious character of the site gives the palaeoenvironmental study a particular interest, reinforced by the good condition of the archaeological levels, especially the oldest one. Thus, the environmental data of the various fittings can be discussed by comparison with the archaeological data. The stratigraphic sequence which covers nearly six centuries, gives also the opportunity to study these elements transversally in terms of rupture and continuity. Besides, it was hoped that the analytical results would not only complement and confirm (or not) the ones of the archaeological study, but would also identify missing elements of facilities or worship.

Some of the results of palaeoenvironmental studies of this site have been preliminary published (Brunaux et al., 1999 – annexe V: 276-279; Fechner 2000; Fercoq du Leslay, 2000; Fechner et al., 2004a). The present paper develops the analysis of the results of all disciplines, relying on partially published elements, that are enriched with unedited palynological results (M. Court-Picon) and soil analyses of the levels inside the enclosure (Fechner et al., 2008) as well as a recent critical review and synthesis of malacological results acquired by F. Deligne and proposed by S. Loicq.

### *I.2. History of the excavations*

The history of the research on this site, excavated almost without interruption from 1966 to 2003, merges with the history of the French archaeological research. It was discovered and photographied by R. Agache, pioneer of aerial archaeology, during the winter 1962-63 (pict. 1) and interpreted then as the remains of a roman villa (Agache, 1964: fig. 202). In 1966 and 1967, the first excavation campaigns by the team of students in archaeology at Paris University (Sorbonne), directed by A. Ferdière, showed that the roman installations were not agricultural but rather had a religious nature (Ferdière, 1966). From

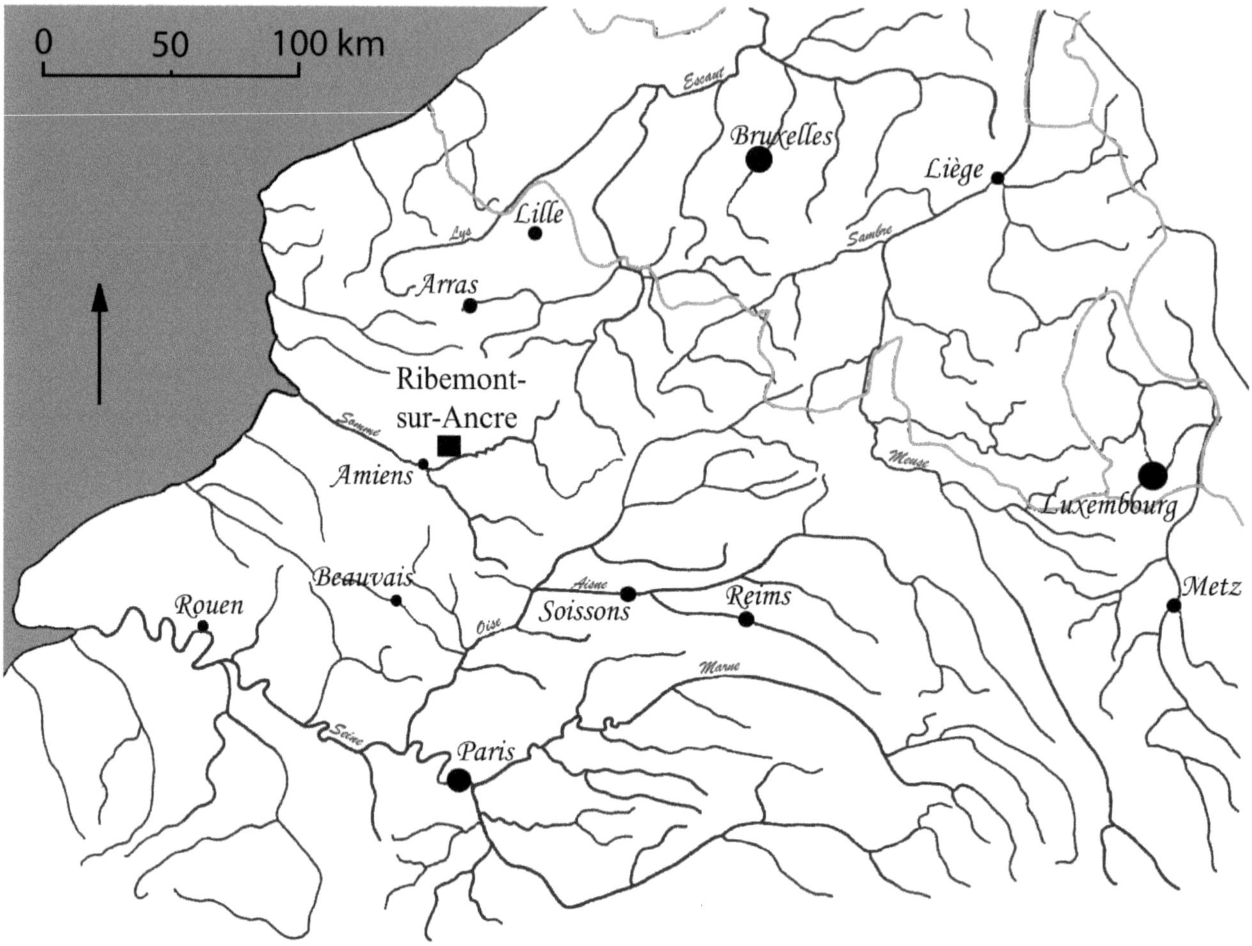


*Figure 1. Localisation map of Ribemont-sur-Ancre in northern France*

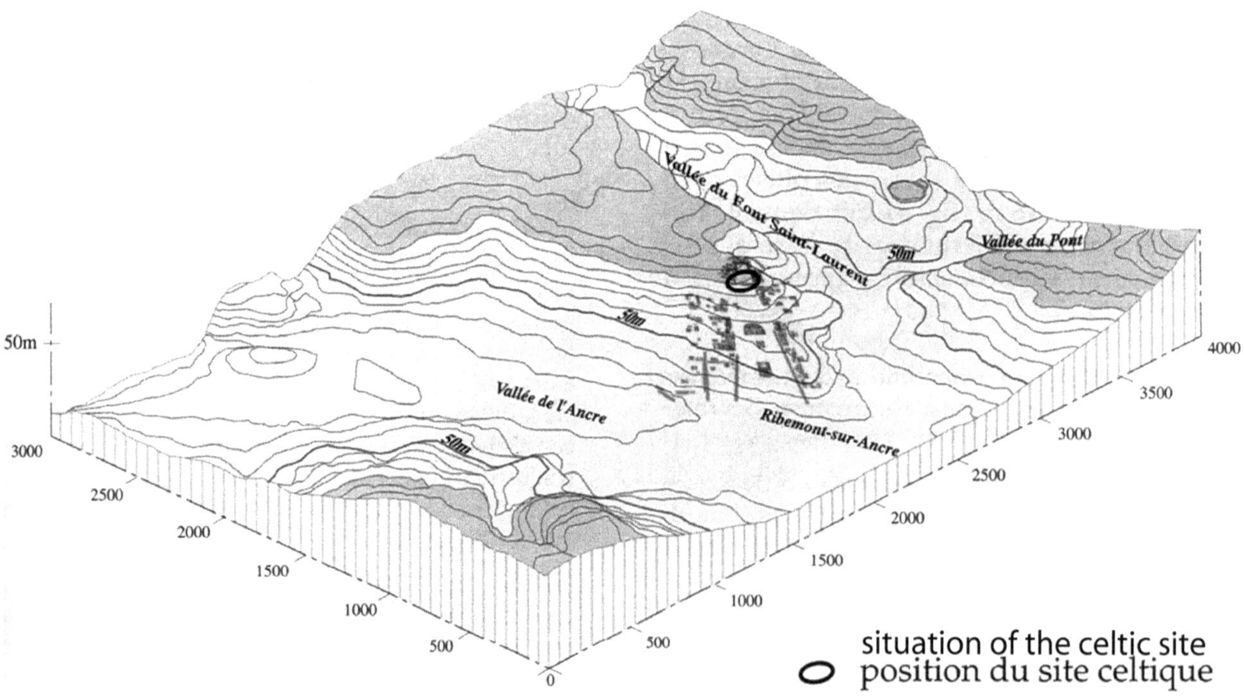


*Figure 2. Topography of the Ancre's valley with the position of roman and celtic remains*

*Picture 1. Aerial view of the site, 1967. Pict. R. Agache.*

*Picture 2. General view of the monumental complex. I Temple, II Theatre, Baths (IInd-IIIrd century AD). Pict. R. Agache.*

1968 onwards, a new team of archaeologists from Picardy University, directed by professor J.-L. Cadoux, resumed the excavations. The religious nature of the site was proved in 1970 when progress of the works revealed the foundations of a roman temple, classical in appearance (Cadoux and Massy, 1970: 469, pl. II; Cadoux, 1971: 43). In the same time, R. Agache, carrying on with his aerial fly-overs, discovered that the temple was part of a monumental worship complex with theatre and bathing complexes, built in a series of stepped courtyards and surrounded by buildings belonging to a settlement (Agache, 1978: 404-411) (pict. 2). The theatre (1971-73) and therms (1974-75) were then excavated and the settlement partly explored (Cadoux, 1984).

At the end of the 70's, the identification of a Celtic sanctuary in Gournay sur Aronde leads to a change in the research. Indeed, at Ribemont, a ditch, oriented south-west/north-east, observed in 1970 under the monumental foundation of the *pronaos* of the temple, contained materials similar to Gournay finds. As a consequence, it appeared that it could be part of the enclosure of a Celtic shrine that was at the origin of the roman sanctuary; but this theory still had to be tested. In 1982, an excavation was practised nearby the temple and revealed a new Celtic ditch (north-west/south-east) containing Celtic material, especially weapons and human bones as in Gournay. But the most impressive discovery was a square construction made of humans and horses bones (Eo-293, pict. 3)), laying on the ground near the ditch (Cadoux and Lancelin, 1987). This structure has not yet found any comparison elsewhere. Following campaigns, up to 1987, showed various deposits of human bones and weapons along the ditch. In 1990, a new team directed by J.-L. Brunaux (CNRS) was in charge to resume the excavation. New deposits of human bones and weapons were then discovered and excavated altogether with two new sections of the ditch. These new traces of a ditch formed, with the two previously identified, square enclosure. Two additional enclosures were also identified for the Iron Age and a new temple for the Gallo-roman period. Thanks to these discoveries, a new chronology and a new interpretation were proposed.

Three main structures enclosed the early fitting dating from the third c. BC. (fig. 3). At first, a square enclosure that dominates both from the topographical point of view and by its nature; this is where the most spectacular deposits are located (fig. 4) and also where the first roman temple was built. In their state of the first century B.C., the structural features are in the form of, sometimes considerable, deposits of bones of horses and especially men, with weapons, lying directly on the antique ground. Some are in the same position as they had been deliberately placed (pict. 4), others in a position indicative of their original position (pict. 5), others scattered on the ground (pict. 6). Much of the artefacts were scattered and buried in the backfilling of hollows, mainly in the ditches (pict. 7).

Forty meters south of the first fence, a ditch delimits a circular area in which the surface soil was extracted from the start. There, artefacts are in the state of discharge in the cuts.

A small shallow ditch, obviously a foundation trench, surrounds the last enclosure, trapezoidal in shape. The ground, badly damaged by roman works, was only partially preserved; its surface is covered with scattered bones and weapons.

The study of the strictly archaeological traces has shown that a battle was at the origin of the foundation of the site (Brunaux *et al.*, 1999: 201). The three enclosures described above were interpreted respectively as follows:

- the first one as a military trophy.
- the second one as a burial ground for warriors of the winners' side.
- the third one as a place for assemblies and commensality.

The first buildings of the roman sanctuary were established in 40-30 BC on the same spaces as defined at the origin, at the expense of the solidity of buildings; walls built above the old ditches slumped and had to be repaired or rebuilt. However, the constructions of the second century (*ca.* 170 AD) were partly free of ancient limitations and the new monumental temple was built downhill, at the rear of the original system of enclosures, at the price of a considerable contribution of embankment which reached almost 8 m at its lowest point. This monument had however a problem with subsidence, as its front colonnade had been erected at the place of a former ditch.

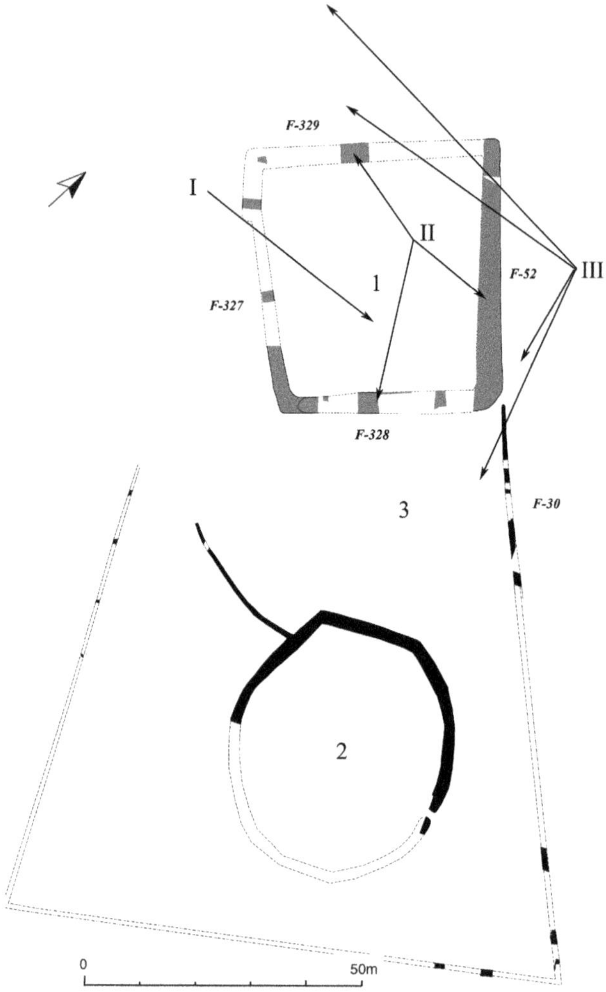


*Figure 3. The Celtic site with the localisation of the main structures of the III^rd c. BC.*
*1 quadrangular enclosure.*
*2 circular enclosure.*
*3 enclosure in shape of a trapezium.*
*I soils inside the enclosure*
*II ditch*
*III outside the enclosure*

The superposition of the layout of fences that successively define the functional spaces of the sanctuary imposed its logic to the organization of the excavation; it also determines the site analysis, since each enclosure can be understood both in relation with its integration into the overall scheme and in view of its evolution during the history of the sanctuary.

An important feature of the site is its longevity, which shows a continuous evolution from the early third century BC to the third century AD; functional spaces that structure the religious space show superimposed structures, sometimes slightly shifted (fig. 5). This long history is also important for the topography and stratigraphy of the site; indeed, while Iron Age enclosures were located on a relatively pronounced slope, the construction of the first roman temple in the second half of the first century BC, required a prior levelling in order to create a vast flat platform. This work has led to the destruction of soils in the western half of the

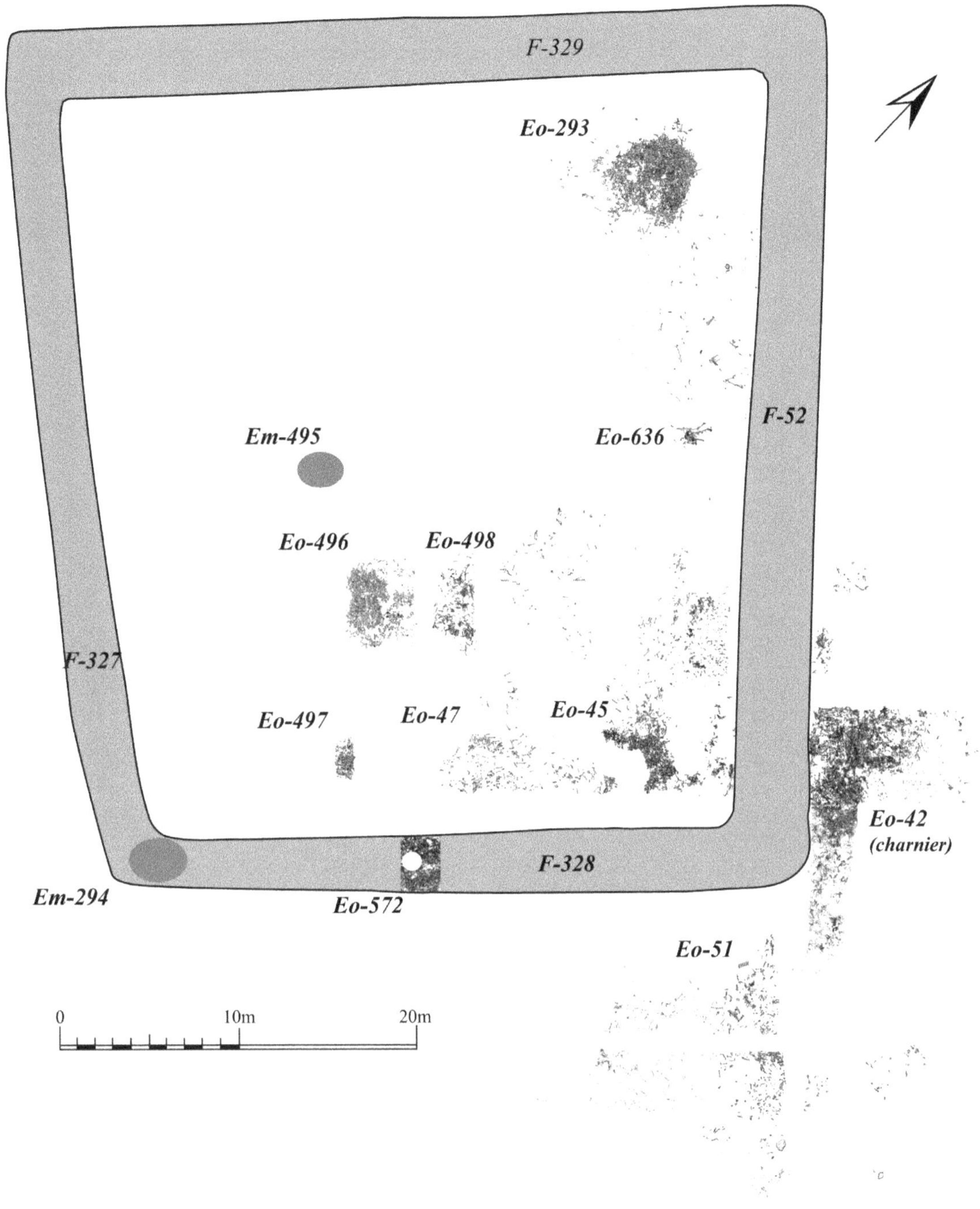


*Figure 4. Localisation of the main features with bones*

site, while structures located lower in the eastern part were sealed by materials of the embankment that preserved them, this case being exceptional in our regions for a rural site.

### *I.3. History of palaeonvironmental studies*

#### I.3.1. The first palynological analyses

It is only recently, as it is the case for most of the sites of historical periods, that the study of palaeoenvironmental context has been incorporated into the dynamics of the research on Ribemont-sur-Ancre. This work starts here during the campaigns 1985-86, with first series of palynological analyses, carried out by Prof. A.-V. Munaut (University of Louvain-la-Neuve) on samples taken from the filling of a ditch of the quadrangular enclosure (F-52). The results, disappointingly, concluded with a low anthropogenic impact for the entire study period and considered the prospect of a broader environmental study as unfavourable (Munaut, 1991)[1].

#### I.3.2. Soil science and malacology

In 1990 the arrival of the team directed by J.-L. Brunaux opened new perspectives. A level previously interpreted

[1] Concerning such layers that are not rich in pollen, (…) "it should be noted however that such work requires a significant investment in time that is only justified to resolve specific problems" (Munaut, 1991: 116).

*Picture 3. Eo-293. A square construction of ca. 2 x 2 m. using human and horse bones as construction materials. The lower layers are in situ (centre of the picture). All around, scattered bones result from the dismantling of the upper part of the feature. Pict. J.-L. Cadoux.*

*Picture 4. Eo-293. Detail of the construction. Pict. J.-L. Cadoux.*

as the remains of an embankment of an "earth rampart", was identified by G. Fercoq du Leslay as the remains of a palaeosol. This interpretation was confirmed by analysis, in 1995, at the laboratory of soil science at the University of Ghent (J. Becze-Deak, University of Ghent). Meanwhile, an initial study on the soil of a particularly spectacular deposit (the bone construction Eo-42), was conducted as part of a university work under the direction of M.-A. Courty (CNRS) (Canseco, 1996). Then a systematic study of soils and the most significant fillings began in 1996 by K. Fechner, B. Bécu (Université libre de Bruxelles, ULB) and R. Langohr (Ghent University). The palaeoenvironmental study was then supplemented by the first malacological and anthracological studies undertaken since 1998 by F. Deligne (ULB). Finally, a palaeoparasitological study was conducted by M. Le Bailly, S. Harter and F. Bouchet (University of Reims) (2002, unpublished).

### I.3.3. New stratigraphic studies

In 1993, the filling of the Celtic enclosure was the subject of a new understanding. In the eighties, it was thought that there was a process of a natural and regular filling interrupted by a new digging of the structure, followed again by a new phase of natural filling and finally quickly filled up by men; now we read a series of discontinuous phenomena. It consists of a first erosion of the ditch walls due to runoff, causing a significant change in the profile of the structure, followed by a long stabilization phase

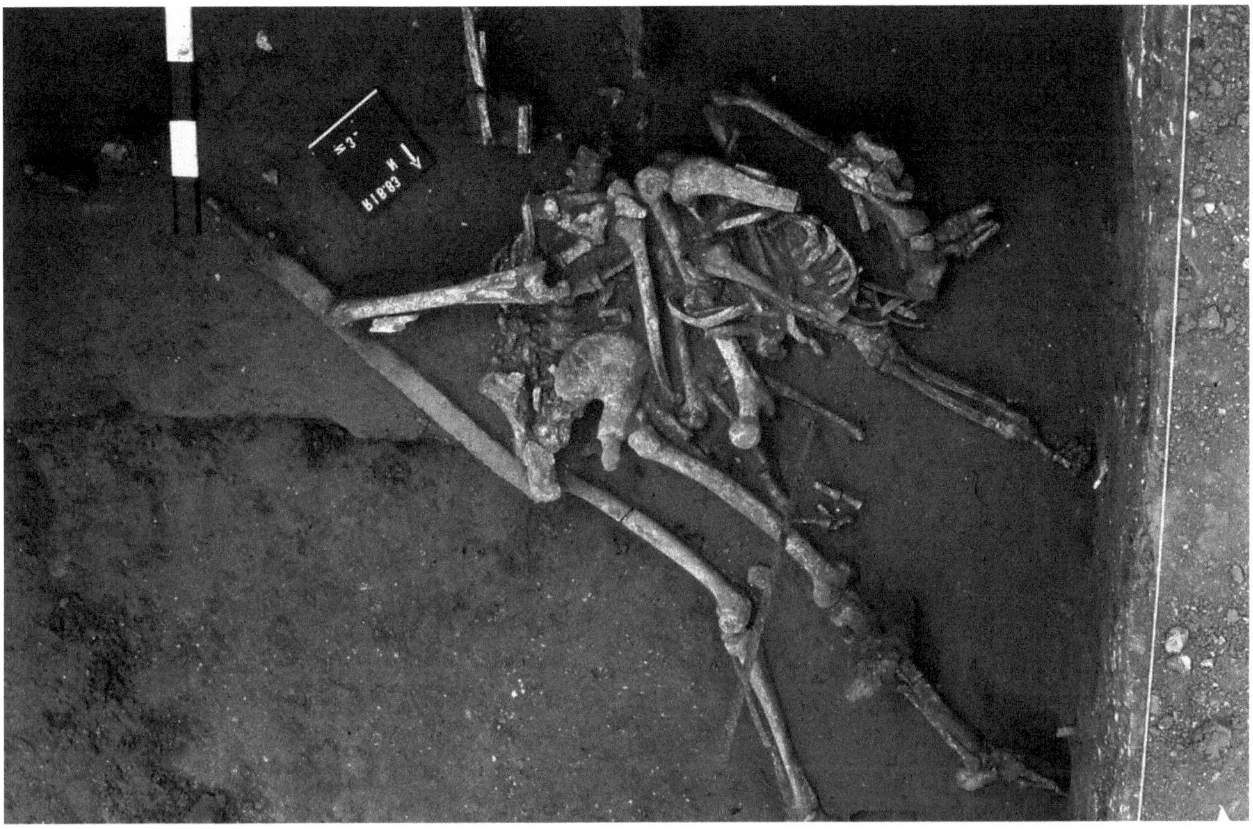

*Picture 5. Eo-636. Bones belonging to three individuals with weapons.*
*One individual is almost complete, with the exception of the head, taken after death. The bones are in connection but the chest appears to be detached from the lower body. The arrangement indicates a fall while the body was already largely decomposed, may be partially mummified ? Pict. J.-L. Cadoux.*

*Picture 6. Eo-51. Bones scattered on the antique ground. Pict. CAD Ribemont.*

*Picture 7. Eo-572. Human bones thrown into the ditch F-328. Pict. CAD Ribemont.*

extending over almost two centuries, and finally by a rapid levelling with a massive fill. The installation of a fence (32/1-Pa) and its restoration (32/2-Pa), unidentified in the eighties, came to disrupt all of these levels; this mistake has affected the results of the first palynological study (pict. 8-10) (Fercoq du Leslay, 2000).

#### I.3.4. New campaigns of palynological analyses

These new interpretations have revived the interest for an environmental approach; a new pollen analysis protocol was established in 1996 (A.-V. Munaut - Université Catholique de Louvain). It is only very recently that this palynological study has been taken over by M. Court-Picon (Ghent University / ROOTS) after the unfortunate passing away of Prof. A-V. Munaut. Therefore, in this area, we do not have as wide a collection as would have been desirable for a site of this size (67 tests for 21 profiles). The planned analysis of a reference sequence, which has been retrieved from natural deposits in the bottom of the valley, less biased by the post-depositional and stational impact, should make it possible to reconstruct the microregional and extra-local vegetation before, during and after the settlement of the sanctuary, and to identify the impact of the site on its evolution.

## II. Materials and methods

### *II.1. Context*

The quadrangular enclosure whose soils are particularly well preserved is the subject of most of the analyses. The enclosed surface is about 1 600 m$^2$ for the Celtic period and 2 500 m$^2$ for the Roman period.

The contexts are defined by their topographical position relative to the oldest structures:

- Inside the square enclosure.
- In its ditches.
- Outside the edges of the square enclosure.
- In the circular enclosure and its ditches

The good soil conservation also helps to clarify the context in matters of stratigraphy. Stratigraphic events and structures seal levels, whether they are constituted by *in situ* elements (as sets of bones) for the older levels or by architectural elements for the most recent levels.

The data quality depends of the potential of the different samples studied, in comparison with the constraints of the different analytical approaches; it also varies according to different contexts depending on the topographic and stratigraphic position of the samples, of the nature of the substrate, and various disturbances caused by the duration of the occupation, the intensity of human activities and the nature of the phenomena and transformations which have affected the site.

### *II.2. Sampling strategy and quality assessment*

Once the contexts were defined, the objective of sampling was to collect in different parts of the studied area, both in

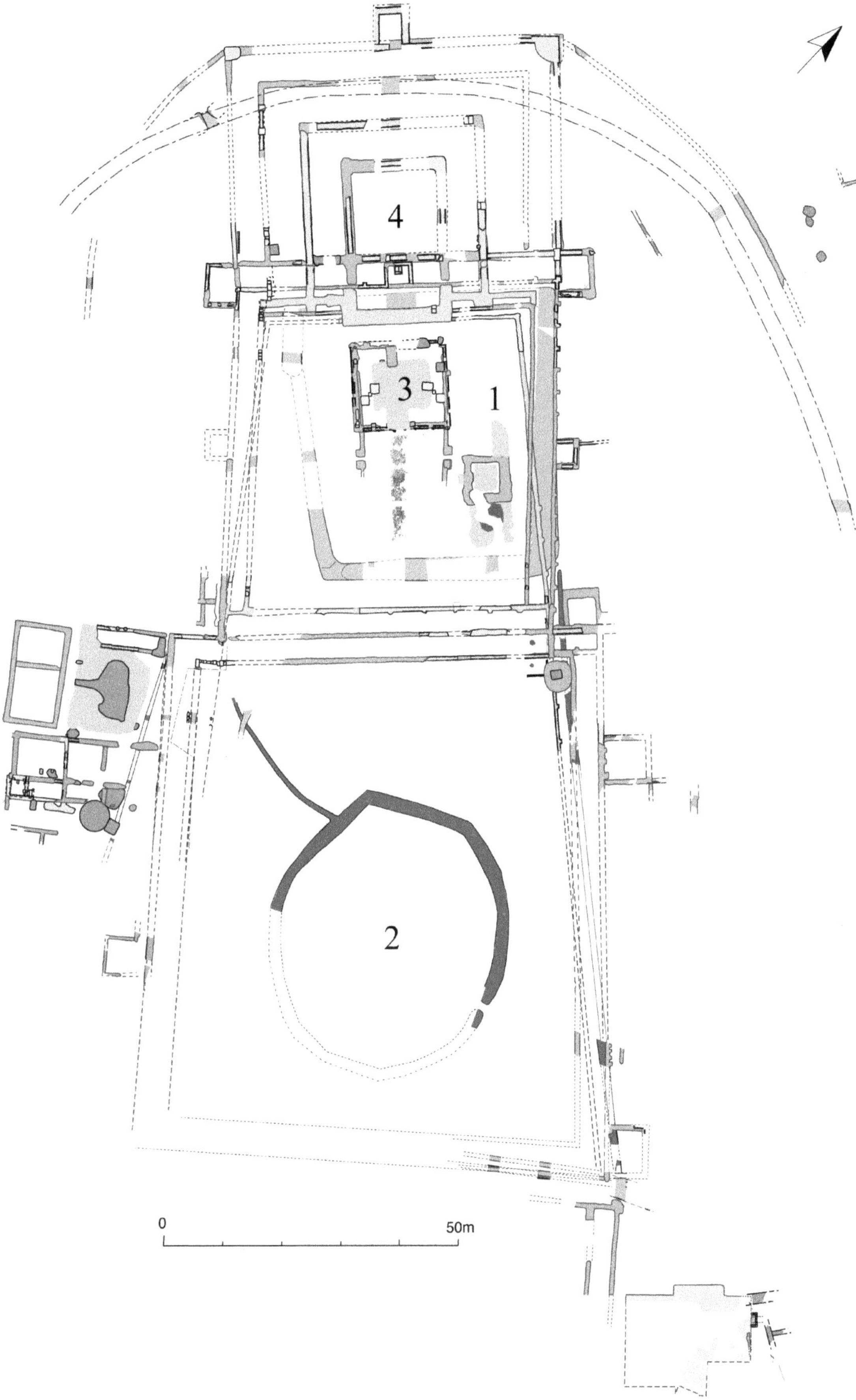


*Figure 5. The roman sanctuary*

*1 sacred enclosure*
*2 esplanade*
*3 roman temple Ist and beginning of the IInd c. AD.*
*4 roman temple from the end of the IInd to the end of the IIIrd century AD.*

*Picture 8. Upper layers of the backfilling of ditch F-52. In the centre of the picture, the edge of the chalk layer, a soil of the I[st] c. BC to the I[st] c. AD., shows the linear limit of the wall M-133 (removed) and circular traces, remains of posts belonging to the wooden fence (Pa-33). Pict. CAD Ribemont.*

vertical profiles covering all recognized periods and locally, in recognized and accurately dated sedimentary complexes. Samples were collected in the natural soil, in fill ups made by humans, in fillings of hollow structures and in sediments associated to sets of bones. The objective was to document a maximum of areas on the widest possible time span.

As a first step, the sampling campaigns of the various involved specialists were carried out separately at different stages of the research. However, given the complexity of the interpretation and the correlation of the first results, a change of strategy has been decided and samples for different tests have been taken jointly and at the same location, in a systematic way.

### *II.3. Sampling*

The review of the malacology data was conducted by F. Deligne and S. Loicq, with the collaboration of O. Decocq, an experienced biologist[2]. They all insist mainly on the low

*Picture 9. Planum at the base of the ditch F-52. The post-holes are difficult to see and situated at the level of the arrows. Pict. CAD Ribemont.*

*Picture 10. Section of ditch F-52. The filling of a post-hole is visible in the top right corner. Pict. K. Fechner. The scale is indicated by a camera lid (ca. 5 cm).*

2 It was also to be noted that the species encountered in the ancient spectra are not the same as the ones found in the different present-day environments (see present-day reference samples 1, 2 and 3, from representative areas that surround the excavation site; Deligne, 2000).

amounts of malacological remains of the site of Ribemont. This weakness is easily explained by the context of non-calcareous soils (the preservation of shells is only due to the calcareous upfill on top of the layers) and by the low amount of analysed sediment (for most occasions samples intended for the soil studies). They also propose to speak, as the palynologist, of trends rather than certainties in the results of the analyses. The same approach was applied to the soil and pollen study (Fechner *et al.*, 2004a), where only data confirmed by at least two of the three disciplines have been taken in account.

### *II.4. Distinction between soils and fills*

The nature of the archaeological layers is crucial for palaeoenvironmental studies. *In situ* surface horizons in the soil profiles and levels of stabilization in natural or anthropogenic fills of archaeological structures can provide the image of the environment at a given moment in the history of the site. Again, the confrontation between different approaches may prove fruitful and reveal specific environments (e.g. bioturbations) that allow to put into perspective the results of each analytical approach and can often solve problems of apparent incompatibility.

The palaeoenvironmental interpretation of fillings is more complex; with regard to the pollen analysis it generally lacks reliability (mixture due to an external input) and was therefore generally not practiced, with the exception of the stabilization level of the surface within the hollow features. The lack of palynology in the sedimentary environment of protohistorical and historical outdoor sites is partly explained by this situation. Indeed, although the archaeological excavation of these sites frequently provides access to sedimentary sequences, they are known to be poor, abundantly reworked, with low palaeoenvironmental potential and difficult to correlate. The validity and scope of the interpretation of palynological data from the analysis of such deposits are still highly controversial and subject to discussion. Often very poor in pollen, even sterile, these sediments provide a sporo-pollen content that is generally very corroded and raises the question of possible effects of differential preservation, phenomena of rearrangement and possible infiltration in the pollen assemblages (Sanchez-Goni, 1994a; Sanchez-Goni, 1994b).

Soil analyses are less problematic and they especially allow to identify phenomena related to the implementation of these fills; here, it allowed us among others to highlight and interpret natural processes (bioturbation, planting, colluvial deposition, collapse, runoff), both as regards *in situ* horizons and fillings of any kind. This, in addition to a better understanding of the ecological dynamics of the site, allowed us to refine the interpretation of pollen analysis, by the comparison of the results. Some aspects of these soils and fillings are related directly to microenvironments due to human activity that could be mapped (pollution by phosphorus, indications of organic matter fittings...).

### *II.5. Distinction between Celtic and Roman periods*

Two independent sets of problems emerge, depending on the nature and originality of the preserved material. They can be summarized in two main phases:

- A phase which covers the Celtic period from the foundation, early in the second third of the third century BC, until removal of original installations and initial Roman processing, immediately after the conquest of Gaul. It is characterized by the presence of the former largely preserved soil surface and of some well preserved structures (some of which are still partially *in situ.*
- A Roman phase ranging from first Roman structures, in the years 40-30 BC, to the abandonment of the site during the fourth century AD. It is exceptional that the soils are protected, as the architectural structures belong to a more conventional type and a better known appearance. Indeed, with the exception of one sample from one of the few Roman soil preserved, layers consist of fill and are of little value in the context of a palynological, malacological or soil studies; overall palaeoenvironmental approach did not appear to offer special interest and was not warranted in a systematic way. The sampling for these layers was done in order to compare with preliminary samples.

When the evidence of continuity between the two major stratigraphic phases was identified, reflection was focused on the key period of the middle of the first century BC, which is the most important to understand the site.

## III. Results and discussions

For the localisation of the structures mentioned for the Celtic and Gallo-Roman phases, we refer to figure 3 and 4 for the Celtic period and figure 6 for the gallo-roman phases.

For the localisation of samples, on can refer to figure 7.

### *III.1. Elements of problematic: palynology*

In the new palynological study (by M. Court-Picon, table 1; see also Fercoq *et al.*, 2009: table 1), the threshold of 10% pollen of trees (and shrubs) is never exceeded, whereas the previous study, conducted by A.V. Munaut, encountered much higher rates (see also note 1). The observed pollen grains and spores are also relatively strongly corroded in the new study, showing a poor state of conservation, probably attributable to the nature of the sediment, not reducing and often quite dry. The corrosion of pollen grains is an argument for an *in situ* origin of these grains, so as they are considered to be contemporary with the sediment.

Three possible interpretations are:

- the pollen grains were corroded preferentially in the profiles chosen for this study, or the old samples were

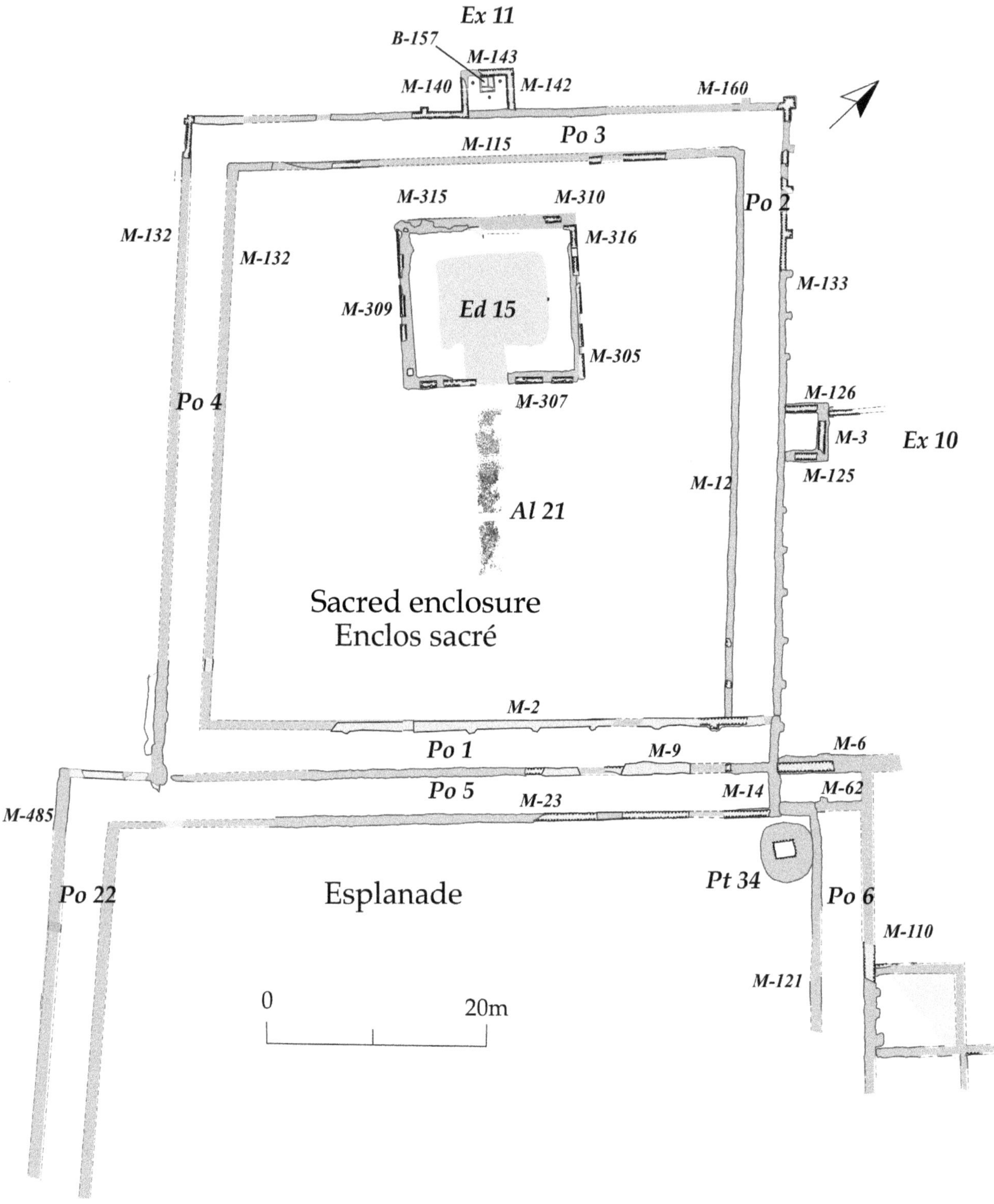


*Figure 6. The main features of the sanctuary from the I[st] and beginning of the II[nd] c. AD.*

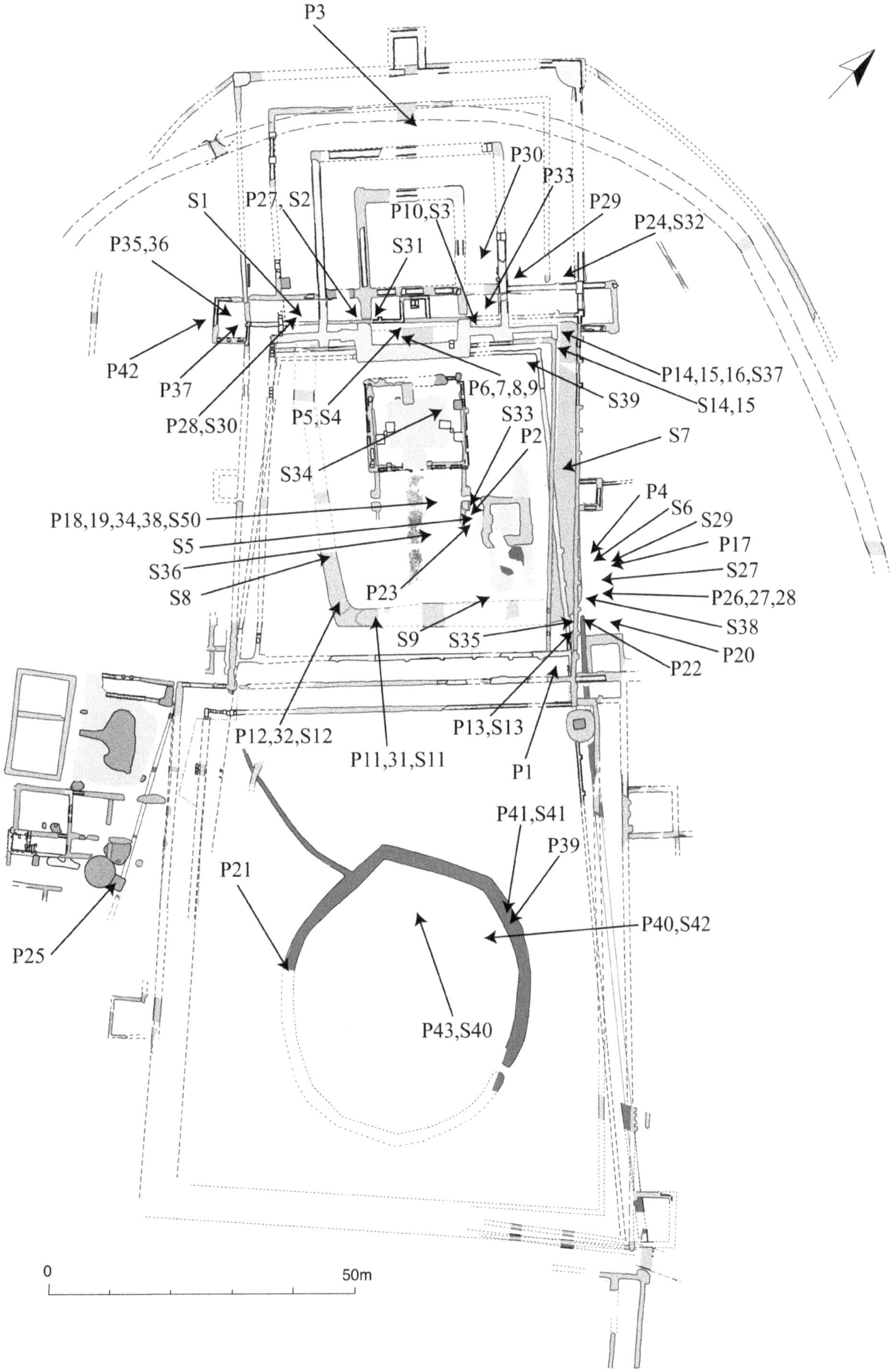


*Figure 7. Localisation of all the treated samples*

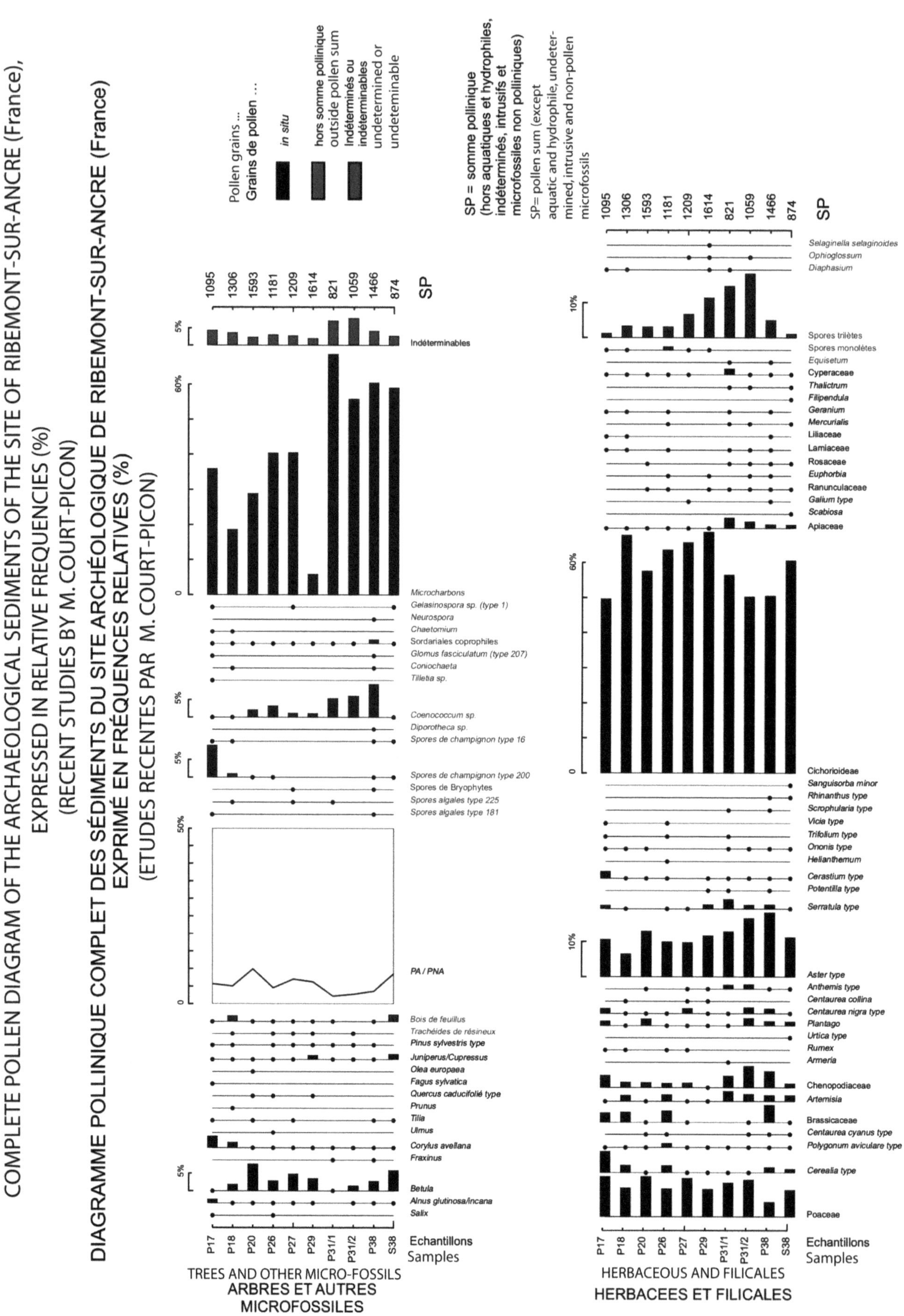


*Table 1. Table representing the most recent palynological data, by M. Court-Picon.*

*Picture 11. Level of erosion at the bottom of the ditch of the enclosure. The alternating of sediments issued from the different geological layers which have been cut by the ditch is clearly visible. Pict. K. Fechner.*

polluted by fresh material because of different sampling method (e.g. without a deep cleaning of the section)

- there has been some drying and oxidation due to long storage of the new samples. Similar phenomena have already been noticed by the palynologist, especially for the analyses carried out in sections at the archaeological sites of Blicquy and Vieil Evreux, without reducing environments and which also show excessively open habitats.
- finally, trees and shrubs would effectively be very rare in the entire site and only be very present in the samples that had been treated before.

The first assumption is eliminated as the necessary precautions have been taken at the time of sampling. The latter two options prevail without possibility to choose one or the other, which leads to use the results of the 2008 analysis with caution. In addition, the encountered tree species are dominated by birch and hazel which are easily transported by wind. Nevertheless, it should be taken into account that the structures that sheltered the elements of the trophy for the older period or monuments and walls of the closing of the more recent period constituted barriers to the dispersal of pollen and were thus the cause of many micro-environments. For all these reasons, concerning the samples processed in 2008, the discussion has been limited to talking only about trends in the local presence of more tree and shrub species.

## *III.2. The third century BC*

### III.2.1. Environment prior to the founding of the site

The oldest horizon is represented by fragments of humiferous soil horizon trapped by the collapse of the walls of the ditch, mainly as a result of frost[3]. This position provides the formation of this characteristic horizon with a relative date.

In the level of collapse, ranging from 50 to 90 cm of depth depending on the studied sections of the ditch, blocks of sediment removed from all the natural levels of the soil crossed by the digging of the ditch are alternating (pict. 11). The fragments of the soil surface, sealed by early collapse (see III.3), have retained the characteristics of the original soil; their study makes possible to identify the characteristics of the landscape prior to the founding of the site and to correlate its analyses with some of the buried surface horizons that are in situ.

#### *Soil science*

The pedological study (S3, S7, S9) reveals that these profiles present a "brown leached soil" of Late Glacial origin with neutral or acid pH, whose upper horizon has been transformed by ploughing (dark horizon with abrupt base) at an unknown date (some Early Bronze Age sherds

[3] Theoretical processes described in Langohr (2000) and Fechner et al. (2000).

*Picture 12. Soil type I in profil S3. Pict. K. Fechner. Legend: 1: humiferous upfill or colluvium; 2: abrupt horizontal limits; 3: homogenous humus horizon (ploughing); 4: micro-biological mottling, 5: base of ploughing horizon cutting uneroded natural ("E") horizon.*

could also be residual). This ploughing precedes the digging of the ditch of the enclosure of the first sets of bones. Its top is accompanied or followed by a slight thickening of probably anthropogenic origin.

In these profiles, the buried surface horizons are defined as part of a "soil type I" (pict. 12). This soil type is also the original state of the other soil types encountered on the site ("soil types II and III").

Starting from a situation which is assumed to be little differentiated, K. Fechner (Fechner, 2000) was able to distinguish 3 types of soil formed from the Celtic occupation in the third century BC (see also similar differentiation on Belgian loess, Langohr, 2001). He characterizes the soil type I in the following manner: the horizons of this soil contain neither limestone gravel nor Roman artefacts. The boundaries of the horizons are still regular and fine accumulations of secondary calcium carbonate (pseudo-mycelium) are well developed and preserved. It is perfectly preserved, buried under a fill made of sediments that also contain no limestone. The upper horizon of this soil has the characteristics of a ploughing horizon (in the field: homogeneousness, abrupt and horizontal base). The reference profile is defined from the characteristics of this soil type.

Representative of the soil type I, profiles S3 and S34 showed a slight enrichment in phosphorus and iron in the Bt horizon, which is distinct from later pollution from a chemical point of view (phosphorus with calcium) and probably due to organic manuring of this ploughing.

The fragments of ancient soil in the collapsed back-filling of the ditch probably indicate that ploughing is not just one event, but reflects a prolonged period of farming; they also have the particularity of being the only soil samples collected on a site that is free of any bioturbation, in contrast with type II and III soils, which implies that the phenomena of bioturbation took place after the foundation of the trophy (Fechner, 2000).

***Palynology***

Palynological analyses performed on samples of the initial sequence of filling reveal an environment marked by the presence of a well developed herbaceous heliophilous layer, dominated by grasses (Poaceae) and Compositae of the *Crepis* type.

The analyses reveal some diversity of the pollen assemblages, according to the topographical location of samples. Samples collected on the north-eastern section of the ditch (F-52) show a low rate of tree pollen, of about 10%, mainly represented by birch (*Betula* sp.), hazel (*Corylus avellana*), and to a lesser extent by tree growing on the edge of wetlands or watercourse, such as alder (*Alnus* sp.; Munaut, 1991 - sample 1985/P26 [4]). The recent analysis (P31-2) on the south-eastern section of the ditch (F-328), near the south-eastern corner of the enclosure, shows a comparable rate of openness, where only birch is quite well represented. The analyses performed on the fragments of soil collected in the north-western section of the ditch (F-329) close to the change of slope, at the rear of the site (P6, A.-V. Munaut - analysis 1996-1997 - Fercoq du Leslay *et al.*, 2009: table 6), reveal the apparent presence of a more dense tree cover (tree pollen exceeding 30%), always dominated by hazel and alder and with a better representation of birch (over 10%). Lime, absent in other samples, appears here although slightly represented (1.5%), while pine, elsewhere less than 1%, exceeds here 5%. Tree assemblages are dominated by trees and pioneer heliophilous shrubs (*Betula*, *Corylus* and *Pinus*), that pollinate when they are very young and at the state of thicket. These trees, especially pine, are also species with high pollen production and distant dispersal by wind. These tree pollen recorded in our assemblages could partially come from thickets that were relatively distant from the vicinity of the ditch; on the other hand the presence of lime, the strongest recorded in Ribemont, and the more discrete presence of beech (*Fagus sylvatica*) and walnut (*Juglans*)

[4] The number of the analyses conducted by A.-V. Munaut in 1985 and 1986 is duplicated by those of recent years. In the text, the reference to the first analyses is systematically preceded by the year to avoid confusions.

implies that some representatives of mesophilous species grew on the slope north-west to the rear of the site, perhaps with groves of birch and hazel.

Concerning herbaceous plants, pollen diagrams from the two phases of analyses also share common characteristics and punctual differences. One constant is the significant presence of Ferns (from about 10% to nearly 20%) and Grass (rate exceeding 10%). High to very high values attested for Asteraceae are not surprising, since these plants have pollen grains among the most resistant and therefore best represented in the pollen spectra of very disrupted environments and/or from sediments that do not favour the conservation of pollen grains. Thus it is dangerous to argument on the base of these very high percentages and to interpret them as the image of local and extra-local vegetation.

Among the Asteraceae, A.-V. Munaut identified the genus *Cirsium* (thistle group) with values below 5%, and especially the *Crepis* type (dandelion group) that exceeds the value of 50% in the sample P2 (1985) and reaches the value of 15% in the sample (P6).

The most important difference is the strong representation of cereals in the sample P6 where they reach the value of 7%, whereas they are totally absent or weak represented in the rest of the samples. The distinctive feature of cereal pollen grains is their low capacity of dispersal; it is logically possible that the north-western section of the ditch was dug nearby, or perhaps even directly in a field or a former field dedicated to the cultivation of cereals. These results are comparable to the interpretation mentioned above of soil profiles S3 and S34, respectively to the north and the centre of the quadrangular enclosure; unfortunately these profiles have not been subjected to palynological analysis. According to K. Fechner (2000), other profiles studied also showed that ploughed horizon originally, but suffered severe alteration by a more or less dramatic bioturbation after its formation; this involves that the ploughed horizon covered the entire studied area at the beginning.

On the other hand, both in sample 1985/P2 and sample P31-2 (2007), *Artemisia*, Chenopodiaceae, *Plantago* sp., *Centaurea nigra* type and *Anthemis* type are at their maximum. In the sample P31-2, the presence of *Serratula* type, *Coenococcum* spores and of Sordariales (coprophilous fungi) is to be noted. In these samples, the high rates of agropastoral indicators, related to animal faeces and/or developing on dung (Van Geel *et al.*, 2003), show a relatively high pressure by cattle breeding in the immediate vicinity.

These results are consistent with the series of analyses by A.-V. Munaut on samples collected in 1986 (1986/P2 to P7); However, these samples, even if they are comparable with those of 1985 concerning the herbaceous (strong to very strong representation of *Crepis* type with 35 to over 85%, and the presence of ferns with 3 to 30%), present a significant variability in terms of tree spectra whose representation varies from 5 to over 20%. The fluctuation of the values of pine (from 0.9 to 23%) is the cause of this surprising phenomenon. These results are surprising for samples expected to come from the same level and could be indicative of local bioturbation due among others to the impact of shrub roots (see below) or of anthropogenic disturbance, such as the establishment of a fence (Pa 32)[5].

***Malacology***

Malacological data of this lower ditch fill (S11) are too poor and not significant (insufficient volume of sediment), except at the depth of 188-195 cm in the south-western corner of the ditch. Here the sample consists of a majority of Pupilla muscorum, which indicates a landscape of rather dry grassland. However traces of stagnant water, probably due to rains or short periods of un-watering, are possible locally, as indicated in a runoff area at the bottom of the ditch for instance (diagr. 1).

***Interpretation***

In the third century BC, the environment is apparently wide open for a very long time; fields were cultivated for some time and probably separated by hedges (hazelnut, birch). The north-western section of the ditch, seems to be in the close proximity to a wooded area which should be located beyond the enclosure, on the north slope and down the dry valley called « la Renardière ».

#### III.2.2. Environment contemporary of the foundation of the site

Another horizon of the third century BC has also been analyzed from the palaeoenvironnemental point of view, but this one is contemporary to and slightly later than the foundation of the trophy (sealed by a rapid collapse of the walls - Fercoq du Leslay, 1996). It appears in the form of thin laminar layers deposited at the bottom of the ditch as a result of erosion of the walls under runoff (fig. 8).

***Palynology and malacology***

This level has not been observed in all studied profiles; it is absent in the north-western section of the ditch, that is gently sloping. Elsewhere its thickness is also variable, but not exceeding 2 cm. On the other hand, it reaches a maximum of 5 cm thick in the north-eastern corner of the enclosure, which is the lowest point of the ditch. Palynological sample P31-1, studied in 2007, shows an open environment (almost complete absence of trees and shrubs, less than 5%) subject to fire (many charred particles) and therefore quite disrupted on the surface (*Coenococcum*). The general picture is that of a fairly moist lawn heavily colonized by ruderal species (Cyperaceae, freshwater algae

[5] As the post holes disrupt the backfilling of the ditch, sometimes to the depth of the level discussed here, and had not been identified during these excavations, it is difficult to go further in the interpretation of this analysis; they should be completed by new samples near the eastern angle of the enclosure.

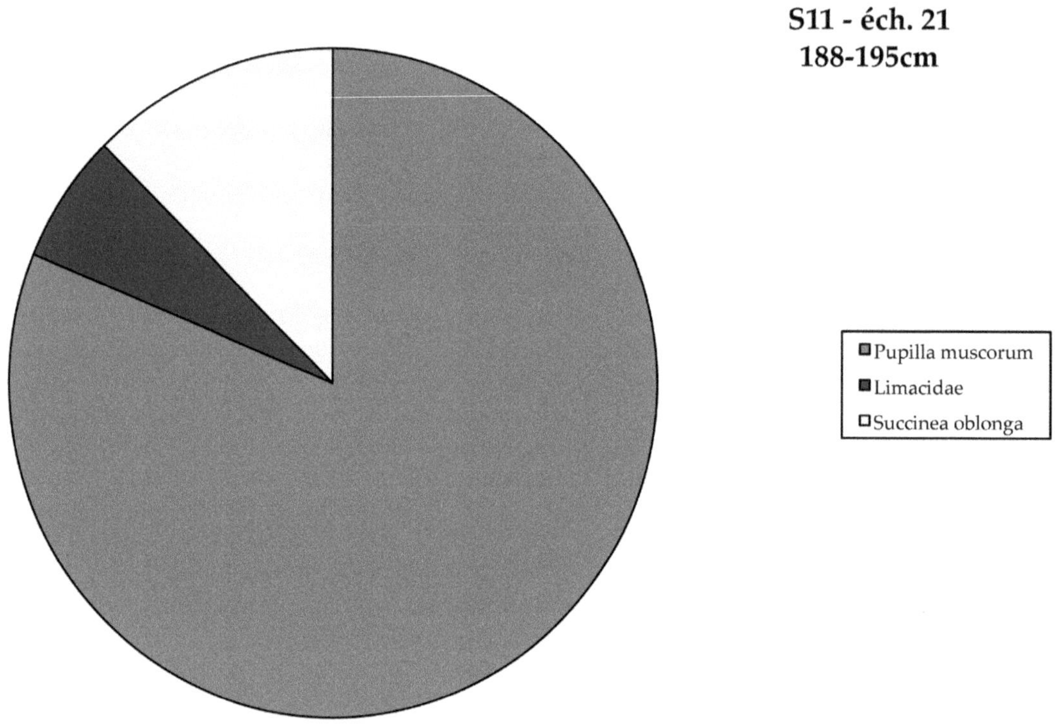


*Diagram 1. The legend is to be followed clockwise. The largely dominant snail species of this lower part of the ditch is* Pupilla muscorum.

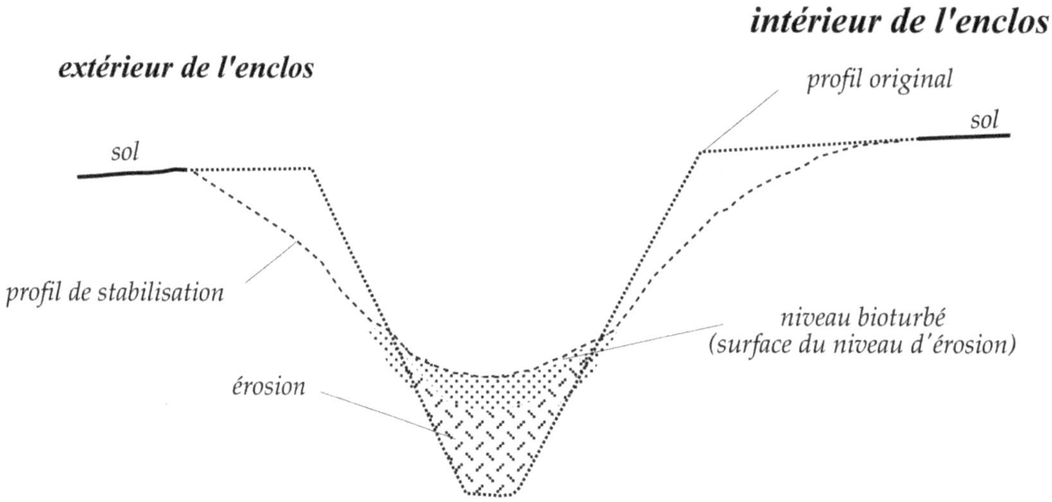


*Figure 8. Stratigraphy of the bottom of the ditch (F52)*

*Picture 13. General view of the section of the ditch F-327. Pict. CAD Ribemont.*

type 225, *Equisetum*, Thalictrum, Apiaceae and ferns / *Artemisia*, Chenopodiaceae).

Despite the poverty of the studied sample, the results of the malacological analysis (S11) are consistent with the palynological data and recall a wide-open environment with the possible presence of a few isolated trees.

***Interpretation***

The response of three individualized phenomena can be detected through these results. The first is a layout of the site after its complete cleaning: eradication of groves, cleaning by fire, intense trampling. Then, the local environment particular to the bottom of a ditch is characterized by pockets of stagnant water. Finally, the first human activities take place in an already very open site; the low representation of anthropogenic pollen markers is probably due to a weak agropastoral impact and/or to the briefness of the represented phenomena (about a few weeks ?). This briefness would explain why the local human impact associated with the founding of the site is imperfectly detected.

### *III.3. From the third century to the middle of the first century BC*

This long period is represented by the ancient surface horizon preserved both inside and outside the enclosure. These soils retain the memory of all human activities that have been made between the foundation of the site and the "cleaning" prior to the works of the first Gallo-Roman temple, early in the second half of the first century BC. The echo of the environmental situation on the eve of Romanization emerges strongly in the analyses of soils levels; this phase is also well represented in the backfilling of the ditch.

#### III.3.1. The ditch (pict. 13)

Early in the IIIrd century BC, a powerful phenomenon of freezing affects the feature and is, at least partly, the cause of the collapse of the upper part of the ditch. Fragments of undisturbed sediment of the various geological strata cut by the ditch sides alternate in the backfilling (cf. pict. 11). The soil study of several sections also shows the impact of frost and or compression on the walls (cracks filled with calcium carbonate parallel to the ditch walls, in the natural soil). Unlike in the site of Kontich (district of Antwerp, Belgium), a comparable deep ditch of a quadrangular enclosure dated to the end of the Iron Age (La Tène finale) (Annaert, 1993; Fechner and Langohr, 1996), there is here no clear evidence of collapse due to frost. Fitting with timber is probable for the phase during which the ditch was open. The profiles of the ditches show that a typical equilibrium profile (stabilization of the ditch walls) has been reached following the erosion of the middle and upper parts of the walls, after the destruction of the supposed timber walls.

The lower filling of the ditch is formed by the superficial part of the materials of this third century collapse. The surface horizon was gradually transformed by pedological phenomena (S4, S11, S12, S37) that will continue in some places up to 70 AD, time of the embankment and of the construction of a portico in the S4 profile (see next section). This phenomenon corresponds to a level of stabilization

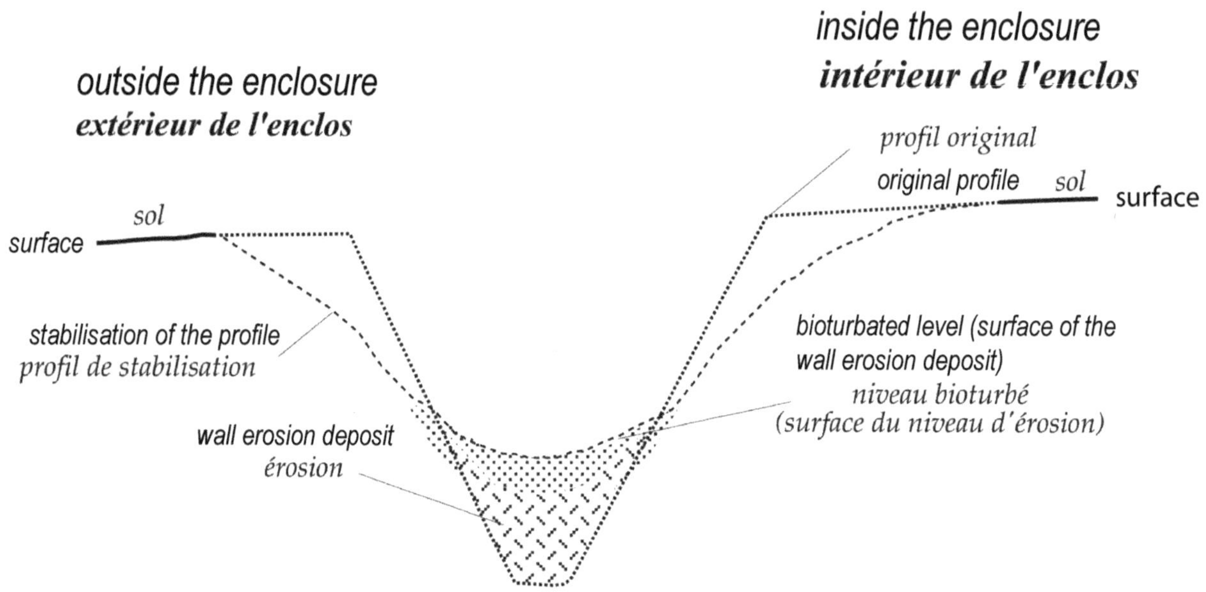


*Figure 9. Interpretation of the ditch's erosion phase of the enclosure of the III^rd c. BC.*

at the top of the bottom fill of the ditch, level that was colonized by wildlife and plants. The ditch thus was open and had not been rearranged by a possible re-cutting, as had been proposed in the past (fig. 9).

*Soil science and malacology*

The wet nature of the ditch bottom is well illustrated by the malacofauna of sample S37 in the northern corner, which is the richest one (diagr. 2). The presence of typical forest (9 *Discus rot.*) and wet (e.g. *Vitria cryst*) species is to be noted, suggesting the closed environment of a riparian forest. Pedological data of S37 reveal a strong bioturbation due to the impact of roots, which confirms the malacological study. Therefore, the arguments for the presence of a shrub cover (thicket) are strong (Fechner *et al.*, 2004a). Unfortunately, samples for pollen analysis could not be studied.

The sample S4 H1A also shows these characteristics, with the dominance of forest species (*Carychium*). It confirms the image of a wooded and relatively humid local habitat, also suggesting the possible presence of a riparian forest, but with nearby drier sectors and possibly more scattered trees. Here, human presence seems less marked than anywhere else.

In the southern corner, the malacological sample (S11) consists mainly of mesophilic species (which are not characteristics of a specific environment). But, in this sample, species of open stations and representatives of semi-forest (Discus rotundatus) are also present; they are characteristic of different places, from dry to wet, under stones, rotting wood and plant waste. The presence of a single representative of the species Pyramidula rupestris (species rarely observed in archaeological samples) is also noted; this species is characteristic of calcareous stones, old walls, and also of dry wood (diagr. 3).

***Palynology***

The pollen analyses also present different characteristics depending on the studied samples, which reflect differences in the local environmental conditions (1985/P3 and P4, 1996/P7). Samples collected in the north-eastern section of the ditch again suggest the presence of a more or less wooded area, dominated by pioneer heliophilous shrubs such as birch and hazelnut, which, altogether with alder, are characteristic of wetland stands. These samples also show a clear trend of decline for these species which are gradually replaced by pine. The herbaceous cover is dominated by ferns (values exceeding 60%), while Poaceae are decreasing but remain in amounts close to 10%. The appearance of *Carex* (Cyperaceae) in the upper level of the sequence (15 cm) indicates a relative humidity of the local environment. *Plantago* sp., poorly represented, is only present in the lower level (30 cm). The diagram corresponding to the north-western section of the ditch shows an even lower tree cover, dominated by birch and hazel; their presence tends to be confirmed over time, but with values that remain very low, and pine is very poorly represented (less than 1%). The herbaceous cover is dominated by representatives of the *Crepis* type that show a slight decrease (from 89.6% to 75.4%), while the ferns show a slight increase (2.5 to 6.1%). *Plantago* sp. is totally absent.

***Interpretation***

The results of the palaeoenvironmental study of the ditch can be summarized as follows: the structure is open, but colonized by vegetation and shows evidence of moisture

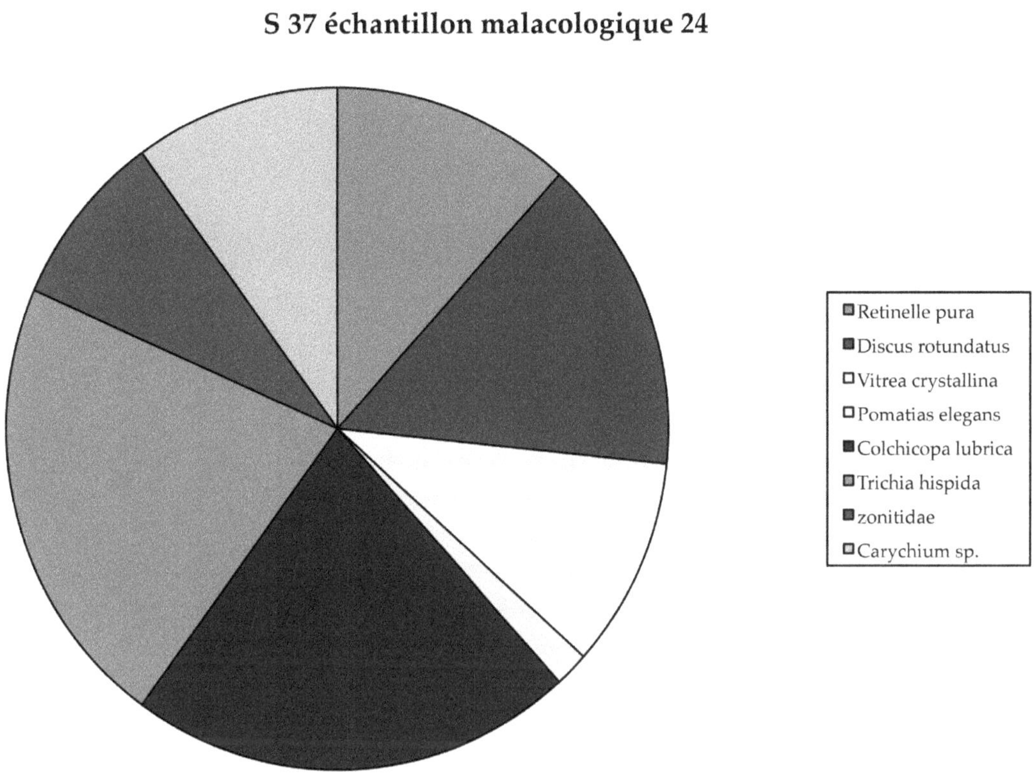


*Diagram 2. The legend is to be followed clockwise. The dominant snail species are Colchicopa lubrica and* Trichia hispida.

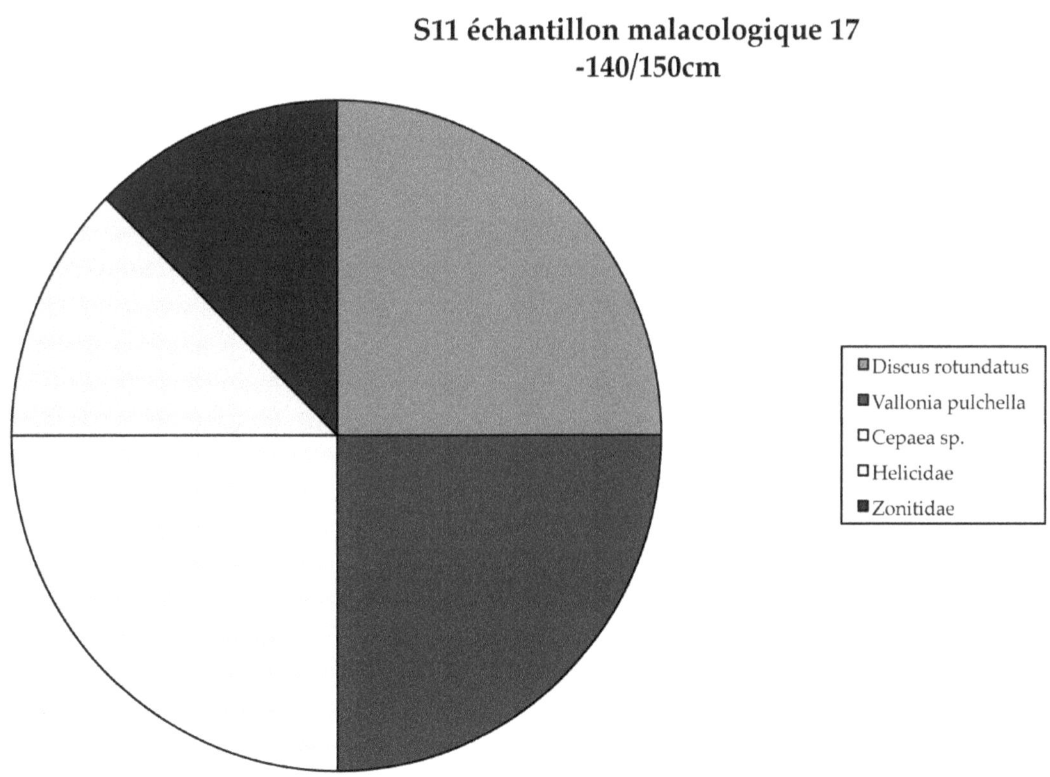


*Diagram 3. The legend is to be followed clockwise. The dominant snail species of this part of the ditch (slightly higher fill than the one of diagr. 1) are* Discus rotundatus, Vallonia pulchella *and* Cepaea sp.

due to poor drainage; punctually, thickets occupy the surroundings of the ditch. The extra-local environment corresponds to a lawn (almost complete absence of trees and shrubs), that is little influenced by man and its activities (poor representation or absence of *Plantago* sp.). The presence of hazel and birch probably refers to the existence of a nearby grove, most likely within the enclosure as indicated by the results obtained for the soil study. The presence of small quantities of a mollusc that colonizes old dry wood could probably agree with the closeness of aging timber frame buildings.

### III.3.2. Soils inside the enclosure

The analytical results are very different depending on the topography; they clearly reflect a significant anthropogenic impact, the structures and the practices that have marked this period of two centuries and have heavily influenced their environment, whether they have generated tangible evidence or not.

#### *Chemical analyses and tests*

The mapping of phosphorus in the Celtic surface horizon (fig. 10-11) shows a clear dichotomy between the profiles in the enclosure (S9, S33, S34) rich in phosphates, and outlying profiles of a few meters distant (S2, S3, S31, S32) that are poor (see Fechner, this volume, for the discussion of the data). The same low rate is found in the fragments of the surface horizon collapsed at the bottom of the enclosure ditch (S7, S9). The phenomenon of phosphate enrichment is therefore subsequent to the foundation site and is not linked to earlier agricultural activity.

The correlation between calcium and phosphate is also an important indicator. In a few cases with particular field morphology (soil type I, observed in S34 and S3, described above), the rather high levels of phosphates are not correlated with calcium but with iron. These are indicative of a vertical movement of phosphates with clay, below a former polluted surface. Here, in concordance with field characteristics, the source of phosphate could be an activity related to cultivation of the land and site maintenance, probably related to agricultural activity prior to the foundation. The good correlation between rather or very high levels of phosphates and calcium in most samples both inside and outside the enclosure is likely to be indicative of the original presence of bones on the surface horizon above it, whether the bones have been left behind definitively (S6, S38, S50) or removed (S4 and S33a-c) (Fechner *et al.*, 2004a).

The phosphorus test map for Celtic levels seems to support the hypothesis that the southern sector of the enclosure is dedicated to the handling/deposition of bones. Indeed, all samples rich to very rich in phosphorus are situated here without exception. The mapping of the analyses of total phosphate is much less dense but provides more accurate and reliable results (Fechner, this volume). It was found on other sites that this type of analysis extracts both more and other types of phosphorus than the tests. These analyses also show high rates of total phosphate in some profiles of the northern half (S4, S33 and S34), but it is more likely that they correspond to a Roman pollution (see below for the profile P38).

#### *Soil science*

Samples taken at the centre and on the eastern periphery of the enclosure present structural differences with the fragments of ancient soil trapped in the lower filling of the ditch. They are related to soils defined as soil type II[6] and III[7]. Soil type II shows only slight bioturbation phenomena which could be due to the presence of organic matter or more likely to the presence of calcium (and collagen ?) present in the bones (Fechner, 2000).

It appears that some areas were only once covered with bone deposits, which are only reflected by the high phosphate content in underlying soils.

The study of profiles collected in four points close to each other inside the enclosure (sample S33a-d), is also indicative of a particular environment. The soil is slightly bioturbated, producing soil type II (33c, 33d), to strongly bioturbated, resulting in soil type III (33a, 33b). This bioturbation is due to roots whose most obvious impact (uprooting) was observed during the excavation (in plan and section, pict.14).

In contrast to the bioturbation affecting the top of the fill of the ditch, the whole of this phenomenon can be attributed to the Celtic era. Indeed, the surface horizon of profile S33 was sealed as soon as the middle of the first century BC[8].

It is useful to recall briefly the results of the multi-proxy study of these bioturbation phenomena (Fechner *et al.*, 2004a). The bioturbation observed since the establishment of the trophy is not due to intensive grazing, but to a locally open environment with shrubs during the period of the use of the enclosure. In one of the profiles, along the outside of the enclosure, malacological and soil studies highlight

[6] Soil type II (Fechner, 2000): This soil shows a bioturbation that is already more pronounced than in soil type I. Small fragments of chalk are concentrated in vertical earthworm's biogaleries. It is poorly preserved and can be sealed by deposits of human bones, themselves covered with Roman upfills that are more or less rich in fragments of chalk. This soil was crossed by vertical galleries, probably due to the activity of worms, which has at least partly been influenced by the Roman level. The base of the ancient buried humus horizon is still horizontal but interrupted locally by bioturbation. The animal that caused these more or less horizontal biogaleries is possibly different from the earthworms that created the vertical galleries.

[7] Soil type III (Fechner, 2000): This soil differs from soils I and II by the presence of fine occasional limestone fragments in the matrix, mainly observed in thin sections. Soil type III, situated below generally rich in chalk Roman upfills, is entirely bioturbated over 30 cm depth. It is characterized by an irregular boundary between the humus horizon and the top of the underlying horizon, which is the result of the mixture of the two horizons.

[8] Another place characterized by a potentially contemporary shrub environment is the northern section of the ditch (F-329/S4, S37, P5). According to the relative dating, it is likely that part of the bioturbation of the ditch, including the lower level of S4, corresponds to this former phase, but we do not have any absolute evidence of this.

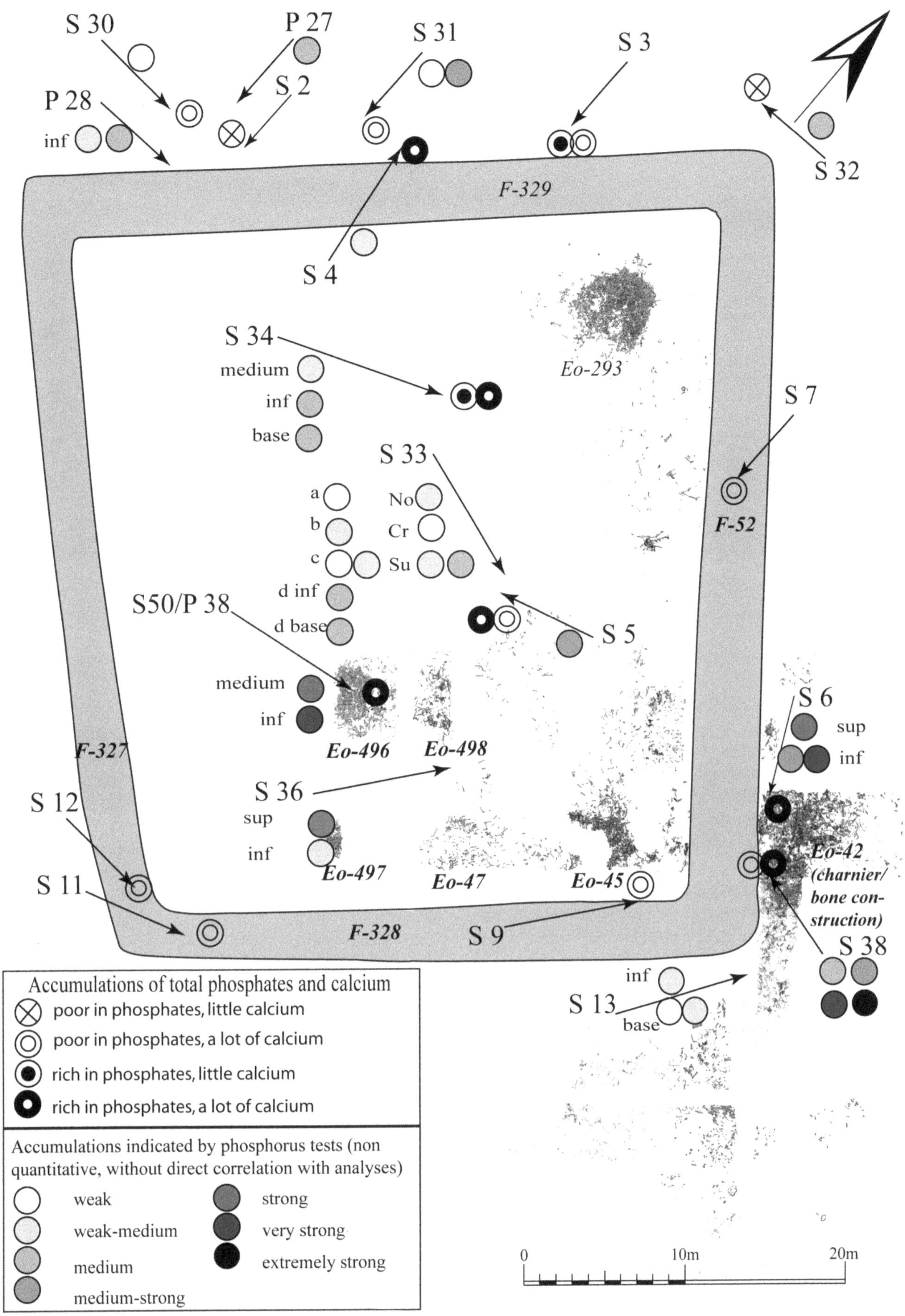


*Figure 10. Celtic and early Roman surface horizons: cartography of phosphate analyses (correlated with amounts of calcium), confronted with more systematic phosphorus tests .*

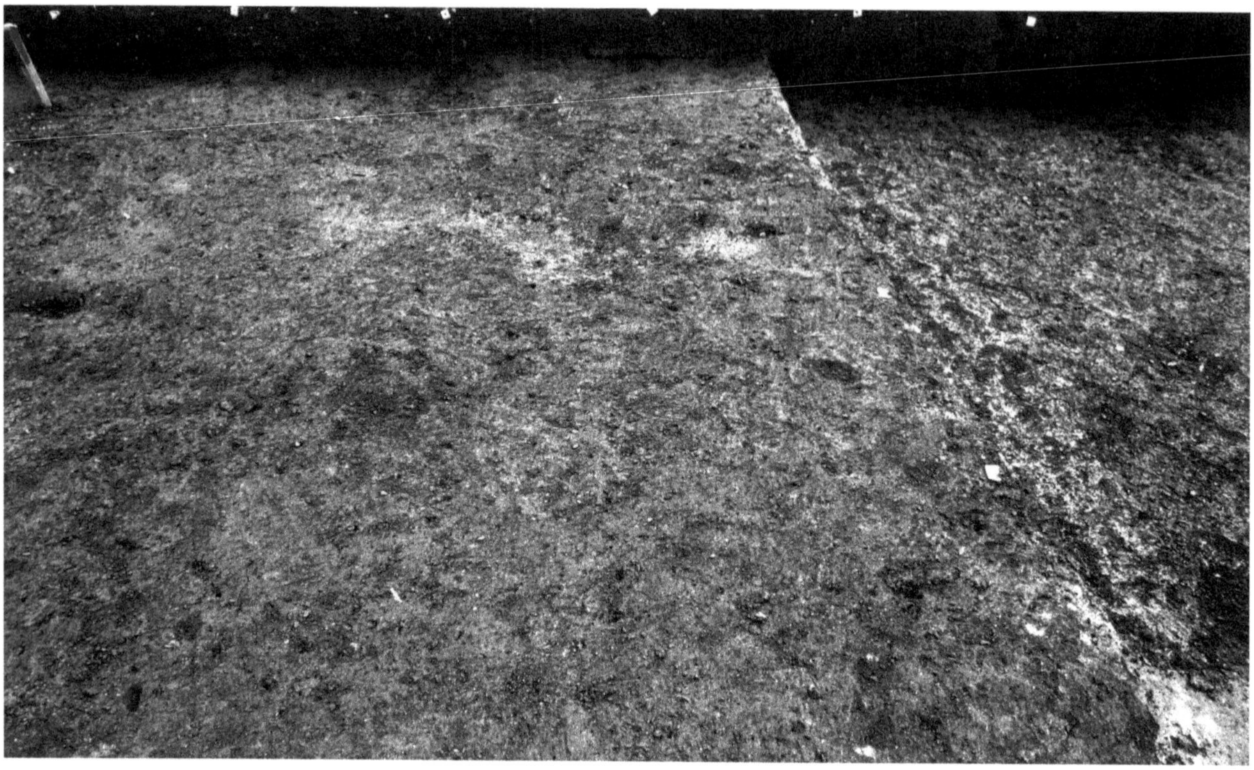

*Picture 14. Surface of the subsoil in the centre of the celtic enclosure. It was disturbed by the roots of trees belonging to the supposed sacred wood. Pict. CAD Ribemont.*

a series of two phases of this nature, both within the range of the third century BC to about 70 AD.

Within the enclosure, many punctual observations reveal the presence of a shrub cover, but it is probably very localized, because these evidences vary from one profile to another, even within the same profile (S50/P26/P18, S33a-d). These points are all located in the central or the northern half of the enclosure or on the immediate periphery of the same northern side (S4/P5; S37). The contamination by bones within the southern part of the enclosure and at its south-eastern periphery could be complementary with the bioturbated areas of the northern half and indicate comparable phenomena.

The sample S34 allows us to relativize and to show the discontinuous nature of certain data from previous samples. It was taken a few meters from the previous (S33) but has different characteristics because of its fossilization by the embankments brought for the construction of the first temple's foundations. Here, one can only note the absence or extreme limitation of bioturbation (soil type I) (see pict. 12). It is a ploughed soil of the same nature as the one observed in S3, whose thickness is accentuated by a former colluvial deposition that is probably related to a repeated ploughing. We have here the example of a Pre-Celtic surface horizon that has hardly been modified by the impact of humans or plants of the Celtic period.

***Malacology and palynology***

In general, the malacological results obtained for S33 show an open and wet environment with locally drier areas. The environment is then probably marked by a small shrub cover and bushes on calcareous soil. Overall, all periods combined, the sequence of samples taken from S33 is characterized by a continuation of grasslands (pastures, gardens, maintained private yards), without seeing the typical forest species that naturally colonize the environment. This phenomenon of prolonged grassland conservation is only possible through human intervention and the controlled management of areas that shows the maintenance and regular attendance of the enclosure over time.

The malacological spectra for S33c and S33d are characteristic of a low-shrubby environment and show a passage from an open to a partially wooded environment. Meanwhile, the pollen analysis (P2) presents relatively high values for trees and shrubs (more than 60%), which are dominated by birches (over 55%). However, birch being a heliophilous and anemophilous pioneer tree with high capacity of pollen production and dispersal, these high values of tree pollen could reflect the presence of a birch thickets at a larger scale (extra-local or regional picture). The herbaceous layer is dominated by grasses and Filicales (ferns); the lacking of pastoral pollen indicators (e.g. Plantago sp.) suggests the absence or keeping away of cattle (little or no trampling).

The malacological results of profiles S33c and S33d (soil type II) are different from those of profiles S33a and S33b (soil type III), that are respectively very poor and without molluscs. The moisture content is high for samples S33b

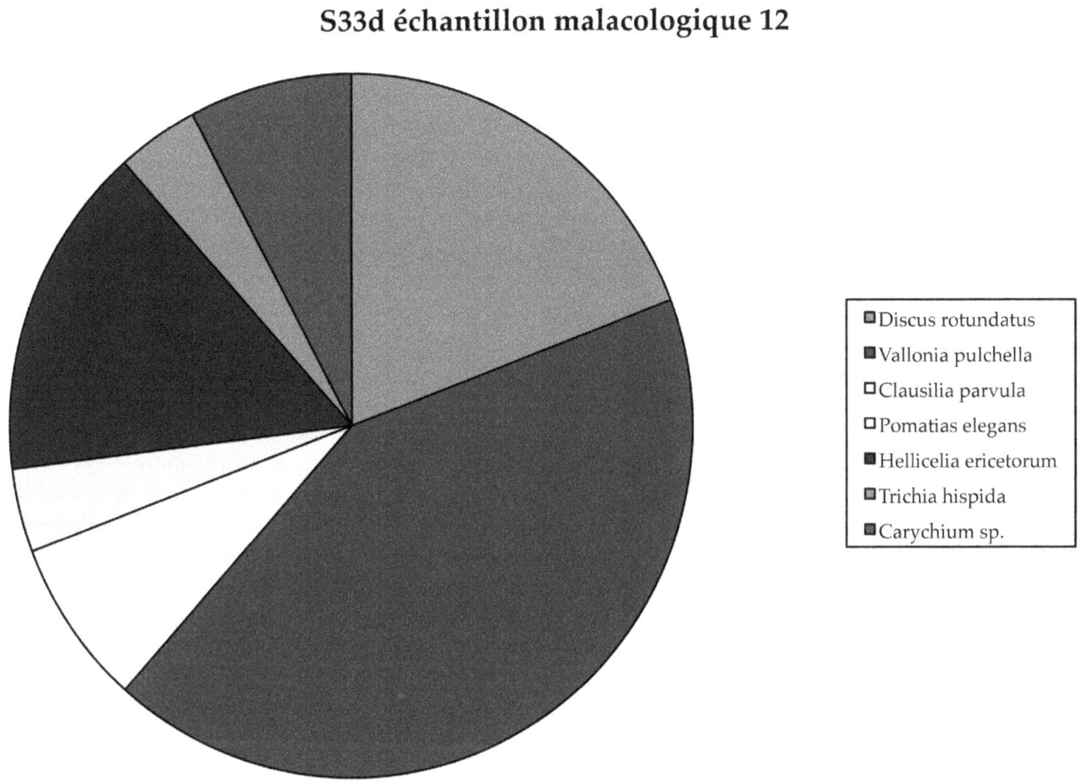


*Diagram 4. The legend is to be followed clockwise. The dominant malacological species is* Vallonia pulchella, *followed by* Discus rotundatus *and* Hellicelia ericetorum.

and rather high for S33c which is mainly wet, whereas the samples S33a and 33d (diagr. 4) are clearly open and dry, with few shrubs (presence of xerophilous taxa and 3 *Discus rotundatus*).

***Interpretation***

From the third century to the middle of the first century BC, palaeoenvironmental data and soil characteristics reveal, in the centre of the sacred enclosure, a fitting made of plants which is only an interface in the stratigraphy, but whose highlight is a major component for the knowledge of ritual practices and their evolution. We are here in the presence of a grove that colonized the central area of the enclosure (naturally or through human action). The local clear presence of phosphates and calcium carbonates (profiles S4, S33A-D, S42) reveals the ancient presence of human remains that have not been found during the excavation. The density of trees is relatively low and its impact is very punctual both on the soils and the fauna. Human presence is evident but relatively weak, and probably related to a punctual event. The presence of birch and hazel, typical of pioneer stands, could be the witness of a will of humans to preserve these thickets, for reasons that can not be advanced here (religious or economic, as source of wood, leaves or fruit offerings, food for cattle or men, manufacture of tools and/or poles and planks, symbol). Given its position at the centre of the enclosure, close to the implantation of the first Roman temple it would be legitimate to see the manifestation of divine presence (sacred wood) (Brunaux *et al.*, 2003: 67) (for the nature of sacred woods in european Antiquity, see de Cazanove and Scheid, 1993; see also example and discussion on artificial sacred woods in Gillet *et al.*, 2006: 191-5).

### III.3.3. Outside the enclosure: the surface of the bone feature Eo-42 ("charnier", pict. 15)

This sequence is dated by the archaeological study to the period from the middle of the third century BC to the middle of first century BC. It corresponds partly to a built feature containing human remains that is very likely to have been tightly closed (no signs of predator attack on the bone) (Brunaux *et al.*, 2003).

***Soil science***

Micro-morphological, pedological and pedo-chemical studies were conducted on the profiles covered by a large mass of human bones called the "charnier", located outside the south-eastern corner of the square enclosure. The objective was to understand the nature of facilities associated with this heap, distinguishing between central (S38), outlying areas (S6) and the edge of the enclosure ditch (S13).

The pedosedimentary characteristics associated with this features are relatively abundant and appear useful for a

*Picture 15. Detail from Eo 42, "charnier": this deposit contains the body of a hundred individuals. Pict. CAD Ribemont.*

better definition of the environment of the bone disposal. Three different situations are observed in four places of the feature.

- On one hand, in the northern outskirts of the feature, only a small accumulation of dark matter has been observed under some bones (in S6); there is no marked redox at the top of the underlying soil at this location.
- On the other hand, inside the feature (study by G. Canseco in 1996 and in the sample S38, a horizon of less than 5 cm is completely leached of its phosphorus, iron and manganese, these oxides forming a continuous and hard crust at the base of this horizon.
- Finally, on the southern outskirts of Eo-42, a thin and less pronounced leaching is present, but without crust or continuing precipitation of iron (S13).

From a pedological point of view, the cause of this redox process called oxydo-reduction is a lack of oxygen which must necessarily be associated with the presence of humus; the process is activated when a certain level at one of these two factors is reached, producing an imbalance of charge that the solution of iron or manganese will compensate.

The environment that triggers this process remains to be determined, but seems to involve a thin compressed and/or waterlogged horizon and, perhaps, a large amount of organic matter. This environment is quite unusual in these soil conditions and must be closely associated with activities related to the feature. That makes its thorough characterization important for a better understanding of this structure.

Pedology shows the establishment of these remains over a previous ploughed horizon, thickened by a slight sedimentation during ploughing. Its base has suffered a slight bioturbation perhaps linked to the presence of corpses or skeletons; it corresponds to soil type II, slightly reworked by bioturbation during Celtic period (Fechner, 2000). In S6, a very thin anthropogenic filling, probably linked to the digging of the ditch, could locally be underlying the human remains. In the same profile, the low accumulation of dark matter under certain bones is not marked in thin sections, which means that it is not mineral but organic matter, very little traces of which are remaining.

The soil type II in S6 and S38 presents an important local bioturbation, but no direct impact by intense rooting or tree falls and trunks was found. An animal, different from vertical earthworm galleries, causes these small and more or less horizontal traces at the base of the humiferous horizon. It also has a very thin and punctual horizontal accumulation organic carbon on the surface, possibly formed under direct and indirect influence of the presence of human corpses (S6; see Canseco, 1996; Fechner *et al.*, 2008). The leaching is complete and the characteristic of a complete lack of oxygen in the presence of some organic matter (Fechner *et al.*, 2004b). To explain this, a strong compression is one possibility, weakly expressed in the thin section, but a complete saturation by standing water seems to be the main cause here, as for instance in a shallow micro-depression filled with water or mud.

A thick crust of iron and phosphate, with phosphorus confirmed both in analyses and in thin section, is found a few centimetres below the level of the bones. It is separated from it by light grey, strongly leached and slightly compressed sediments (S38).

***Micromorphology***

The micromorphological study of G. Canseco (1996), conducted in the central part of Eo-42, highlights a surface stabilisation and superficial bioturbation of the surface before the coming of human remains (Canseco: 34-5 – pict. 16-1 to 16-4). The hypothesis of a confined environment and an organic cover of the kind of a wall in leather would explain the high redox (Ibid: 35), but it seems difficult to prove this in the current state of the research. Chemical analyses confirm the impressive and unusual leaching within the 5 cm under the bone layer and show the high rate of total phosphate in the crust located below (3460 ppm). A sharp limit between the soil without oxygen and the oxygen-rich soil is present which, for the underlying soil, can not be explained otherwise than by an extreme waterlogging or compression in the presence of some organic matter, even under a floor or another fitting made of organic matter. In thin sections, the phenomenon which caused the lack of oxygen (in S38, closed to the sample studied by G. Canseco) could be characterized in more details by the organic chemistry analysis of ingredients of non-mineral nature accumulated in places below the bones (e.g. in S6, where the redox apparently did not erase the traces of other phenomena). This important new approach might contribute significantly, in the future, to the complete understanding of the process of the setting up of this bone feature.

***Palynology***

The pollen spectra associated with this phase have a high variability according to the sample locations (P4, P27, P29, S38). The profile P4 shows, from bottom to top, a drastic fall of the tree cover (from 24% for the sample of the lowest part of the profile to more than 5% in the sediments located between the bones); there is a faint echo of the presence of birch (3%), while the herbaceous cover is mainly composed of Asteraceae (*Crepis* type) in strong increase (up to 71%). *Plantago* sp. is weakly represented (less than 2%) while the Poaceae values are constant and slightly above 10%. Samples P27, P29 and S38 were taken from the layer which supports the set of bones. P27 shows a slightly humid environment (presence of freshwater algae, type 225, and moss spores, which might support the mentioned pedological and micromorphological information. There are some ruderalized species, in connection with the presence of herds (Chenopodiaceae, *Centaurea nigra* type).

P29 shows a landscape dominated by species characteristic of lawns (over 20%) and by ferns (10%); these lawns do not seem to have been affected by fire (very little charred particles), and are colonized by thickets with juniper (2%) and birch (4%). The sample S38 is very rich in pollen and indicates the presence of nearby crops (cereals and weeds), although in small amounts. It presents ruderalized open lawns related to the presence of herds (*Artemisia*, Chenopodiaceae, *Plantago* sp., *Urtica* type, *Aster* type, Sordariales), but weakly directly impacted (little or no soil erosion). In the vicinity, the presence of birch trees and the abundance of juniper are to be noted. The presence of cereals (about 2%) and simultaneously of arable weeds (accompanying crops), such as Brassicaceae, *Polygonum aviculare* type and *Centaurea cyanus* type, suggests relatively important and/or close agricultural activities, rather than a deposit of grains and/or spikelets of cereals, for instance as offerings. In contrast with P29, this sample below the level of bones presents many charred particles with spores of carbonicole fungi (*Gelasinospora* sp., *Neurospora*, *Chaetomium*), which suggests the use of fire to clear the last thickets in the vicinity.

***Malacology***

In S38, malacology has especially shown the lack of carnivorous species (associated with the presence of flesh) and the probable association of molluscs with an open environment prior to the establishment of human remains (Deligne, 2001).

***Palaeoparasitology***

The palaeoparasitological study of 5 samples taken at the level of soil underlying the bone feather revealed the presence of three parasites in the studied material, allowing specifying that the body with their intestinal contents were present. This result seems to prove the *in situ* decomposition of corpses in the Celtic trophy. In addition to the conclusions concerning the anthropological field, the discovery of parasites in the studied material provides additional data on diet and health of populations. The first components identified are the eggs of *Taenia* sp. (pict. 17). Tapeworms are, most of the time, responsible for abdominal pain, vomiting, diarrhea and a weight loss of the affected individual. Parasitosis is contracted by ingestion of poorly cooked meat of beef or pork, containing the encysted larvae of the parasite.

The second type of eggs found in the samples studied is the egg of *Fasciola* sp. (pict 18), the large liver fluke (Bouchet *et al.*, 1998). These parasites cause liver enlargement and cirrhosis. (hepatomegaly) of the patients. This parasitosis is contracted by the ingestion of poorly washed raw salads, carriers of infective forms of parasites: watercress (*Nasturtium officinale*), dandelion (*Taraxacum dens leonis*), corn salad (*Valerianella olitoria*), chicory (*Cichorium intybus*) or Bear's garlic (*Alium ursinum*).

Finally, the third and last type of egg highlighted at Ribemont-sur-Ancre is the egg of roundworms of the genus *Ascaris* sp. (pict. 19). They are hematophagous and cause

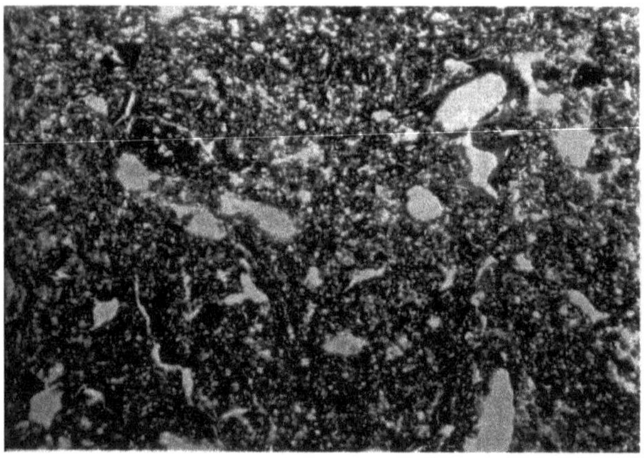

*16-1. Leached soil corresponding to the virgin soil 130-95 cm. Abundant clay coatings micro-layered embedded in the basal body and around grains. Presence of a ferruginous nodule strongly impregnated with abrupt edges. (x 1.6.) (according to G. Canseco).*

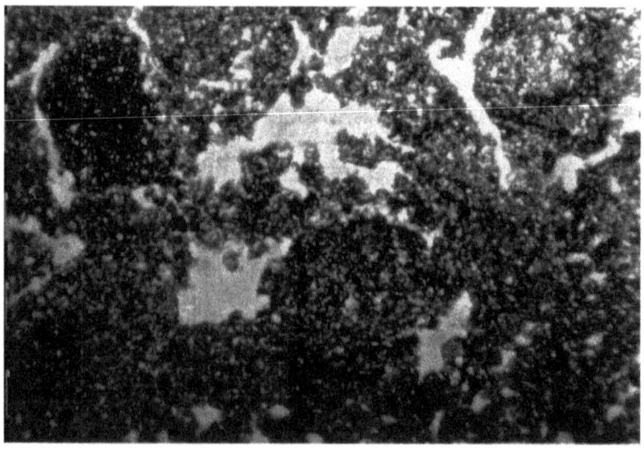

*16-2. Eluviated horizon corresponding to the Latenian archaeological level, 95-80 cm. Presence of ferruginous nodules amorphous, moderately impregnated. (x 1.6) (according to G. Canseco).*

*16-3. Palaeo-surface of the "charnier", in situ, formed by a facies of compacted silts with micro-layering, indicative of micro-colluviation and runoff. It presents abundant ferruginous features. (x 1.6) (according to G. Canseco).*

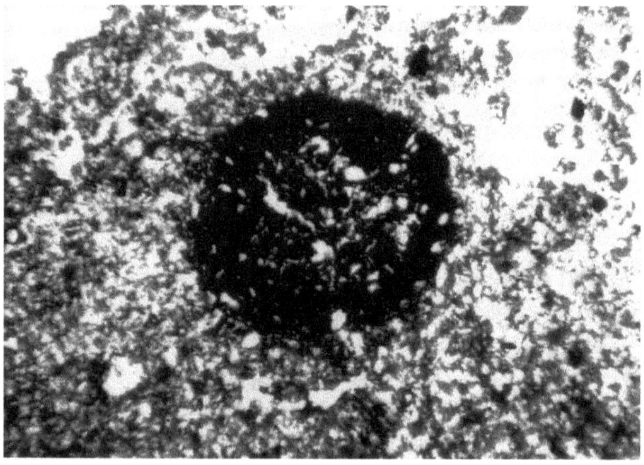

*16-4. Palaeo-surface of the "charnier". Ferruginous nodule showing an almost pure impregnation, constituted by ferruginised organic matter. (x 3.5) (according to G. Canseco).*

*Picture 16. Details in thin section of Eo-42, called "charnier"*

minor illnesses as some abdominal discomfort. However, in case of hyperparasitaemia, they can cause serious intestinal occlusions. Ascarosis does not reflect a particular nutrition, but gives information about a lack of hygiene of the populations (dirty hands, badly washed food).

### Interpretation

Palynological and malacological samples of S6, gathered on the outskirts of the construction, are not directly indicative of the building but probably correspond to a sheltered space at the foot of the latter. On the other hand, the sample S38 was collected within the building and it has very specific characteristics that indicate an artificial hydromorphy and compression.

### Perspectives

The detailed restoration treatment of part of the weapons associated with this structure with a micro-sandblaster, after observing with the binocular microscope, has allowed the identification of putrescible materials protected by corrosion. The presence of fragments of wood, leather, skin and clothe adherent to the metal was observed. The treatment has also made possible to identify remains belonging to insects (pict. 20). These findings open the prospect of a future archaeoentomological study. This is an area of particular interest because the presence of insects is in general related to the phenomenon of decaying corpses and hence, to the conditions of their treatment at the time of the foundation of the site.

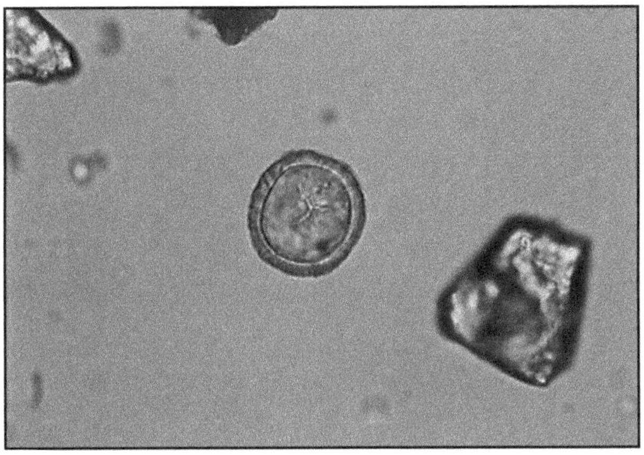

*Picture 17. Embryo-phorus of Tænia sp. 35 x 36 µm (x 400) (according to F. Bouchet).*

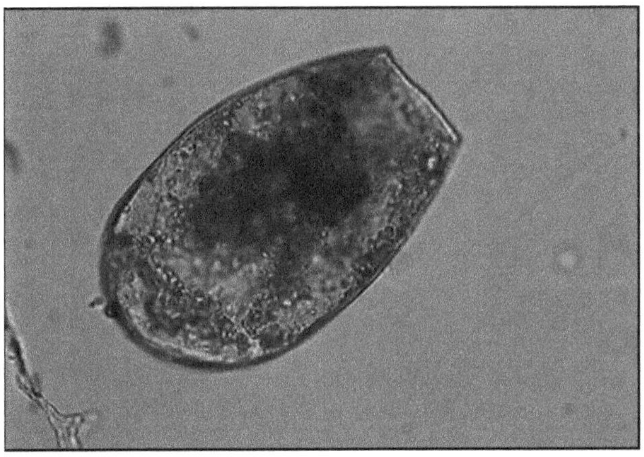

*Picture 18. Egg of Fasciola sp. 124 x 72 µm (x 400) (according to F. Bouchet).*

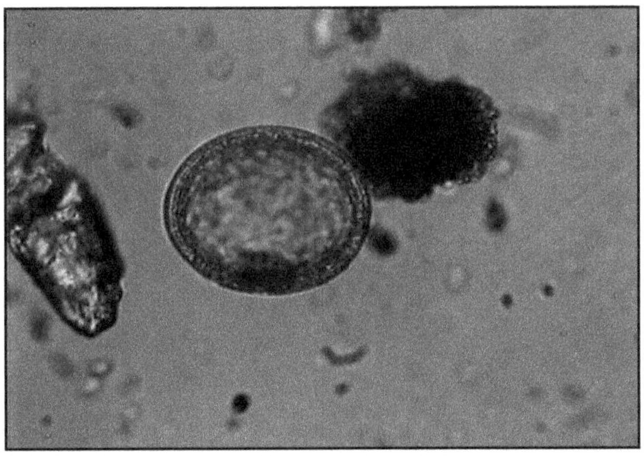

*Picture 19. Egg of Ascaris sp. 64 x 58 µm (x 400) (according to F. Bouchet).*

### III.3.4. Outside the enclosures: the soil at the southern periphery of the bone-feature Eo-42

As shown by pollen analyses, the sample P1 (pict. 21) corresponds to an open environment, slightly invaded by ferns (10 to 15%), with a grove of birches that grows nearby (14%); on the other hand, P13 has revealed an environment of thickets with nearly 60% of tree pollen marked by an absolute domination of birch (over 40%), together with hazel (10%) and pine (less than 10%).

For P20, we note the presence of birch in the vicinity (less than 10%). The environment is still quite open, grasses representing over 10%, Asteraceae nearly 15%; P20 is situated on the immediate outskirts of Eo-42 and close to P13, but situated in a free space left between the closing systems of the sacred enclosure and of the esplanade delimited by a trapezoidal enclosure.

The area has probably not been submitted directly to fire, as spores of carbonicole fungi (living on burnt wood and showing the local nature of the fire) are absent from the pollen assemblage. The environment is quite disrupted, as indicated by *Coenococcum* (fungus representative of erosion facies) and considering the lack of Monoletes fern spores, nearby flocks or droppings (Chenopodiaceae, *Plantago* sp., *Centaurea nigra* type, Sordariales).

This marked difference in samples taken a few meters from each other is probably due to the presence of obstacles (building or closing system), P13 is located at the junction of the ditch that limits the enclosure and the fence (F30) which surrounds the esplanade; thus a passage is materialized between these two structures that are not connected. The image given by plants in the sample P13 probably reflects the environmental conditions inside the enclosure, confirming the observations of the soil preserved within the enclosures (P2, S33), whereas the vegetation of the esplanade is highlighted by P1, and the environment outside the enclosure by P20.

### *III.3.5. The circular enclosure (pict. 22)*

According to the pollution of the underlying soil in the circular enclosure, the latter was also probably covered with bones at the origin (S40 and S42). The type of molluscs found in these samples could indicate the presence of past fittings or constructions made of stone (flint ?) in this enclosure. However, as other causes, the presence of dead wood also might explain this malacological assemblage.

## *III.4. The first Gallo-roman fittings, second half of the first century BC*

In the years 40-30 BC, the site is subject to important transformations. The ancient installations are dismantled; the space is cleaned by fire[9]. The western half of the Celtic site is levelled, while materials extracted during this operation were spread on the surface of the eastern part.

[9] A thick layer of charcoal was identified in the whole ditch.

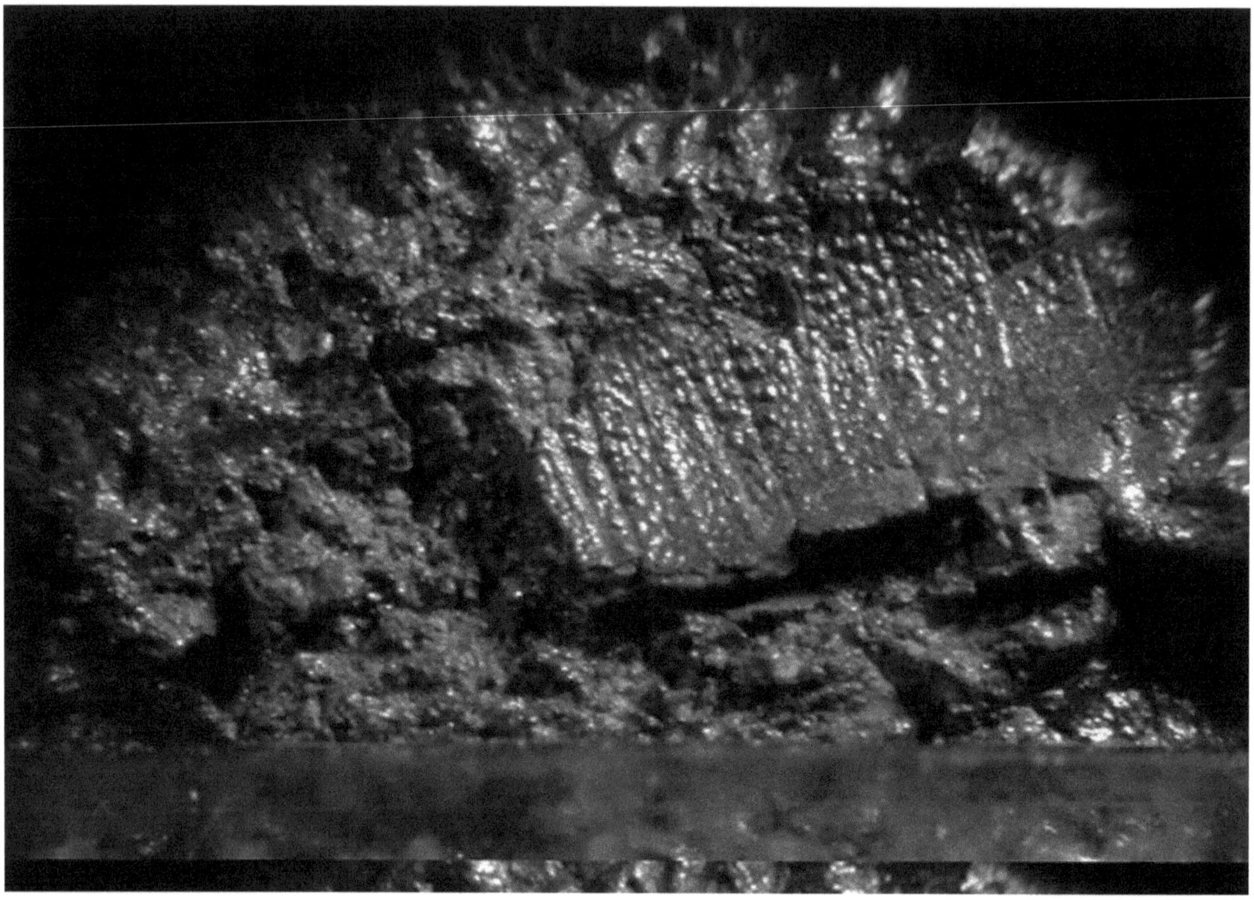

*Picture 20. Pupa of a fly preserved by the corrosion of iron, on the surface of a scabbard (Eo-42). Pict. M. Leroux.*

*Picture 21. Location of the sample P1. The stratigraphy is well preserved. Two roman soils (chalk lines) show two stages of construction (respectively 50 and 70 AD.). They cover a layer of the Ist century BC. At the bottom of the picture (centre) a spear butt shows the location of the original ground (IIIrd century BC.).*

*Picture 22. The circular enclosure with the hollow altar (in the centre). Pict. CAD Ribemont.*

***Palynology***

Samples P2, P18, P26 and P38, corresponding to a level of fill under the set of bones Eo-496, are associated with the levelling of the Celtic site. Pedological analyses have also been undertaken for P38 (S50) (Fercoq du Leslay *et al.*, 2009: table 8).

The presence of numerous carbonized particles in the palynological samples, with spores of carbonicole fungi (*Gelasinospora* sp., *Neurospora*, *Chaetomium*), confirms the use of fire for cleaning. Compared to the Celtic and early Gallo-Roman levels (sacred wood ?), the palynological results of the sample P2 reveal a drastic fall of trees from nearly 60% for to average values of 7 to 15%. The birch (around 5%) is still present, accompanied by pine (less than 2%) whose presence is more discrete and certainly from long distance origin. On the contrary, the willow (less than 2%), characteristic of wetlands, is of local origin. The environment is wide open with Asteraceae (80%) dominating the assemblages. P26 shows a fairly ruderalized (*Artemisia*, Chenopodiaceae) and disrupted (*Coenococcum*) environment. These taxa grow generally nearby plots cultivated for cereals or may be imported with cereals and weeds into the site after a harvest (presence of significant quantities of *Cerealia* type, *Polygonum aviculare* type, *Centaurea cyanus* type and *Brassicaceae*). Some birches could however be present near the sampling point.

P38 is the sample most affected by the proximity of flocks and/or droppings with a maximum of Sordariales, *Coniochaeta*, *Diporotheca* sp. (coprophilous fungi) in the pollen assemblages. *Glomus fasciculatum* and *Coenococcum*, typical of surface soil erosion, are also present. Rudero-nitrophilous species associated with pastoralism and soil trampling are very well represented (*Artemisia*, Chenopodiaceae, *Plantago* sp., *Centaurea nigra* type, *Serratula* type, *Aster* type). The environment is fairly moist with the presence of *Equisetum*, Apiaceae, spores of Bryophytes (mosses) and algal rests of type 181. As for sample P2, P38 suggests a presence of crops in the vicinity (*Cerealia* type, Brassicaceae, *Polygonum aviculare* type, *Centaurea cyanus* type), and possibly of few isolated birch trees.

P18 reveals the presence of hazel and birch near the sampling point, which is supported by the presence of many fragments of deciduous tree wood in the palynological slides). The environment is fairly dry (presence of fungal spore type 200) and still under the influence of fire (microcharcoal and *Chaetomium*). The space around the area of sampling was certainly shared between areas of cereal crops (*Cerealia* and their arable weeds Brassicaceae, *Polypodium aviculare* type, *Plantago* sp.), and areas of pasture (Poaceae, Cyperaceae, *Serratula* type, *Cerastium* type, *Aster* type, *Centaurea nigra* type, Artemisia, *Rumex*).

*Picture 23. Bone deposit Eo-496. It was set up after the phase of levelling prior to the construction of the first Gallo-Roman temple. Was it to reconstruct an ancient deposit moved during works ? Pict. CAD Ribemont.*

***Soil science***

With regard to the sedimentological analysis, three different samples were treated; they are located inside the quadrangular enclosure, nearby the bone feature Eo-496 (S50/S18-2 and S50/S18-3 within square BB4E in the bone level, S50/S18-4 within square BB4E under the bone level; see pict. 23).

Bipolarity appears between S18-3 and the other two samples. These last two are very close in terms of amounts of clay and sand, of silt ratios, of relationships between chemical elements and in terms of rates of iron, lime and organic carbon[10]. However, if these two sets of samples show differences, they are quite close to each other when compared with other analyses made previously on the site. This lack of real comparison with the samples of the *in situ* soil is quite surprising. In S18, particularly poor sorting of the sands is indicative of a sediment that is strongly reworked or influenced by anthropogenic impacts. Samples that are the closest in terms of amounts of silt are at the top of the *in situ* soil (S4 and S33 south); but these samples are both richer in clay content and much less reworked, as indicated by the rates of sand.

[10] There are little differences in the amount of organic carbon between the three samples.

The characteristics related to the three samples S18-2, S18-3 and S18-4 promote the hypothesis of secondary fillings rather than layers formed in a pre-existing *in situ* soil, such as natural (E and Bt) or old ploughed (Apb) horizons. It therefore appears that the sediment on which the bone feature Eo-496 has been established is most probably an embankment and not the natural soil. It corresponds to a feature that makes the connection between the Roman and the Celtic installations and has to be linked to practices that are posterior to the Roman conquest.

***Chemical analyses***

The only clear exception to the specified gradation between the three samples mentioned above concerns the rates of total phosphate: S50/18-3 and S18-4 show a rate that is slightly higher than 18-2.

This can be explained by the movement of this chemical element, as already seen in S38. The most important information for these three samples concerns the extreme richness in total phosphate in relation to other profiles on the site.

For S50/18-3 and S50/18-4, such richness is only approached by the Celtic level, under Eo-42 in S38, which has spots of phosphorus (visible by eye). For S50/18-2, the rate is slightly lower than that level, but still significantly higher than the other samples of the site (fig. 10).

In addition to this very high rate of phosphate, we note the slightly higher rate of organic carbon for the three samples, compared to those of all other samples of the site. It could indicate the ancient development in this fill of a layer of decaying vegetation or the former presence of a construction made of organic materials. It is also possible that the embankment is derived from one or another of these layers, and has kept some of their characteristics, but in a secondary position.

#### *Malacology*

Malacological data from the study of the sample S36 indicate a fully open environment (grassland) with variable humidity. Through the profile, no change in the environment is observable over time, there is evidence of regular maintenance and of a continuity of meadows that needs a recurring anthropogenic management, as in S33. This prevents any important growing of local scrubland or forest. S34 is taken from the horizon above the bone feature Eo-42. It also shows an open environment of grassland type, associated with a low shrub cover (presence of *Discus rotundatus* which is a semi-forest species, *Orcula doliolum*, forest species) and some trees from various ligneous structures (hedges, shrubs or bushes).

Outside the quadrangular enclosure, the malacological content of the sample S6 contains too few shells that can be determined to be totally reliable, but suggests a more open environment (no forest species), where some bushes remain. We are at that time in the outside periphery of the sanctuary that is a sparsely frequented area.

#### *Interpretation*

In conclusion, in profile S50/S18/P38, both the nearness between the three samples and the absence of comparison with the natural soil profiles analyzed elsewhere in the enclosure, allow us to conclude that this corresponds to an upfill with sediments and not to a development into the Pre-Celtic soil. This information is of great importance for the understanding of archaeological stratigraphy, as it allows to distinguish the *in situ* levels affected by the Celtic occupation and more recent fills.

### *III.5. Third quarter of the first century AD to the end of the second century AD (fig. 12-13)*

#### III.5.1. The environment of the first temple

The ground inside the enclosure, which should correspond to a stabilization level, is unfortunately too bioturbated to study the details of the environmental developments for these periods. Even if two levels of soil may have been identified punctually, they are often not distinguishable, neither archaeologically nor by sedimentological analysis. According to artefacts collected in the archaeological layer within the enclosure, this layer is extremely heterogeneous and represents a chronological mixture from the end of the first century BC to the early third century AD. It is only possible to discern clearly the latest horizons near the buildings.

The north-western section and the north end of the north-eastern section of the ditch also present a stabilization level of its fills (pict. 24-25). Here, the stratigraphy has not been disturbed and shows that the fillings corresponds to successive constructions. These constructions have been stirred by bioturbations within the enclosure, until the construction phase of the *quadriportico*, where the ditch was finally filled up in the second century. Unfortunately, the pollen samples taken for A.-V. Munaut have not been kept.

#### *Soil science*

This long period is represented by the soil inside the sanctuary and in particular by the stabilization horizon in the Roman upfills (S36). The analysis is completed by the analysis of the soil in the *quadriportico*, which is the only well-dated level (third quarter of the first century AD to the third quarter of second century). This soil has been sampled for pollen analyses (P9), although its nature is not particularly adapted to a proper preservation of the pollen grains. It forms the upper layer of an embankment, inside a covered construction and closed to the outside of the temple yard (*Portico* Po 3).

The ancient soils outside the *quadriportico* are preserved, on its north-western side, and up to the wall of the temple courtyard of the second and third centuries. They are well preserved and indicative of the external environment of the sanctuary, in its rear part. Here the soils are sealed by the fills corresponding to the construction of the great temple and his yard between the end of the second and early third century AD (P10, P3).

#### *Palynology*

Inside the enclosure, as no *in situ* horizon was observed, the palynological analysis refers indiscriminately to levels dated from the beginning of the first century to the end of the second century AD.

For these horizons, two pollen samples have been taken in the profile P2 (levels 0-10 cm). From bottom to top, palynological results show a slight increase in tree cover (from 7 to about 15%), with a clear dominance of birches (6 to 14%). The lower level reflects a great diversity of species with some heliophilous, mesophilous and wetland taxa, but in amounts that are still fairly low. The higher level shows diversity of species that is significantly less important, but with slightly stronger representation rates of birch, associated with hazel, lime, charm, elm, oak, alder and willow. These samples should reflect an aspect of the environment within the enclosure as the concerned area is surrounded by a fence. It is likely that the diversity

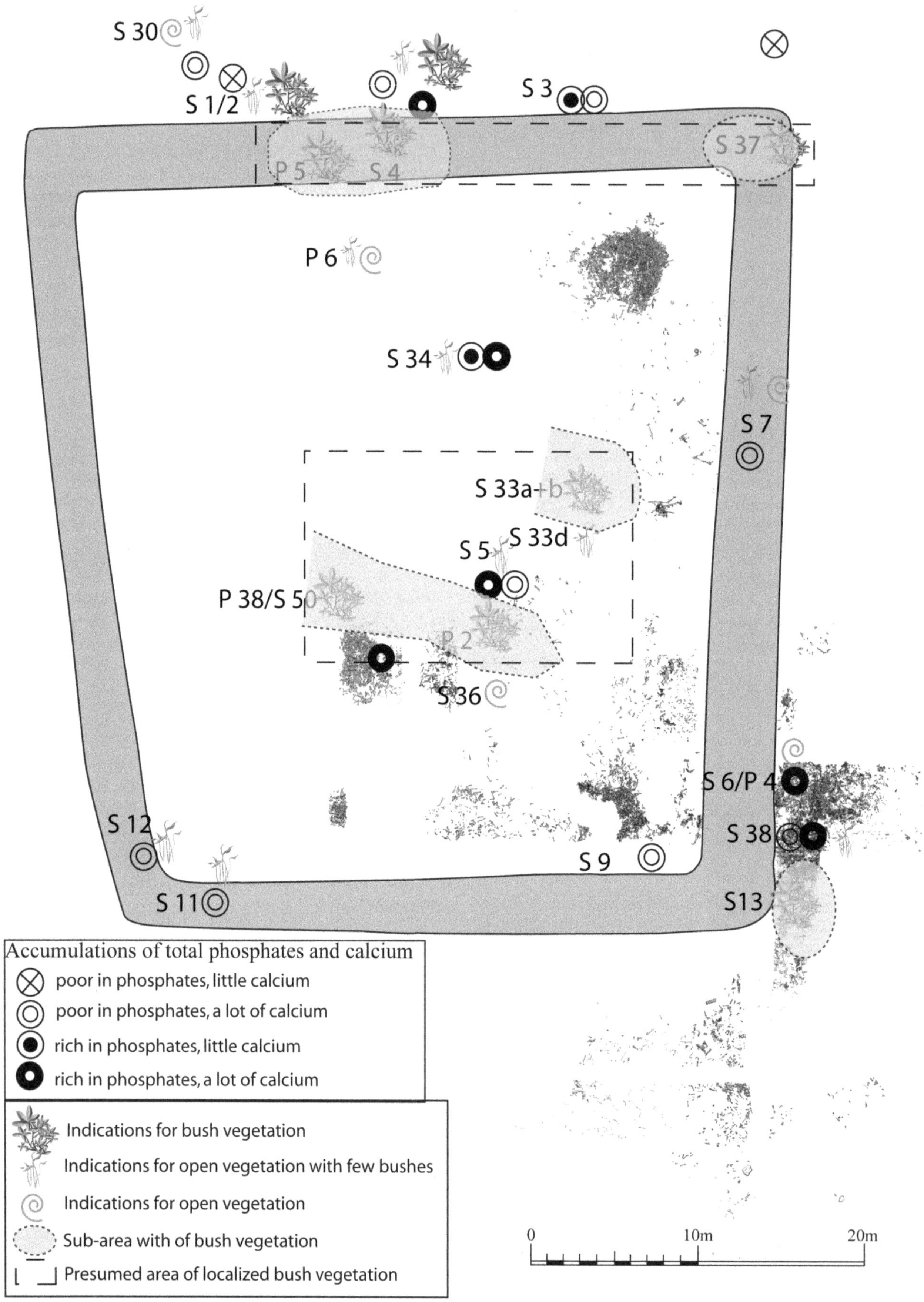


*Figure 11. Celtic and early Roman surface horizons: cartography of former bone distribution confronted with vegetation cover (synthesized from malacological, soil science and palynological studies).*

*Picture 24. The ditch F-329 is partly cut by the foundation of the pronaos of the roman temple. Here, the ditch is situated outside the area enclosed in the beginning of 1st C. and was thus only filled back around 70 AD, during the construction of the quadriporticus (Po-3). Sample P5 correponds to the border of the ditch on the inner side of the enclosure, profile S4 to its equivalent on the outer side, below the foundation of the pronaos. Pict. CAD Ribemont.*

*Picture 25. Section of the ditch F-252*

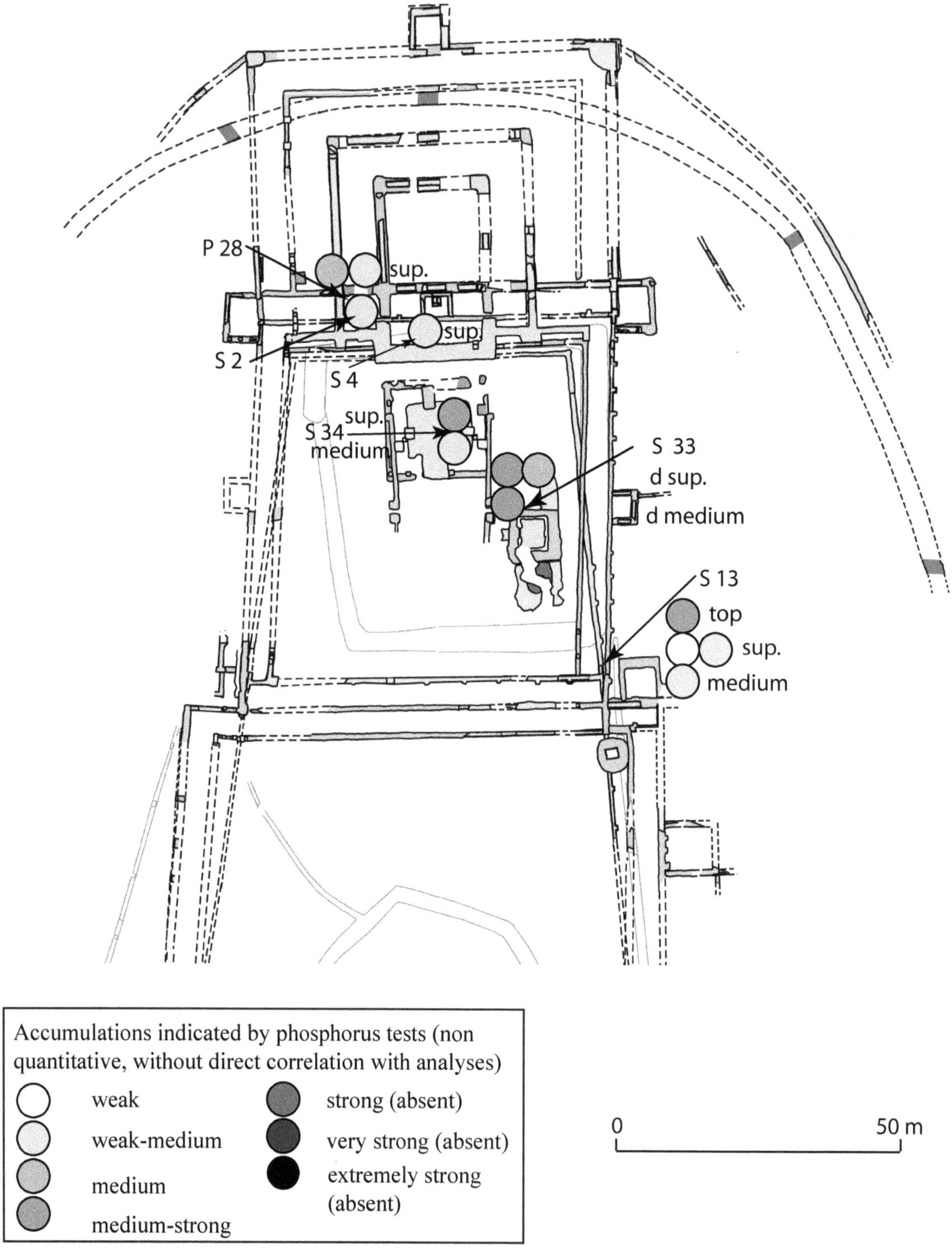


*Figure 12. Cartography of phosphorus tests for the Roman surface horizons (non quantitative, only relative amounts).*

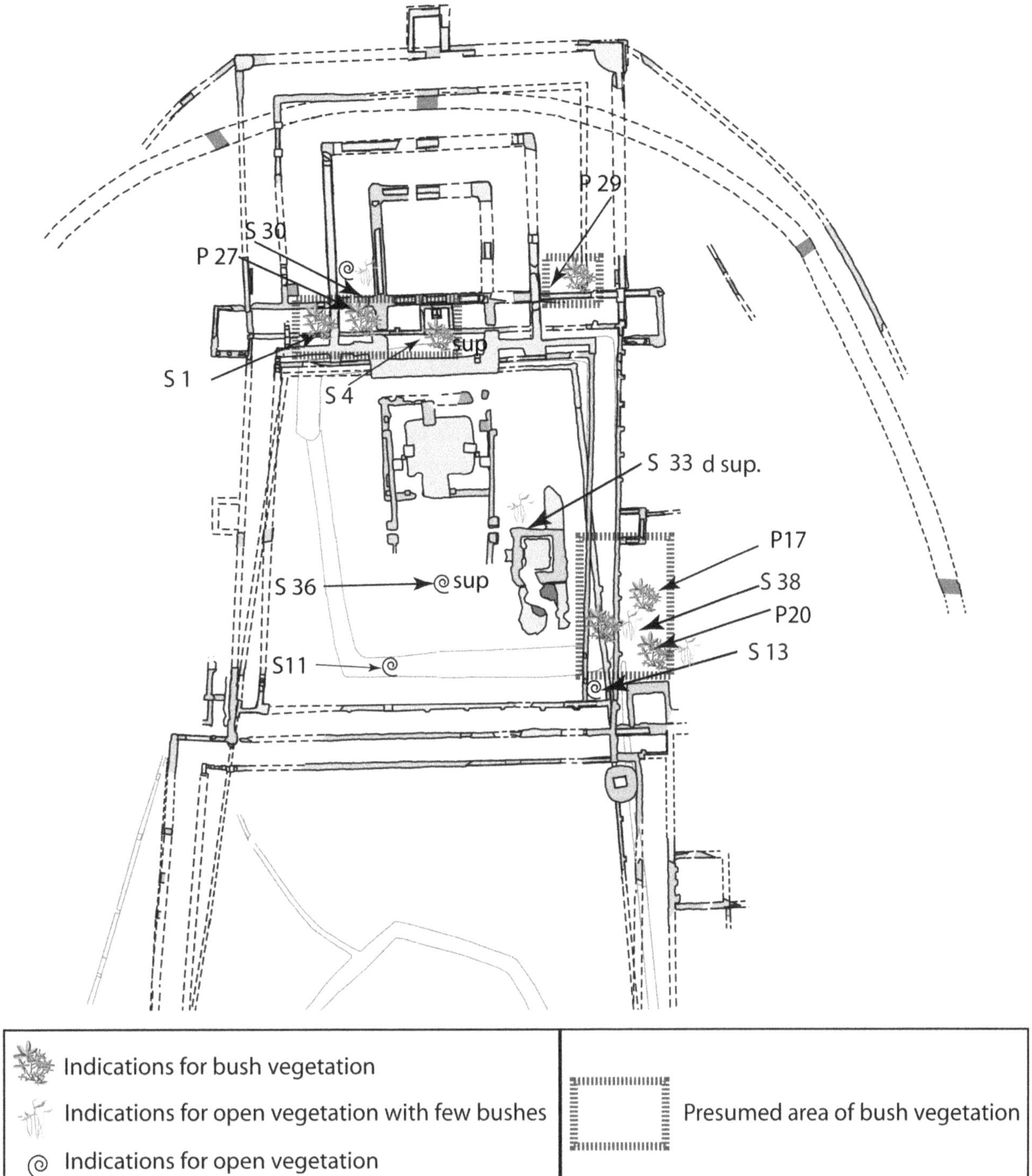


*Figure 13. Synthesis of the results of malacological, soil science and palynological studies for the Roman period, showing localized accumulations of phosphorus and outside samples with bush vegetation*

observed is the result of a voluntary choice (planting) rather than a natural restocking, considering the mentioned human maintenance of the area within the enclosure. We must imagine that the trees are confined within a specific space. Indeed, inside the enclosure, the grassland is the absolute dominant, with *Crepis* type representing alone nearly half of the identified species (45%) and grasses (nearly 20%). There is a faint echo of the presence of cereals (2% in the upper level), which may indicate offering practices rather than the proximity of agricultural activities; in the same level, *Plantago* sp. makes a timid appearance (1.2%), while is was absent in the lower levels.

In the sample from the *portico* (P9), palynology also shows the presence of trees (13.5%) with birch (9%) and hazelnut (2.6%), which is consistent with the data of P2. Pine is weakly present (1.9%) and come certainly from long distance transport. The herbaceous vegetation dominates with an overwhelming representation of the *Crepis* type (nearly 55%); wormwood (about 5%), grasses (over 10%) and ferns are (more than 5%) are also well represented. The presence of sedges (Cyperaceae) could reflect ambient humidity. Note that the population of molluscs (S33) does not seem to vary from the bottom to the top of the sequence, and one must take into account the very high anthropogenic or biological disturbance that affects the ground inside the enclosure for the Roman period.

***Interpretation***

The interior of the enclosure is an open environment that is more or less humid. Although the frequenting does not seem to be very important (punctual frequenting during special events ?), the area is maintained by man. The different sedimentological, malacological and palynological approaches agree to show a second shrub phase within the enclosure (sacred wood ?).

***III.5.2. Outside the enclosure***

***Palynology***

Outside the enclosure, in the immediate vicinity of the *portico* of the first century, the soil is perfectly conserved (S3) and is the reference level for the type I soils. This soil is fossilized by the upfills during construction of the classical temple. The tree pollen values of the two samples taken from this soil (P5 and P10) indicate the presence of a birch grove (from 30 to more than 55%), with hazel (10%) and lime trees (2%). At the bottom of the sequence (first century AD ?), the herbaceous vegetation is still important (45.9%) and the area is a semi-open environment. In the second century level, the trees play an increasingly important role (from 53.9 to 70%), whereas the Asteraceae, mainly represented by the *Crepis* type, are in sharp decline (from 30 to 10%). Poaceae and ferns slightly increase, respectively from 9 to less than 4% and more than 10%. *Plantago* sp. is remarkably absent.

For the period from the first to the second century AD, the nearby area outside the enclosure is a wasteland colonized by a grove (fig. 13) that grows in the vicinity and that tends to grow as time goes on; this area seems to be very little frequented by man

Outside the fence, on the northern side and along the ditch F 252 (P3, fig 14), the vegetation is characterized by a strong representation of ferns (above 70% in all levels analyzed in profile P3) and by Poaceae that do not exceed 5%. As far as trees are concerned, pine is the absolute dominant and constantly progresses over time, but its values remain rather low (representing less than 5% in the bottom of the sequence, it exceeds 10% in the upper profile). The other taxa present less than 1%, with the exception of hazel whose peak representation level barely exceeds 3% at the bottom of the sequence.

***Malacology***

Molluscs (S4 - lower level) show a partially shrub environment, and this vegetation is responsible for the bioturbation associated to soil type III.

The upper level of profile S4 shows a second bioturbation phenomenon whose dating is slightly earlier than the construction of *quadriportico*. The sample S4 H1A (-4 to -14 cm) is a little more reliable than S33 in terms of number of shells and species. It reveals an environment that is particularly wet (*Carychium*), with a presence of more trees than in S33, but without being a forest. Represented shells range from forest to semi-forest species. It could be a relatively open and clear wood, or perhaps a riparian forest. It is also to be noted that H1A is wetter than the layers above it.

***Interpretation***

We are situated in an area that is peripheric to the sanctuary, in a place that seems neither frequented nor used until a late date (second century AD ?); the digging of the ditch F 252 apparently did not substantially change the vegetal environment.

## IV. Conclusions and perspectives

The palaeoenvironmental studies carried out at the sanctuary of Ribemont-sur-Ancre have allowed us to provide information on the nature of the soils selected for the original location of the site, their subsequent development according to their location in different parts of the Celtic site, on particular aspects such as the phenomena associated with the bone feature or "charnier" (Eo-42) and at the bottom of the quadrangular enclosure's ditch, and on the correlation between these phenomena (e.g. between the filling of the bottom of the ditch and the Celtic surface horizon).

The different approaches that were sometimes in apparent contradiction (Fechner *et al.*, 2004a) are better understood in relation with archaeological contexts and dating. The environment prior to the foundation of the site appears as a landscape of cultivated fields and hedgerows, but this should be confirmed by further analyses (e.g. in the base of S34). By contrast, the environment of the Celtic site appears to be of a highly complex nature, with areas of shrubs and trees, with open and constructed areas, and with zones of accumulation of human remains, either maintained as such or displaced in a later phase (fig.15).

After this Celtic period, whose decoding necessitates a multidisciplinary approach, the Roman phases are less rich and more difficult to identify, because of the scarcity of *in situ* soils. They are characterized primarily by an open environment, maintained by humans who seem to have set up an artificial area of shrubs and trees, possibly and partly in continuity with the Celtic practices (?). These places seem relatively little frequented, although regularly

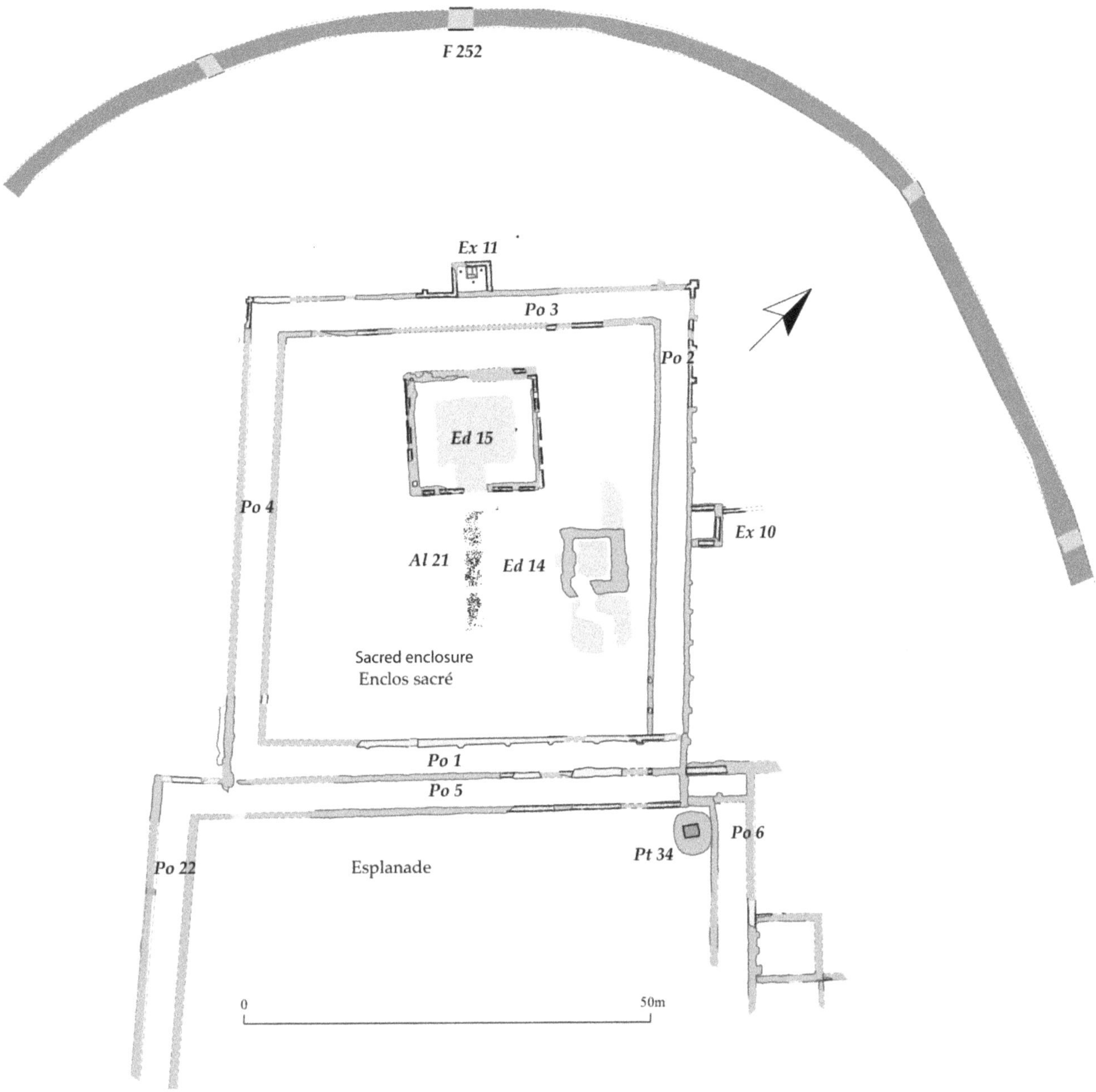


*Figure 14. Localisation of the ditch F 252*

maintained. Outside the site, a natural wooded and therefore not directly exploited area is developed, as opposed to the Latenian and earlier phases when the site was used for livestock and agriculture.

If the conditions of exceptional sealing at Ribemont-sur-Ancre have allowed the detection of phenomena such as a sacred "wood" and the monitoring of its evolution over the lifetime of the sanctuary, some problems remain to be overcome. For malacological and palynological studies, the too small number of analyses is to report and is a handicap that does not allow providing definitive answers. A new sampling campaign should be considered, taking into account three factors: samples of sufficient size for the malacology of ancient soils, fresh columns for palynology and their correlation with a long reference natural sedimentary sequence that has been thoroughly retrieved in the bottom of the main valley a few years ago.

## Bibliographical references


Agache, R., 1964. *Archéologie aérienne de la Somme : recherches nouvelles 1963-1964*, Amiens. (=Bulletin de la Société de la Préhistoire du Nord, bull. spécial n°6).

Agache, R., 1978. *La Somme pré-romaine et romaine*, Société des Antiquaires de Picardie, Amiens, 1978. 515 p.

Annaert, R., 1993. Een *Viereckschanze* op de Alfsberg te Kontich (prov. Antwerpen) : meer dan een cultusplaats, *Archéologie in Vlaanderen*, 3 : 53-125.

Bouchet, F., Bentrad, S. and Paicheler, J.C., 1998. Enquête épidémiologique sur les helminthiases à la cour de Louis XIV, *Médecine/Sciences*, 14 : 463-466.

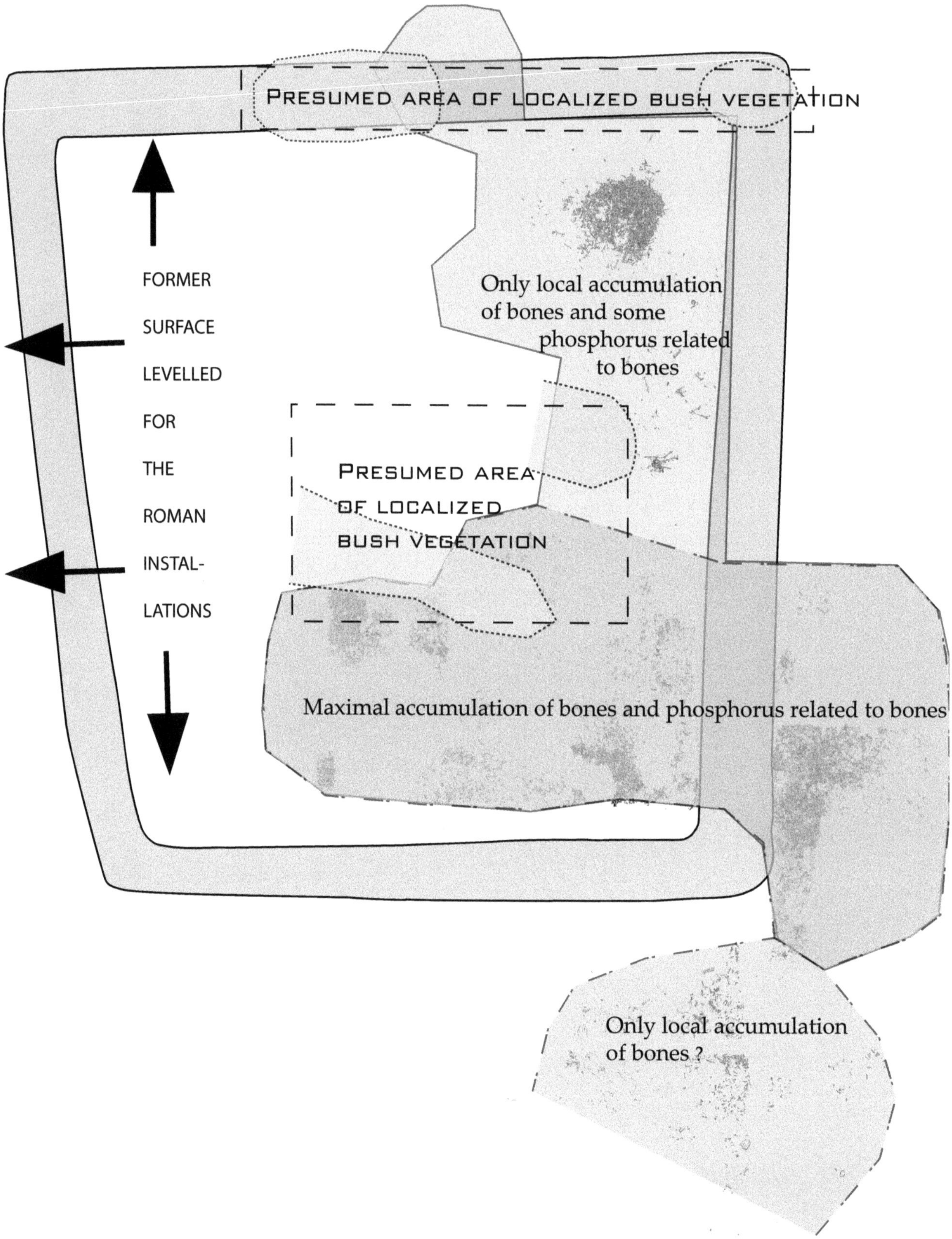


*Figure 15. The results of the environmental studies show traces of practices indisputably linked to ritual practices inside the enclosures, before the Roman period: distinguishing shrub areas (sacred wood ?) and areas of bone manipulation.*

Brunaux, J.-L., Amandry, M., Brouquier-Reddé, V., Delestrée, L.-P., Duday, H., Fercoq du Leslay, G., Lejars, T., Marchand, C., Méniel, P., Petit, B. and Rogéré, B., 1999. Ribemont-sur-Ancre (Somme) : bilan préliminaire et nouvelles hypothèses, *Gallia*, 56: 177-283.

Brunaux, J.-L. (ed.), Amandry, M., Bécu, B., Berard, F., Bouchet, F., Brouquier-Reddé, V., Chaidron, C., Court-Picon, M., Delestrée, L.-P., Deligne, F., Di Vita-Evrard, G., Fechner, K., Fercoq du Leslay, G., Hartner, S., Le Bailly, M., Langohr, R., Marchand, C., Munaut, A.-V. and Wyss, S., 2009. *Les temples du sanctuaire gallo-romain de Ribemont-sur-Ancre*, St. Germain en Laye.

Brunaux, J.-L., Malagoli, C., Bataille, G. (collab.) and Lambot, B. (collab.), 2003. La France du Nord (Champagne-Ardenne, Ile-de-France, Nord, Basse-Normandie, Haute-Normandie, Pas-de-Calais, Picardie). In : Arcelin, P. and Brunaux, J.-L. (eds.), *Cultes et sanctuaires en France à l'Age du Fer : dossier. Gallia*, 60 : 9-73.

Cadoux, J.-L., 1971. Le sanctuaire gallo-romain de Ribemont-sur-Ancre, *Bulletin de la Société des Antiquaires de Picardie*, 2 : 43-70.

Cadoux, J.-L., 1984. Le sanctuaire gallo-romain de Ribemont-sur-Ancre (Somme): état des recherches en 1983, *Revue du Nord*, 66, 260 : 125-145.

Cadoux, J.-L. and Lancelin, P., 1987. L'ossuaire de Ribemont-sur-Ancre (Somme), *Anthropozoologica, 1er numéro spécial* : 91-96.

Cadoux, J.-L. and Massy, J.-L., 1970. Ribemont-sur-Ancre : études, *Revue du Nord*, 52, 207 : 469-511.

Canseco, G., 1996. *Étude micromorphologique du charnier de Ribemont-sur-Ancre, les surfaces aménagées*; Mémoire de DEA, sous la dir. de M.-A Courty, Université Paris I - Paris X, 37 p.

de Cazanove, O., Scheid J. (eds.), 1993. *Les bois sacrés. Actes du colloque international organisé par le Centre Jean Bérard et l'Ecole pratique de Hautes Etudes (5e section). 23-25 novembre 1989*, Naples. (=Collection du Centre Jean Bérard, 10).

Deligne, F., 2001. Analyse malacologique de l'occupation celtique du site de Ribemont-sur-Ancre. In : Fercoq du Leslay, G., Deligne, F., Fechner, K. and Munaut, A.-V., *Les études environnementales sur le site de Ribemont. Ribemont-sur-Ancre*, Rapport triennal 1999-2001 vol. II, Annexe XIII.

Fechner, K., 2000. Comparaison des histoires pédo-sédimentaires des fossés protohistoriques étudiés en Belgique et dans le Nord de la France, *Revue Archéologique de Picardie*, 1-2 : 23-56.

Fechner, K., this volume. Case studies of phosphorus mapping in Late La Tène to Roman enclosures and ponds between Rhine and Seine. In: Fechner, K., Devos, Y., Leopold, M. and Völkel, J. (eds.), *Enclosed and buried surfaces as key sources in Archaeology and Pedology (papers of the EAA congress of Krakow)*, British Archaeological Reports. International Series.

Fechner, K., Deligne, F. and Fercoq du Leslay, G., 2004a. L'environnement de l'occupation celtique du site de Ribemont-sur-Ancre. *Lunula, Archaeologia protohistorica*, 12 : 147-159.

Fechner, K., Langohr, R. and Devos, Y., 2004b. Archaeopedological checklists. Proposal for a simplified version for the routine archaeological record in Holocene rural and urban sites of Nord-Western Europe. In: Carver, G. (ed.), *Digging in the dirt: Excavations in a new millennium*, Oxford: 240-256. (=British Archaeological Reports, International Series, S1256).

Fechner, K., Loicq, S., De Lil, A., Deligne, F. and Court-Picon, M., 2008. *Rapport de l'étude pédologique 2006-2007 et synthèse environnementale sur le site de Ribemont-sur-Ancre*, Research Team in Archaeo- and Palaeo- Sciences (ROOTS), unpublished report.

Fercoq du Leslay, G., 1996. Chronologie et analyse spatiale à Ribemont-sur-Ancre (Somme), *Revue Archéologique de Picardie*, 3-4 : 189-208.

Fercoq du Leslay, G., 2000. L'apport des fossés de Ribemont-sur-Ancre (Somme) à la chronologie et à l'interprétation du site, *Revue Archéologique de Picardie*, 1-2 : 113-146.

Fercoq du Leslay, G., Court-Picon, M., Fechner, K. and Loicq, S., 2009. *Analyse environnementale du sanctuaire gallo-romain de Ribemont-sur-Ancre (Somme)*. Annexe VII p. 255-312. In : Brunaux, J.-L. (ed.), *Les temples du sanctuaire gallo-romain de Ribemont-sur-Ancre*, St. Germain en Laye.

Ferdière, A., 1966. Première campagne de fouilles dans la villa gallo-romaine de Ribemont-sur-Ancre (Somme), *Revue du Nord*, 191: 439-543.

Gillet, E., Paridaens, N. and Demarez, L., 2006. Le sanctuaire de Blicquy - « Villa d'Anderlecht » (Prov. Hainaut, Belgique). Itinéraire d'une topographie religieuse dans la cité des Nerviens. In: Dondin-Payre M. and Raepsaet-Charlier M.-T. (eds.), *Sanctuaires, pratiques cultuelles et territoires civiques dans l'occident romain*, Bruxelles : 181-215.

Langohr, R. 2000. Creusement, érosion et comblement des fossés ; l'approche des sciences de la terre, *Revue Archéologique de Picardie*, 1-2 : 57-65.

Langohr, R., 2001. L'anthropisation du paysage pédologique agricole de la Belgique depuis le Néolithique ancien – Apports de l'archéopédologie, *Etude et gestion des sols*, 8, 2 : 103-118.

Le Bailly, M., Harter, S., Bouchet, F., 2002. *Compte-rendu de l'étude paléoparasitologique du trophée celtique du site archéologique de Ribemont-sur-Ancre (Somme, France)*. Laboratoire de Paléoparasitologie, EA 3308, Associé C.N.R.S. ESA 8045, Université de Reims, U.F.R. de Pharmacie, Reims 2002. Unpublished report, 4 p.

Munaut, A.-V., 1991. L'environnement végétal de deux sanctuaires picards : Ribemont-sur-Ancre (somme) et Saint-Maur (Oise). In : Brunaux, J.-L. (eds.), *Les sanctuaires celtiques et leurs rapports avec le monde méditerranéen. Actes du colloque de Saint-Riquier, 8-11 novembre 1990*, Paris : 113-121. (=Dossiers de protohistoire, 3).

Sanchez-Goni, M.-F., 1994a. L'environnement de l'homme

préhistorique dans la région cantabrique d'après la taphonomie pollinique des grottes, *L'Anthropologie*, 98 : 379-417.

Sanchez-Goni, M.-F., 1994b. Végétation et climat sur le pourtour de la Méditerranée au cours du Pléistocène supérieur. In : Sacchi, D. (ed.), *Actes du colloque International XXIVème Congrès Préhistorique de France – Carcassonne 26-30 septembre 1994 – Les faciès leptolithiques du nord-ouest méditerranéen : milieux naturels et culturels*, Carcassonne : 237-250.

van Geel, B., Buurman, J. Brinkkemper, O., Schelvis, J., Aptroot, A., van Reenen, G. and Hakbijl, T., 2003. Environmental reconstruction of a Roman Period settlement site in Uitgeest (The Netherlands), with special reference to coprophilous fungi, *Journal of Archaeological Science*, 30: 873-883.

# Chapter 5
# Functioning and Hierarchy of Farms in the Gallic Society from the 3rd Century BC to the Roman Period

*François Malrain*
*INRAP Parc d'activité de Passel,*
*60 400 Passel (France), UMR 7041 ArScAn Protohistoire Européenne*
*francois.malrain@inrap.fr*

*In collaboration with Gertrude Blancquaert*
INRAP ZI de la Pilaterie,
11 rue des Champs, 59 650 Villeneuve d'Ascq (France), Geertrui.
blancquaert@inrap.fr

Summary: *The increase of rescue excavations has allowed the discovery of a great number of sites, which have been grouped under the label of "indigenous farms". This term doesn't address the question of their actual function. In order to clarify this situation, the characteristics of a little over three-hundred settlements located in the Parisian basin have been synthesised; and we have analysed in greater depth ten sites from the Oise valley (individual on-site studies). Characterising the sites based on the size and the artefacts from the burials is possible up to the first half of the III rd century BC. From the second half of that century BC on to the roman conquest, a great number of changes occurred. Settlements are spread out over the available land and are almost always demarcated by a ditch, which represents ownership. The farms were ranked in four categories based upon the size of the ditches, the architecture of the houses, the nature of the artefacts and the quality/nature/kind of the consumed meat.*

*Advertisement: in this paper, we only deal with the rubbish in the ditches. The relics of banquets, the manifestations of cults or specials deposits are not examined.*

*Keywords : Iron age, farms, functioning, hierarchy, agriculture, Gallic society.*

## History of research on *ædificia* in the Gallic Wars

In France, there has been a long fascination for the significance of a word found a dozen times in Caesar's text about his description of the Gallic Wars. This term, *aedificium*, has been translated in numerous ways. It has been taken to mean a variety of different structures: houses, buildings, farms, barns and so on, but the absence of a precise definition illustrates the difficulty in accurately determining the reality, despite countless different translations. Arbois de Jubainville first proposed applying *ædificium* to buildings that occupy the same role as the Italian Gallo-Roman *villa* (Arbois de Jubainville, 1890). In 1956, in his handbook on Gallo-Roman archaeology, A. Grenier also stressed that it might be combined with the word for an Italian villa (Grenier, 1970). However, latterly he revised his opinion, and suggested that it might mean instead an isolated agricultural building. E. Thevenot extended the interpretation by proposing that this term included *any* places occupied by influential Gallic landowners, that is to say any of their barns, stables or outhouses (Thévenot, 1966).

In 1976, R. Agache suggested that these dwellings were aristocratic residences playing the role of farms. For this researcher, the *ædificia* had the same role as the villa in Italy. If Caesar did not classify them in this manner it is because they were not built on the same plan, nor with the same materials, but from wood and clay (Agache, 1978).

Caesar however uses identical vocabulary to identify several different categories of settlement and the final definition given to *ædificium* varies from author to author. In fact, there is a close link between the definition given to this word and the evolution of research on the farms. At first, when Arbois de Jubainville put forward his theories, these were drawn from the written history of the Gallic wars since at that time no farm had been excavated. This is why he made a connection with the *villa.* With the advent of aerial survey, these sites were recognised as being similar to those mentioned by Caesar. Today, both the development of rescue archaeology and the increase in numbers of excavations have made it possible to get closer to their true economic and social functions.

So the definition given to *ædificium* can be used to cover several forms of settlement structure among which farms

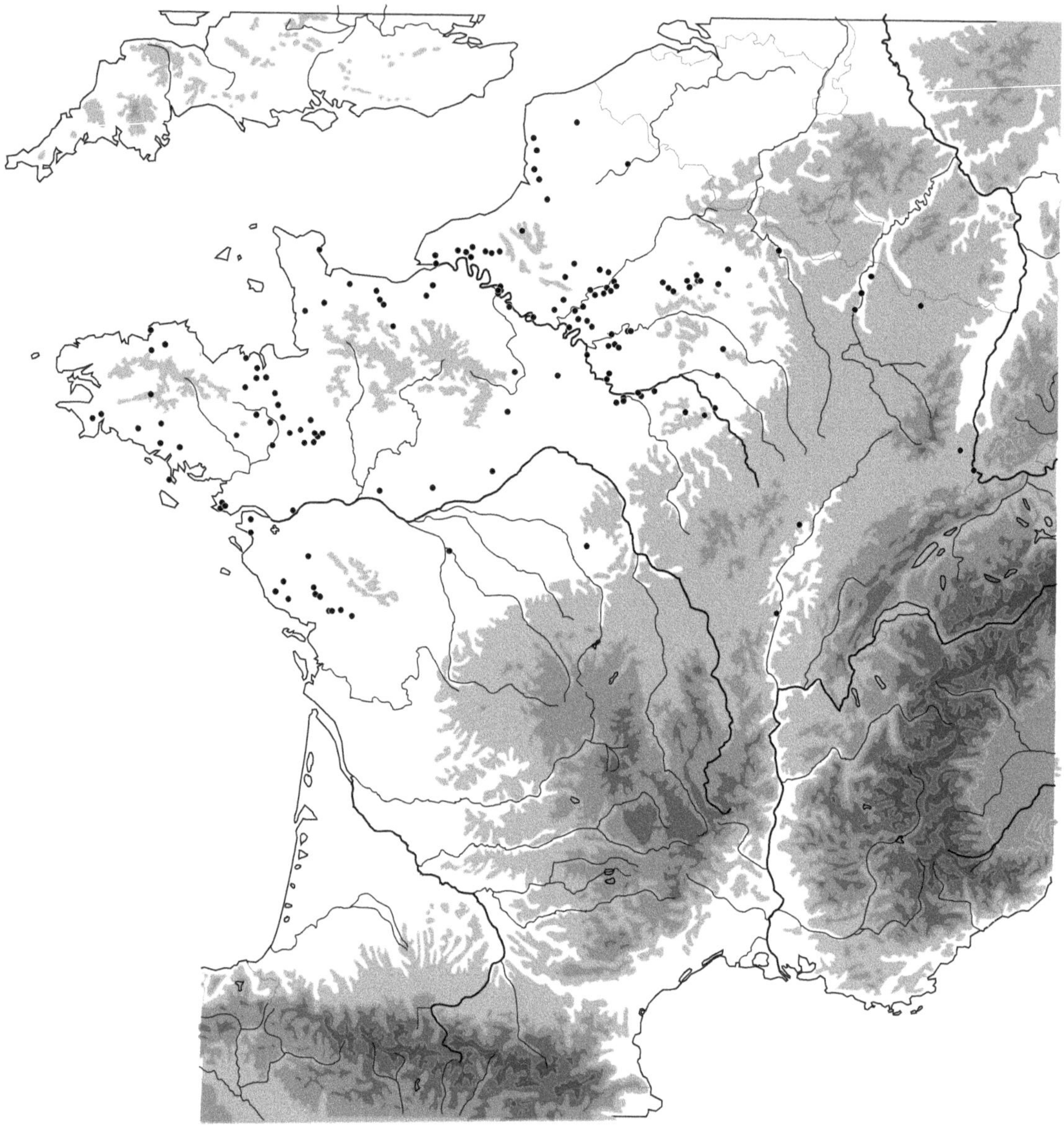

*Figure 1. The three hundred sites listed confirm a fairly dense occupation for the region (Ill. F. Malrain/© Inrap).*

certainly have a place. What needs to be stressed is the strategic importance of these *ædificia* in the unfolding of the Gallic Wars. They are noted as potential supply places as well as locations to supply foraging cavalry and men with meat and grain.

## Settlement patterns

Recently, research on these farms has seen some new developments. In fact, large-scale construction works in the region brought into being the utilization of the term preventive archaeology. The increase in excavations undertaken in the directly threatened zones has made it possible to progress considerably in this field. The three hundred sites listed today make it possible to update considerably the image which we have of the Gallic farms.

## Farms almost everywhere in Gaul

The location of these farms in the landscape confirms the idea of a fairly dense occupation for the region **(Figure 1)**. All the topographically suited situations were utilized: three-quarters of the sites occupy valleys and hill tops; the plains and their slopes are also used. On the other hand, their location changed over the centuries. In Early Tène, between the fifth and third centuries, the majority of sites are confined to the valley bottoms, followed by a progressive colonisation of the higher lands which can be observed as starting in the third century and which is fully developed by early Gallo-Roman times when close to three-quarters of the farms are located on hill tops **(Figure 2)**.

The pollen analyses reveal the presence of widely deforested spaces around the settlements in north-western Gaul (Leroyer, 1997; Munaut,1988 and 1989; Marguerie,

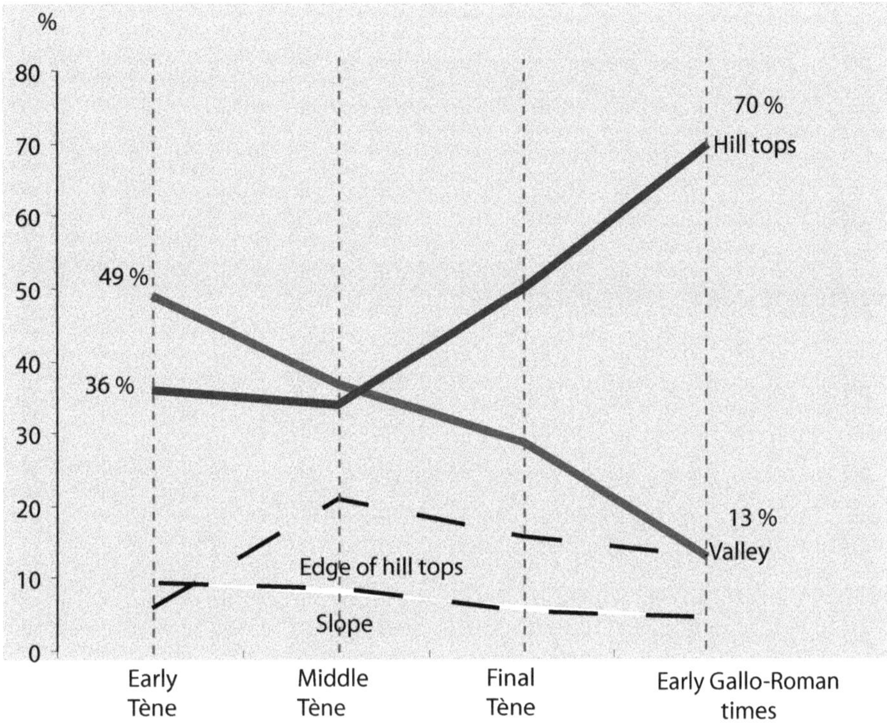


*Figure 2. In early Tène, sites are close to the valley bottoms; in early gallo-roman times, sites are on hill tops (Ill. F. Malrain/© Inrap).*

1992), as well as wide open spaces in the south (Puertas, 2000). The results of the carpological studies confirm the importance of the land under cultivation. (Matterne, 2001 ). The archaeozoological data show that the forest animals had but a tiny part in the diet of the Gauls, because the hunting products were obviously reserved for the happy few. The hunted species were mostly animals living in open spaces as the hare or in spaces with dual purpose as the roe-deer and the deer; the other wild game, the wild boar and the flesh eating animals are far too few to give reliable information on the close environment of the country settlements. The whole of studies emphasizes a landscape where the forest was either not very widespread or far away.

It seems obvious that spaces were cleared to allow people to settle down. The Vermand farm in the Aisne department is a good illustration of it. In order to set up a fence in the ditch delimiting this farm from the 3rd century BC, nearly five hundred trees had to be felled. If we reckon on 50 to 60 trees per hectare, being 30 to 40 cm in diameter (Ophove and Plaisance, 1963, part 2320: 23 ), and if we agree that clearing was necessary to settle down the site, ten hectare of the forest had to be cut down. In another farm, at Chevrières in the Oise, the palynological study of the soil filling in the ditch delimiting the settlement showed that if the forest was still relatively near in the beginning of the human settlement, it receded little by little during the occupation of the site. (Leroyer *in*: Malrain and Pinard, 2006). The impact of a farm on the environment is still difficult to measure. Because of the number of sites revealed by preventive archaeology it can be inferred that the landscape was strongly anthropic. But the clues to define the extent of a farm remain tenuous. In order to determine the impact on the environment , one should know all the settlements, farms and others, their farming area but one should also make sure they all were synchronous, which remains difficult in the present state of research even in the best known regions. The area necessary to the functioning of one farm includes areas suitable for cultivation and others devoted to grazing, wooded parcels and marshy areas which would be close to 60 hectare. (Malrain *et al.*, 2002: 151). This estimation of the area comes close to the area considered on the excavations at Onnaing in the Nord department where on the 260 scraped hectare farms 800 to 1500 meter away from each other could each use about sixty hectare (Buchsenschutz, 2004 ). Similar findings of extent, anthropic impact, as well as diversity of the immediate environment, have been observed on the archaeological project of Dourges in the Pas-de-Calais department (Blanquaert and Deschodt, 2005).

Between the 9th century BC and Caesar's campaigns in the north of Gaul farms underwent a remarkable evolution. If from the 9th century BC to the 4th century AD, they mostly and according to the regions show some structures sometimes delimited with ditches , from the 3rd century onward this trend becomes general with ditched enclosures on most sites.

## Ditches to make enclosures

The enclosures defined by the ditches take on various forms whose basic, curvilinear or rectilinear shape can be simple or complex.

Beginning with the second Iron Age up to the Gallo-

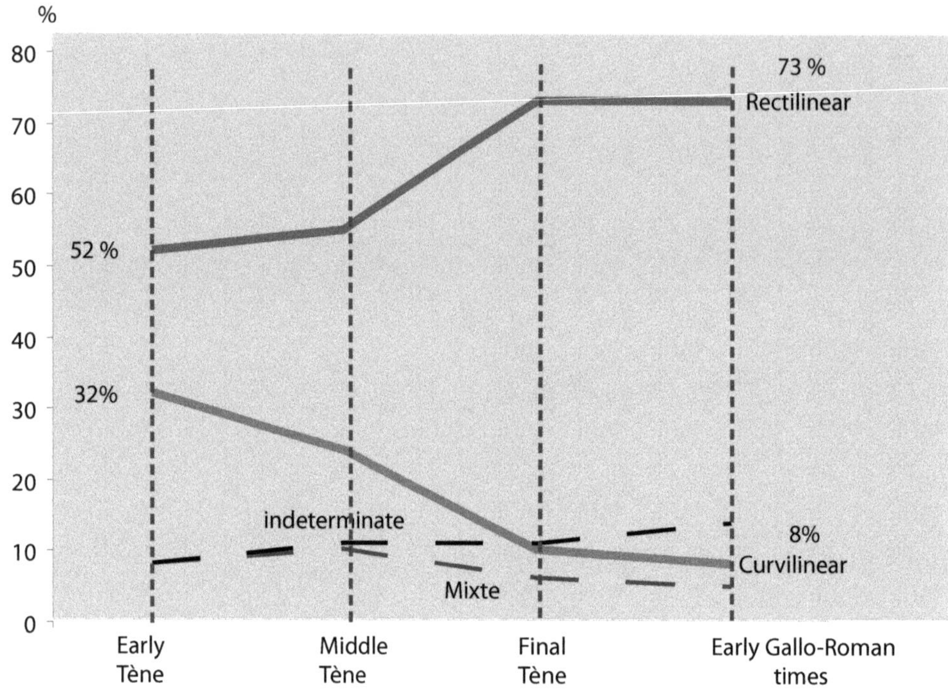


*Figure 3. Beginning with the second iron age up to the gallo-roman period, there is an increasingly strong tendency towards regularity in the shape of the enclosure (Ill. F. Malrain/© Inrap).*

Roman period, there is an increasingly strong tendency towards regularity in the shape of the enclosure. This change manifests itself most strongly by the third century, when about two thirds of the enclosures are defined by a rectilinear layout (**Figure 3).**

Ditches leave clearer traces in the earth; consequently, they are most easily recognisable, to the detriment of other more ephemeral boundaries like palisades or quickset hedges.

The profiles of the ditches reveal the multiple ways in which they were dug; for example the walls can be wide, sub-vertical, vertical, concave or convex and their bases, flat, rounded, or V shaped; in other words no uniformity can be seen here.

The dimensions of the ditches are also very varied. Three types can be distinguished. The first, encompasses the majority of the sites, and is characterised by a not very broad and not very deep ditch (between 0.20 m to 1 m on average). The second includes ditches whose width and depth are deeper than one meter. The third, rarer, form consists of huge ditches whose width and depth can reach, and even exceed, tree meters.

The large heaps of spoil generated from digging the ditches were often used to make a bank the presence of which can be detected in the asymmetry of the fillings of the ditches after their collapse. On the other hand it is more difficult to determine on which side the earth was thrown. However, observation of the plans often reveals, inside the enclosures, an empty space between the settlement structures and the ditches, which could certainly correspond to the negative evidence for a bank.

For example in Picardy, at Ronchères, the soil from the digging of a wide ditch was used for the making of a slope still to be seen nowadays. The linings of the ditch were wooded and deeper digging was noticed near the buildings perhaps to ensure an easy access to water **(figure 4).**

In another case, at Longueil-Sainte-Marie in the Oise, the micromorphological analyses showed that unbaked clay bricks had been raised at the lower part of the slope to hold it in position **(Figure 5).**

During the construction of a farm, the most noticeable feature is the generalised use of ditches to define various spaces. The location of a ditch, a bank and sometimes of other means of delimitation such as palisades or quickset hedges is an ostentatious way of defining the "property". Significant differences in layout between farms may be explained by a desire to show off clearly the status of the occupant. So elements of their plan such as the size of the ditches and their configuration probably result from a desire to mark the properties in different ways.

In this way the social status of the tenant is clearly identifiable on looking at the more or less remarkable improvements he made: the wider and deeper the ditch is, the more imposing the slope is, the more important his status probably was in society.

*Figure 4. The bank raised with the soil from the ditch at the site of Ronchères (France, Picardy, Aisne ) is still raised nowadays (photo. F. Malrain/© Inrap).*

*Figure 5. The unbaked clay bricks in the ditch at Longueil-Sainte-Marie (France, Picardy, Oise) (photo. F. Malrain/© Inrap).*

To this physical definition of the property is added the construction of the buildings necessary for the functioning of the settlement. In the settlement enclosures the buildings are not set out randomly and there is an obvious rational organisation of space.

Generally less than ten constructions are located near to the ditches and parallel to them. At more or less regularly distances from each other they alternate with additional buildings. Sometimes, constructions are isolated from each other by internal divisions of the enclosure; in other cases one finds a succession of enclosures established to meet specific needs. Generally a vast central space appears empty of all features. This situation is the most common, but some settlement enclosures consist only of a single construction positioned in the centre of the enclosed area.

## Spatial analysis

The organisation of the farms and the function of the buildings can be seen from the spatial distribution of artefacts found within the structures, according to their distribution or density by weight **(Figure 6).** In order to achieve this, the sites of the valley of Oise were the subject of an excavation method aimed at recording the precise location of each artefact (Gransar *et al.,* 1997). This method of proceeding has a major disadvantage however, that of the time necessary to carry out the excavation **(Figure 7).** The experiment showed that the effective sampling for the excavation of a ditch consisted in digging by hand one in every four meters. This practice constitutes an effective compromise between a full excavation, which is seldom possible, and random sampling that prohibits the analysis of the spatial distribution of artefacts. It was applied to the totality of the sites; however generally a more extensive excavation also supplemented it in order to check its effectiveness (Malrain, 2000).

The collected material was separated and weighed by category (flakes, bones, heated stones, ceramics, etc). This data was recorded on a spreadsheet and sorted statistically. The linkage of the spreadsheet data with a geographical information system (GIS) made it possible to chart various artefacts.

Domestic rubbish is mainly found in the ditches (more than 80 %). Assumptions about the usage of areas and the purpose for which they were constructed were drawn from the spatial information (charts). Waste is found in the ditches closest to the building, which produced them. Its analysis is a good source of information in recognising the function of structures or activity areas. Beyond a functional attribution of areas from which they result, waste is an indicator of the social status of the occupants. The spatial analysis makes it possible, moreover, to have access to some information concerning the practices of deposition.

## Hierarchical classification

Artefacts show us clearly defined divisions between social classes **(Figure 8).** The majority of the sites have the tools necessary for their functioning such as grinding stones, and loom weights. However, the ornaments found on various sites indicate clear differences. Some are impoverished, while others reveal more prestigious artefacts, notably bracelets made from glass, and lignite, and bronze objects. The distinctions are even more striking when weaponry is taken into account; this class of artefact is quite rare on

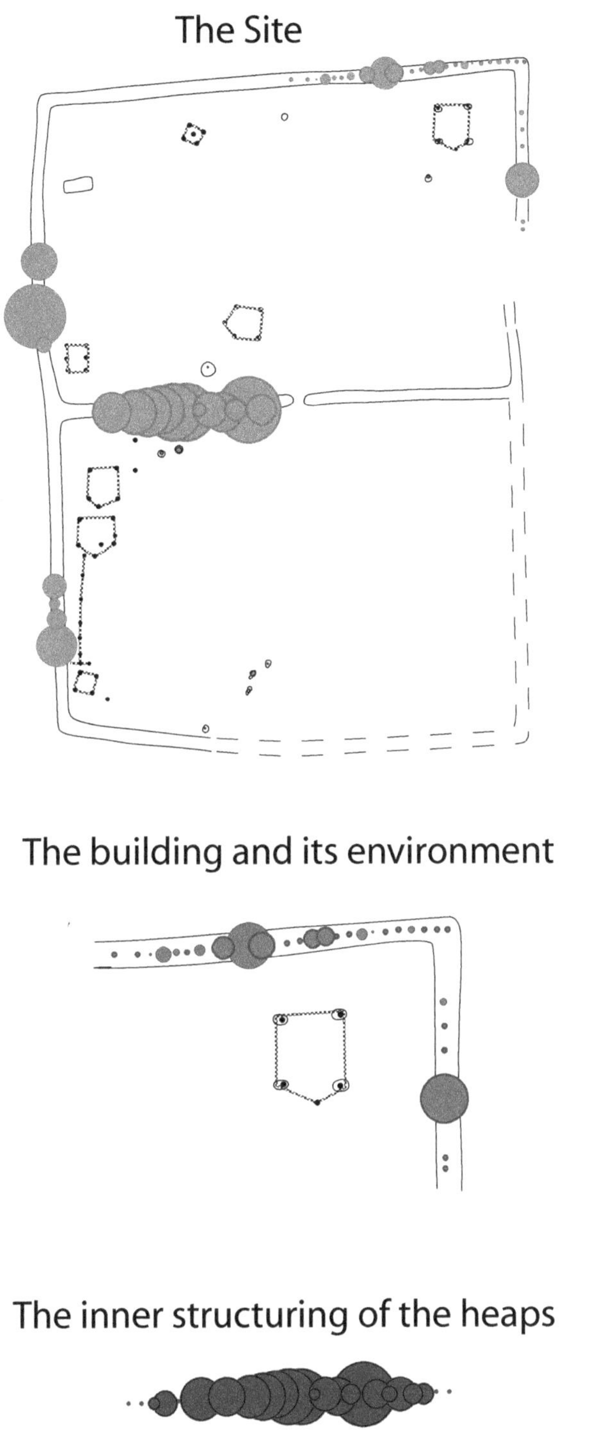


*Figure 6. Methods of spatial analysis of the main parts of the farm (Ill. F. Malrain/© Inrap).*

the farms; therefore its presence, when found is even more remarkable. So, the shield boss and spear-butt, from the site of Herblay as well as the pieces of armament from Maroilles sur-Seine "les Gours aux Lions " point to the high status of the occupants. It is the same for the coinage, also not very frequently found on the farms. Its presence indicates trade and the ability of the owner to get goods, which he was not able to produce. Similarly on certain farms that exhibit signs of wealth we find amphorae.

All of the analysis allows us to establish a hierarchical classification for the farms. The top the hierarchy is represented by the aristocratic farm of Longueil-Sainte-Marie (**Figure 9**). This establishment is characterised by an enclosure with a deep, broad, ditch, with the structures well maintained and set up with a narrow entrance, probably protected by a gateway. The enclosed space revealed two successive constructions. In addition nearby there is a small enclosure of the same period, used as a funerary space. These elements are situated within a larger network of ditches, which enclose the other structures of the homestead. Glass beads, lignite bracelets and bronze objects are a strong indication of the unusual status of this settlement. It can be regarded as a one at the top of the hierarchy. The single building erected inside the ellipsoidal enclosure suggests that it was occupied by only one family that of the Master, in contrast to the other sites, which consist of several domestic units, built in the same space.

Farms of second rank within the hierarchy are characterised by imposing ditches and artefacts, which suggest an elevated status for the occupants. Nevertheless, there is not, as in the top rank, an obvious division of space. However social differences are perceptible between the inhabitants. The house of the Master and those of his assistants are built in the same enclosure, but the artefact and the waste from bones consumed by each household highlight differences of status between the inhabitants. The example at Verberie is a good example illustrating this **(Figure 10).**

Though all the buildings of this farm are built with same materials, they have different plans. The kitchen equipment corresponding to each house shows that, though made of similar types of clay, some appear more sophisticated than of others. This is also the case for a whole range of cooking utensils, which, in addition to some more decorative vessels also included a lignite container. In addition to the quality of the table wear other elements of ornamentation must also be considered such as the fragments of lignite bracelets, a bronze pin, glass beads, two keys and other fragments of metal. The fauna consumed in this house is evidence of a diet of quality meats (Méniel *in*: *Malrain et al.*, 2002).

These signs are limited, but they may be used to interpret the different standards of living in one house as compared to another. The analysis of the functional relationship between the buildings has made it possible to understand the privileged place that this house occupies. It is situated in a central position within the enclosure making it possible for its inhabitants to supervise the whole of the establishment. Also worth noting, is that this house is close to the only storage pit on the site.

For the first time in this category of settlement, a slaughteryard for animals could be observed. In the north - eastern angle of the enclosure, the most important remains of the carcasses of big animals, oxen and horses, were thrown into the ditch. On the opposite side, in the north – western angle, a concentration of stones linked to a building in which a stack had been discovered , has

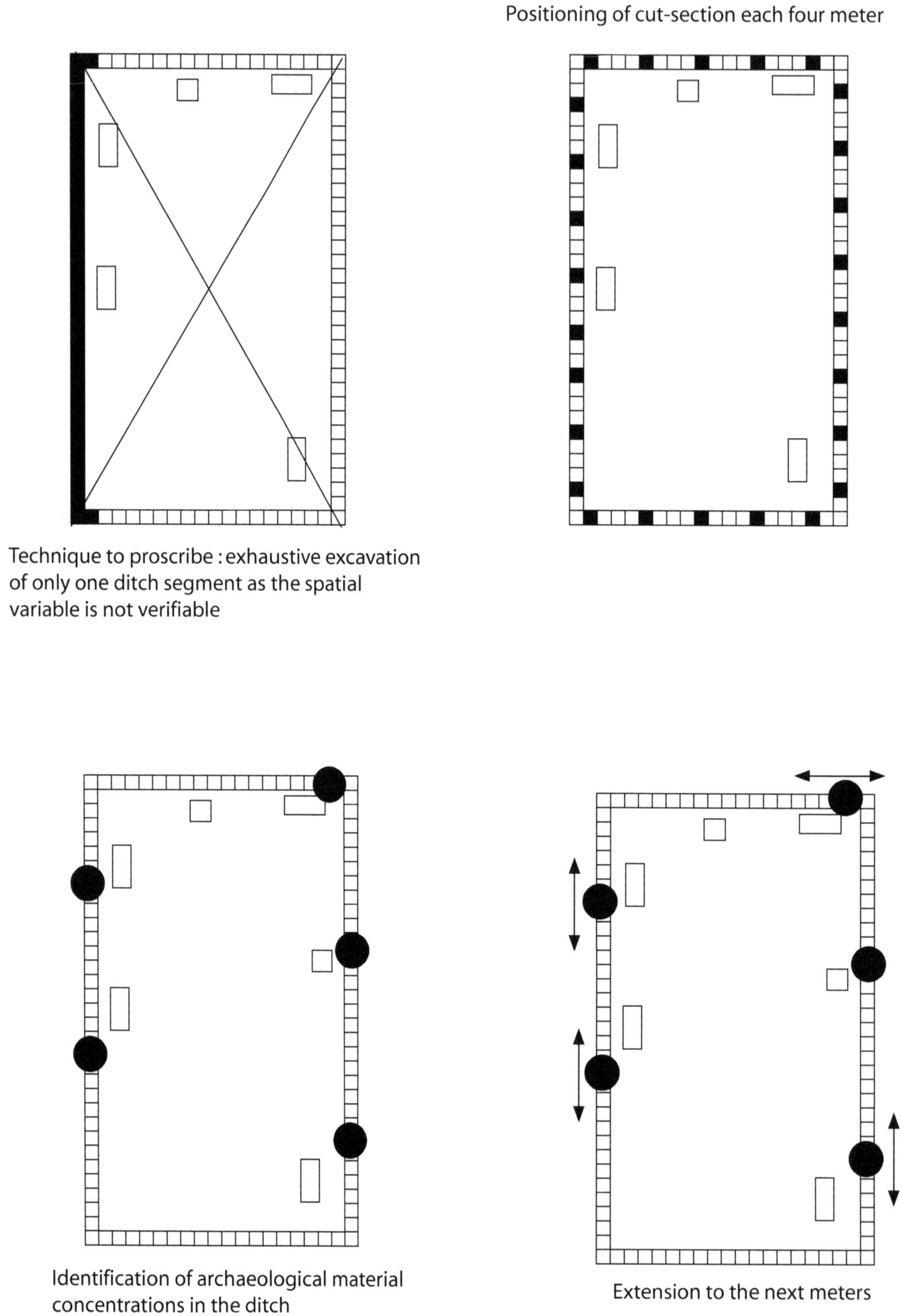


*Figure.7. Method of digging (Ill. F. Malrain/© Inrap).*

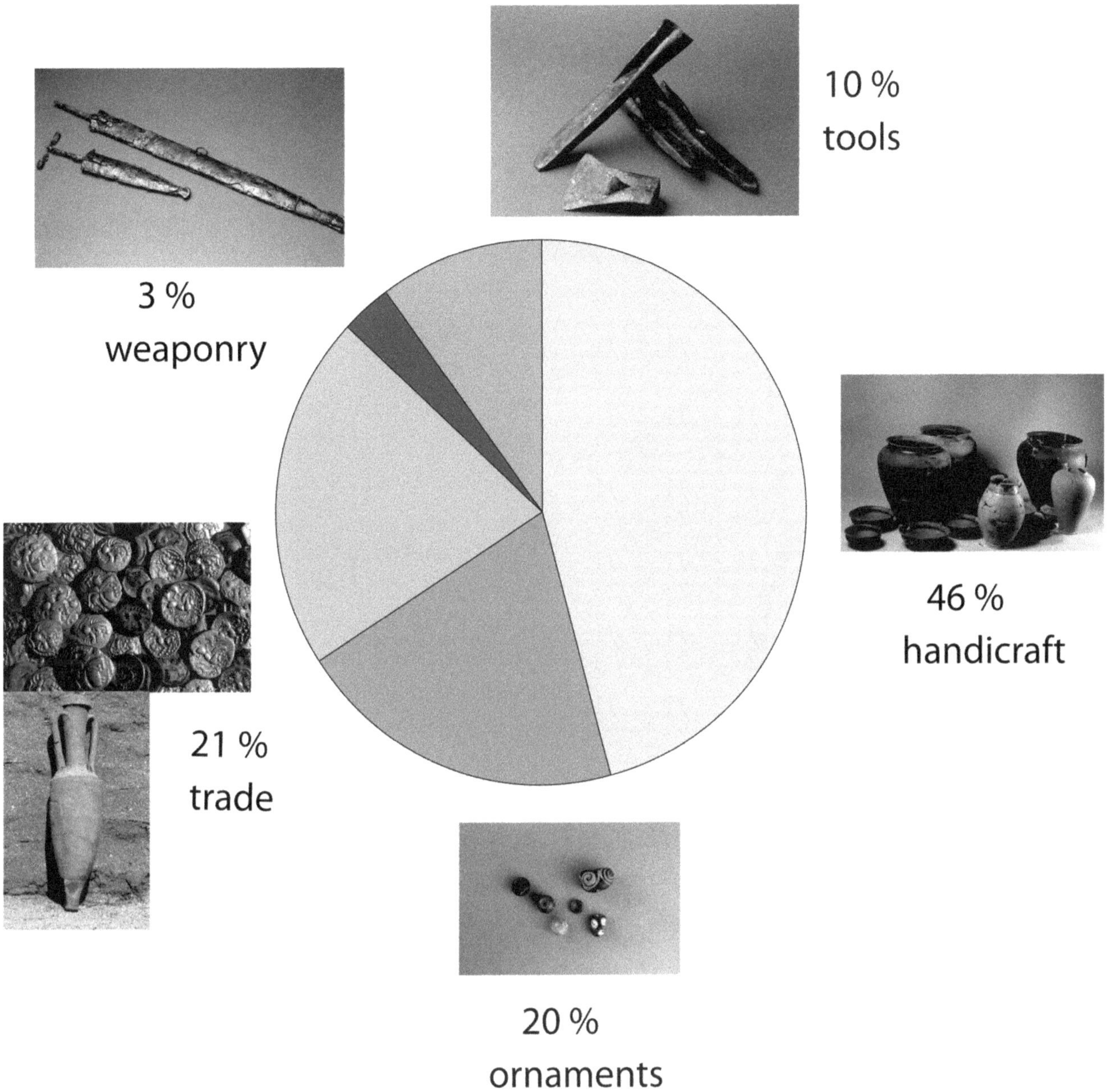


*Figure 8. Artefacts show us clearly defined divisions between social classes (Ill. F. Malrain/© Inrap).*

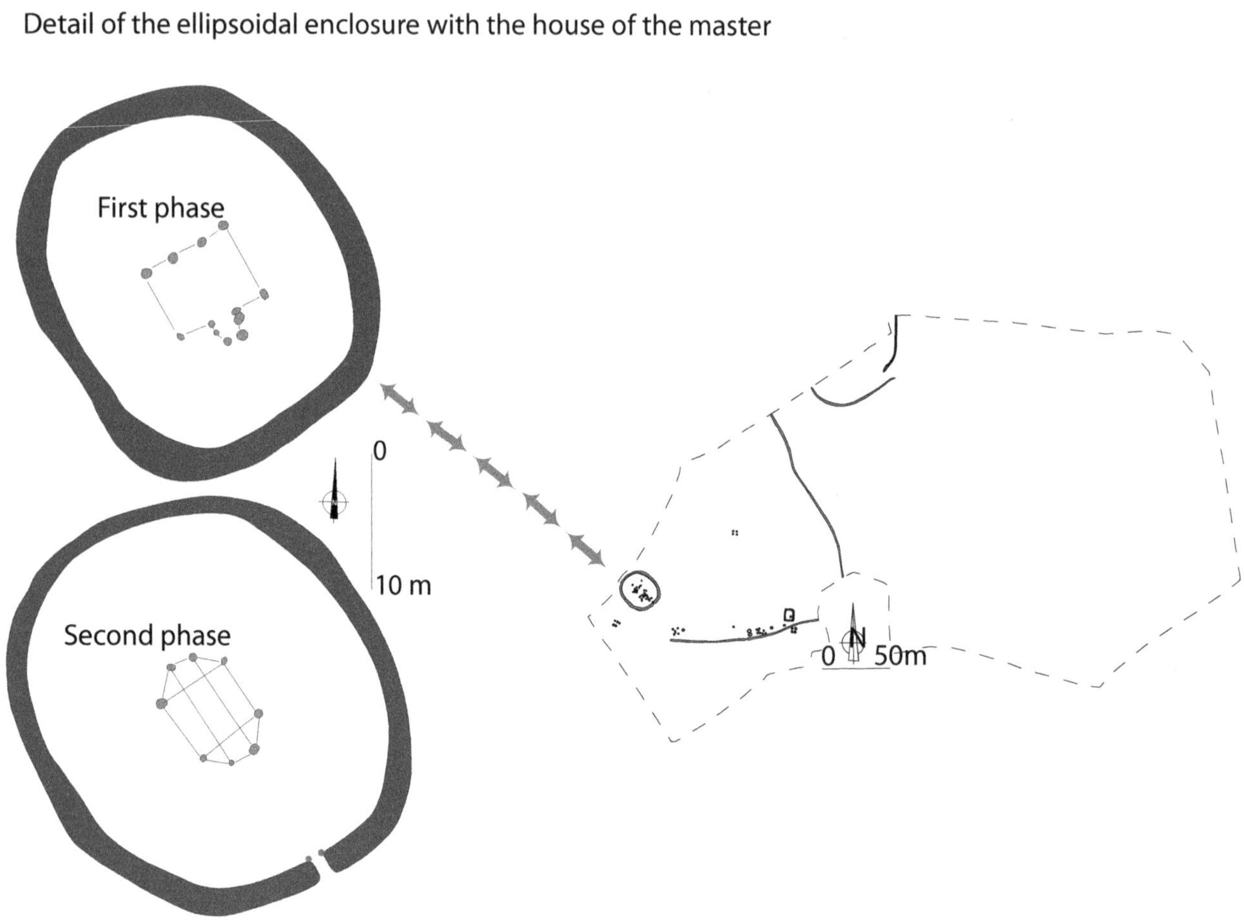


*Figure 9. Settlement of Longueil-Sainte-Marie (France, Picardie, Oise) (Ill. F. Malrain/© Inrap).*

something in common with an outbuilding near a threshing area. Moreover there is a more enigmatic dump of more or less forty mortal remains. More than half of these remains are skull fragments from at least four adults and one youth. Worship rites could explain their presence in front of a building which could have a particular assignement within this establishment.

The type of establishment most frequently found is the farms of third rank, defined by not very wide and shallow ditches. In terms of, artefacts, objects and bone waste these are of low quality. Nevertheless, in some cases, such as at Jaux, the status of the central house was not indicated by its artefacts to be of higher rank than the others, but its situation close to the gate which separates two enclosed areas and the fact that nothing could happen between the inhabitants without their activities being observed due to its proximity, suggests that it could have had a dominant position: it is also close to a storage pit **(Figure 11).**

Lastly, at the base of this pyramidal hierarchy, are the farms of 4th rank **(Figure 12).** They are characterised by some isolated pits containing only a few objects, and only in some cases do they have a building. Their function is not clearly defined. They are seldom discovered because only extensive stripping of the topsoil brings them to light. Several questions arise: were they outposts used to exploit land further away from the principal settlement, or seasonally occupied to allow more efficient exploitation during the important periods of agricultural work (harvests, for example), or perhaps huts used to guard the herds? It is still difficult to answer these questions, but it seems, in the absence of storage structures and storage vessels for provisions and given the small quantity of artefacts discovered, that their populations must only have numbered a few people. The scarcity of artefacts from these structures suggests that they played a satellite role when compared with the more significant occupation sites, unless they belonged to poor farmers.

**Conclusion**

The examination of the various spaces that comprise the farms, living areas, work and funerary places highlights the hierarchy in these sites which specialised in agricultural production. The development of trade, exchange, and improvement of production techniques brought about deep changes in Iron Age society; this can be seen by the way sites are established within the territory as a whole, by a clear delimitation of the property by using enclosure.

These upheavals had significant consequences to which

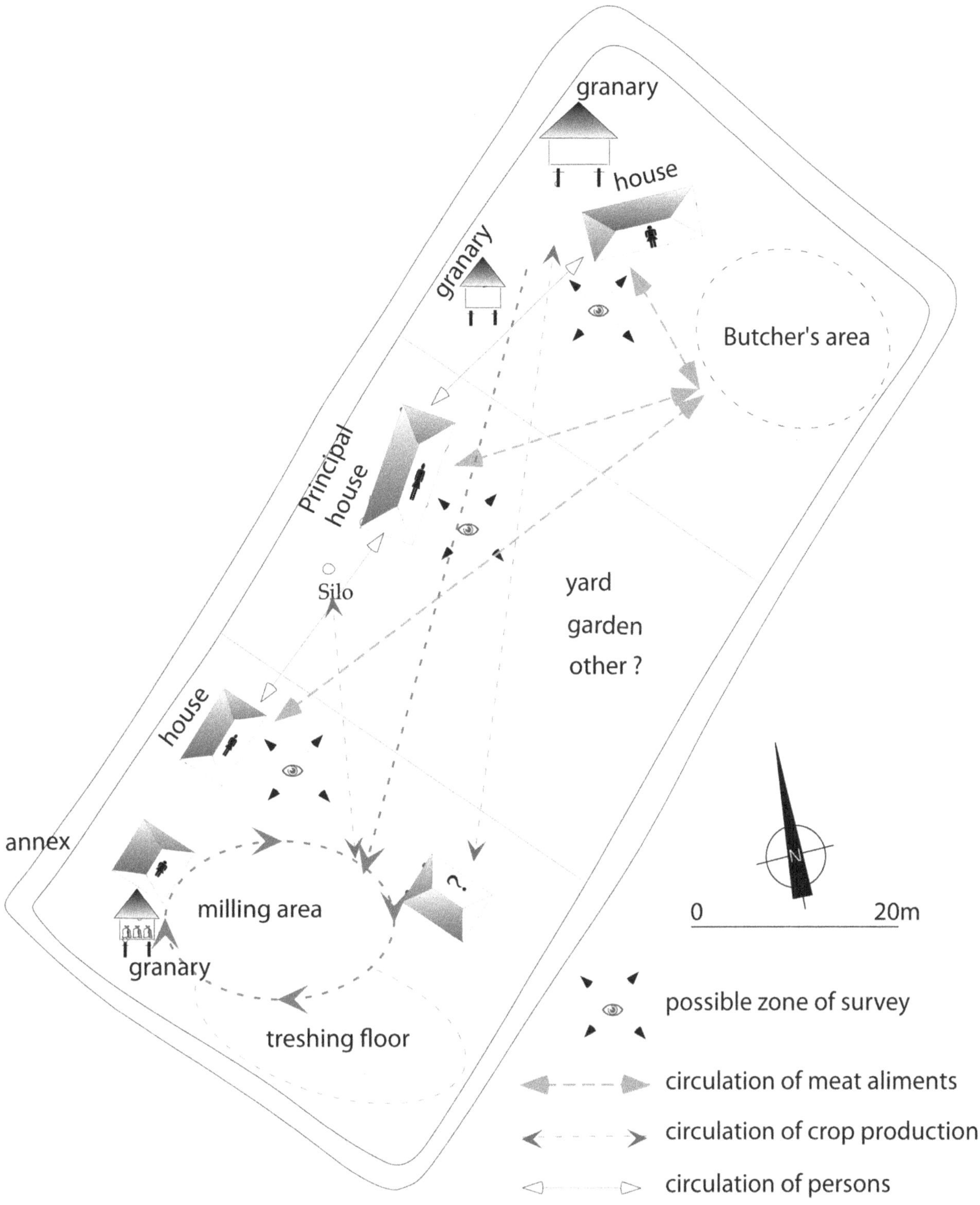


*Figure 10. Settlement of Verberie (France, Picardie, Oise) (Ill. F. Malrain/© Inrap).*

the culture testifies while being based at the same time on both an intensive system of mixed farming and also on monospecific grain production (Matterne, 2000). Appearance of the forge and the generalised use of iron tools such as the ploughshare doubtlessly allowed the "conquest" of new lands such as those located on the hilltops. Mutations within agricultural strains are accompanied by new technical methods to handle them. The rearing of cattle in stalls, whose winter fodder undoubtedly improved considerably with the appearance by the 2nd century of the scythe facilitated by the availability and use of manure.

In another area, saddle querns have come and gone to be replaced by the rotary quern, which simplifies work by considerably reducing the time necessary for grinding grain. In the same way, the intensification of salt production in

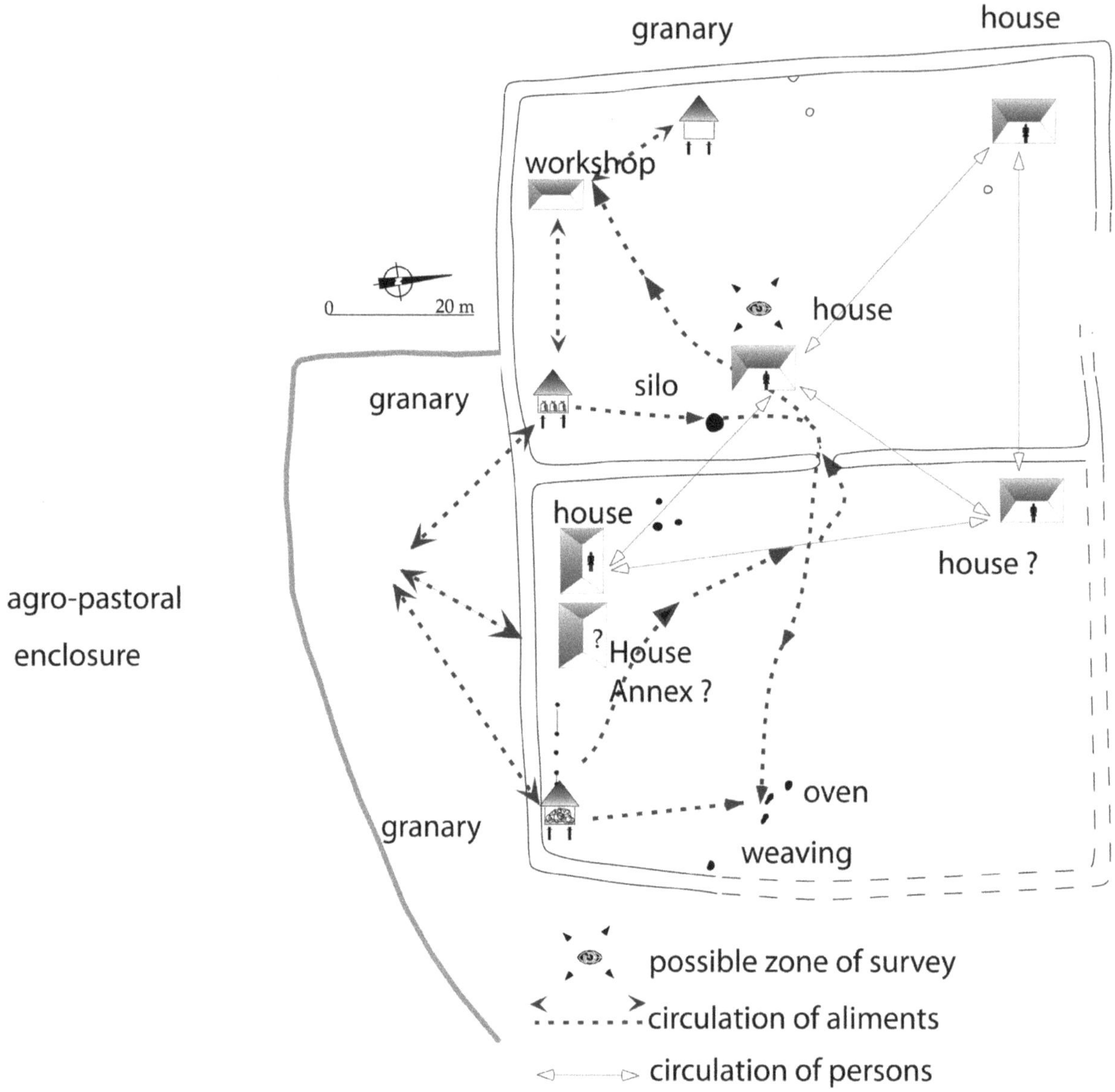


*Figure 11. Settlement of Jaux (France, Picardie, Oise) (Ill. F. Malrain/© Inrap).*

specialised establishments allows for better conservation of food, whether it be meat, fish or certain plants.

Storage structures are adapted to the needs of the consumers inhabiting the farms, but it does not seem that surplus was stored there. A part was preserved at the production site for consumption and cultivation while the surplus does not seem to have been passed back to the farm. So it must be supposed it was conveyed from the production area to other sites. Sites for artisanal productions are set up alongside sites of agricultural specialisation, and the *oppida* are established by the end of the 2nd century. By the beginning of 1st century BC a clear diversity of Gallic settlement types appears. It shows a highly complex stratified society whose farms would have been closely integrated as they must have formed the basis for this formidable economic expansion of Gallic society.

## New research prospects

The tackled item raises new questions. A new research undertaken by the INRAP is being carried out nationally. Its aim is a homogeneous census of the rural establishments from Hallstatt D3 to the early Gallo-Roman period. The whole file will be connected to a GIS (geographical information system). The new project is planning to complete the information with recent data in order to reconsider the old documentation and to develop cross themes. Logically, the operating data base should make it possible to derive several thrusts:

- the spatial approach of the distribution of the sites (topography, geology) ;
- the staging of the sites in order to better grasp the emergence, the interruptions and the decline of the settlements on a regional and national basis;

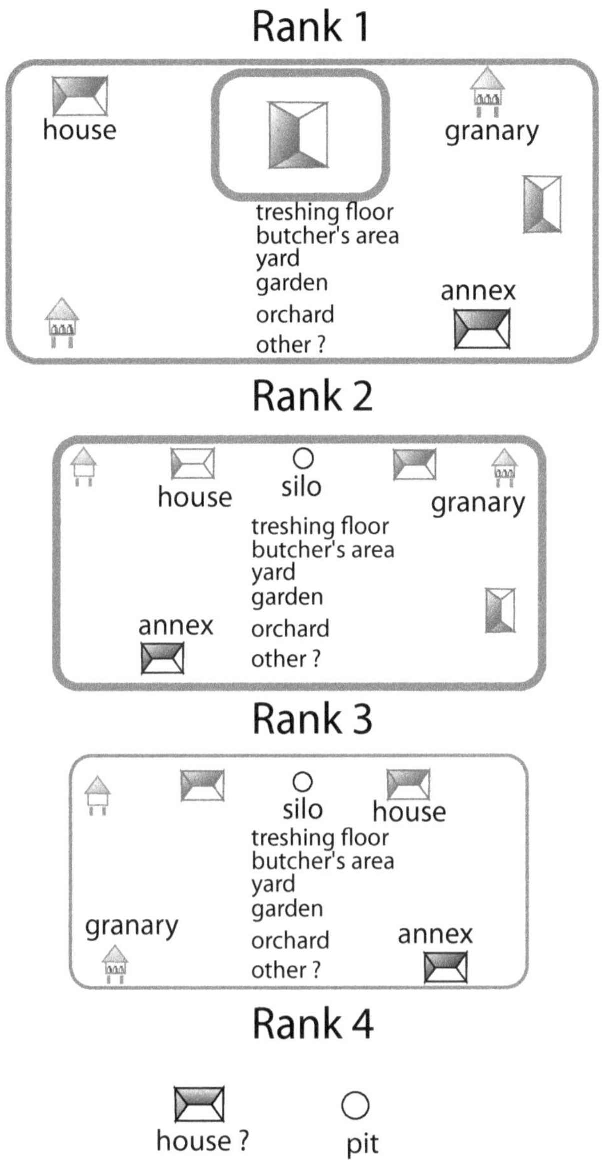


*Figure 12. Pyramidal hierarchy of the farms (Ill. F. Malrain/© Inrap).*

- the analysis of the sites ( structuring and furniture ) to reach a subtler and more varied definition of the forms of occupation and their working, an aspect which must allow to go beyond too narrow a sense of " farm ", and to detect much more complex units ( or networks ) on different scales ( regional, supra-regional, national );
- typology and function of the buildings;
- lots of contributions to the material culture ( chronology, functional categories, economical and social function, hierarchy );
- records of specific studies already taken up ( palynology, carpology, phytology, malacology, micromorphology, sedimentology, geology, pedology, dendrochronology, etc. ) in order to draw up a first inventory and to analyse the contributions of these studies and their development.

This data base is devised as a means for thought and will be open to all researchers.

## Bibliography


Agache, R., 1978. *La Somme pré-romaine et romaine d'après les prospections aériennes à basse altitude*, Amiens. (= Mémoires de la Société des antiquaires de Picardie: XXIV).

Arbois de Jubainville, H. d', 1890. *Recherches sur l'origine de la propriété foncière et de noms de lieux habités en France*, Paris.

Audouze, F. and Buchsenschutz, O., 1989. *Villes, villages et campagnes de l'Europe celtique du début du IIe millénaire à la fin du Ie siècle avant notre ère*, Paris. (= Bibliothèque d'Archéologie).

Blancquaert, G. and Deschodt, L., 2005. Le site de Delta[3] à Dourges : un territoire qui a fait ses preuves, *Archéopages*, 17: 6-21.

Buchsenschutz, O., 2004. *Les Celtes de l'Age du Fer dans la moitié nord de la France, Histoire de la France Préhistorique du VII^e siècle au 1^er siècle avant J.-C.*, La maison des roches éditeur, Tours, 123 p.

Gransar, F., Malrain, F. and Matterne, V., 1997. Analyse spatiale d'un établissement rural à enclos fossoyés du début de La Tène finale : Jaux « le Camp du Roi » (Oise). In: Auxiette,G., Hachem, L. and Robert, B. (dir.); Bocquet, A. (ed.), Espaces physiques, espaces sociaux dans l'analyse interne des sites du Néolithique à l'âge du Fer, actes du 119ème congrès national des sociétés historiques et scientifiques, Amiens 1994, Paris: 159-182.

Grenier, A., 1970. *Les Gaulois*, Paris. (= Petite bibliothèque Payot, 157).

Leroyer, C., 1997. *Homme, climat, végétation au Tardi- et Postglaciaire dans le Bassin parisien : apports de l'étude palynologique des fonds de vallée*. Thèse doctorat en archéologie et préhistoire, Université Paris I - Panthéon - Sorbonne, vol. 1, 574 p.

Malrain, F., 2000. Contribution à l'étude des fermes. In: Marion, S. and Blancquaert, G. (eds.), *Les installations agricoles de l'âge du Fer en France Septentrionale*, Paris: 15-37. (= Études d'histoire et d'Archéologie, VI).

Malrain, F., 2000. *Fonctionnement et Hiérarchie des fermes dans la société gauloise du IIIe siècle à la période romaine : l'apport des sites de la moyenne vallée de l'Oise.* Thèse datylographié. Paris, Université de Paris I, décembre 2000. 436 p.

Malrain, F., Matterne, V. and Méniel, P., 2002. *Les Paysans Gaulois (IIIe siècle, 52 av.J.-C.)*, Paris Errance, 236p.

Malrain F., Pinard E. (dir.) with contributions of Gaudefroy S., Leroyer C., Matterne V., Maréchal D., Méniel P., Pastre J.-F., Pommepuy C. 2006. *Les sites laténiens de la moyenne vallée de l'Oise : contribution à l'Histoire de la société gauloise*, Abbeville, 272p., 2CD. (= Revue Archéologique de Picardie, n°spécial 23).

Marguerie, D., 1992. *Evolution de la végétation sous l'impact humain en Armorique du Néolithique aux périodes historique*, Rennes. (= *Trav. Labo. d'Anthropologie, Préhistoire, Protohistoire et Quaternaire armoricains*, n° 40, UPR 403 du CNRS, Université Rennes I).

Matterne, V., 2000. *Agriculture et alimentation végétale durant l'âge du Fer et l'époque gallo-romaine en France septentrionale*. Thèse de doctorat dactylographiée. Paris : Université de Paris I, 2000, 2 vol. 314 p.

Matterne, V., 2001. *Agriculture et alimentation végétale durant l'Âge du Fer et l'époque gallo-romaine en France septentrionale, Archéologie des Plantes et des Animaux I*, Montagnac.

Munaut, A.-V., 1988. La forêt gauloise dans le Nord de la Gaule Belgique. Enquête palynologique préliminaire, *Revue du Nord*, 70, 276: 5-21.

Ophove, H. and Plaisance, H., 1963. Sylviculture. In: Keilling, J., Martin, M. and Casalis, J. (dir.), *Technique Agricoles, Encyclopédie Agricole Permanente.*, Paris: 1-35. (= Publication de l'Association pour le développement et la vulgarisation des Techniques Agricoles, fasc. 2320).

Puertas, O., 2000. Stabilisation des activités agricoles dans la delta du Lez a partir de l'analyse pollinique, In: *Els productes alimentaris d'origen vegetal a l'edat del ferrode l'Europa Occidental: de la produccio al consum.*, Girona: 43-49. (= Derie monografia, 18, Museu d'Arqueologia de Catalunya Girona).

Thévenot, E., 1966. *Histoire des Gaulois*. 4e éd., Paris. (= Que sais-je?, 206).

# Chapter 6
# Holocene Soil Development and Pedogenic Clay Mineral Transformation in Luvisols with Different Development Ages at the Viereckschanze of Poign (Central Bavaria, Germany)

*Alexandra Raab (1)*
*raab@wzw.tum de, phone +49 (0) 8161 / 71 – 2501, fax +49 (0) 8161 / 71 - 2502*

*Jörg Völkel (1)*

(1) Special Field for Geomorphology and Soil Science,
Research Department for Ecology and Ecosystem Sciences,
Center of Life and Food Sciences Weihenstephan,
Technische Universitaet Muenchen,
D-85350 Freising

## Introduction

The Celtic square enclosure of Poign, situated ca. 10 km southwest of Regensburg (Central Bavaria, Germany) (Fig. 1), was the focus of various landscape history investigations extending over several years (Völkel, 2006; see also Leopold, this volume). The studies include soil erosion processes induced by agricultural land use during Celtic times (Völkel *et al.*, 1998, 2001) as well as the reconstruction of a Latène age land-use model based on multivariate analyses of geoarchives in the surroundings of the Viereckschanze (Leopold and Völkel, 2005, 2007a; Raab *et al.*, 2005). In this paper, we present one aspect of these studies, the Holocene soil development and clay mineral transformation in Luvisols with different development ages at the Viereckschanze of Poign.

Clay minerals are very sensitive indicators for soil forming processes (Birkeland, 1999; Schaetzl and Anderson, 2005). Initial pedogenic processes can be proven by the identification and characterisation of clay minerals using the X-ray diffraction (XRD) technique. Therefore, clay mineral analysis can be a very helpful tool for geoarchaeological approaches because even short lasting surface exposure during different prehistoric periods may be identified.

Within the area of the Poign Viereckschanze, three stages of soil development in relation to age can be differentiated due to the characteristic architecture of the earthwork (Figs. 2, 3).

a) The original and mainly undisturbed soil developed from loess and preserved from erosion in the interior of the Celtic square enclosure (profile 7038-21).
b) The soil on the top of the rampart, which formed on the earth wall after its construction (profile 7038-22a).
c) The soil which became fossilized under the Latène earth wall (profile 7038-22b).

As the age of the ditch-earth rampart system is known, the time intervals for the soil formations are ascertainable. For the Luvisol in the interior of the earthwork, the soil development starts post-sedimentary and lasts during the Late Glacial and the Holocene. Pedogenic processes of the soil formed on top of the rampart start after the construction of the earthwork about 2000 years ago. And, in the case of the soil under the rampart, the fossilization of the soil caused an interruption or at least a drastic decrease of the pedogenic processes about 2000 years ago and reflects the level of soil development until Celtic times.

By comparing these three different soil stages we aim to characterise the soil properties before, during and after the Celtic land-use period in a prehistorically long and intensively settled landscape of Central Europe, the so-called Altsiedelland around Regensburg (Torbrügge, 1984). By using the Celtic square enclosure as a geoarchive, we can check the hypothesis that acidification of soils in Central Europe is a comparable recent process during the Holocene and would be a mainly man-induced phenomenon of the industrial age.

## Studied Site

The Viereckschanze of Poign is situated ca. 10 km southwest of Regensburg (Central Bavaria, Germany) (Fig. 1). It lies on the top of a weakly inclined (0-1°) and east-exposed plateau (Fig. 2). North and south of the plateau there are

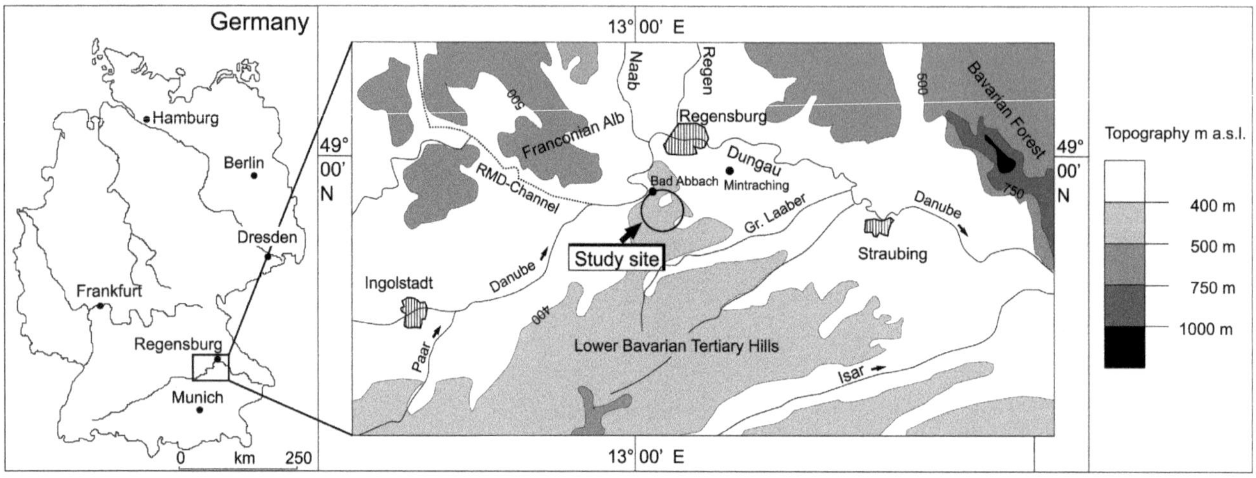


*Figure 1. Location of the studied site Viereckschanze Poign, situated ca. 10 km southwest of Regensburg and 3 km east of Bad Abbach in the Lower Bavarian Tertiary Hills. RMD-Channel = Rhein-Main-Donau-Kanal.*

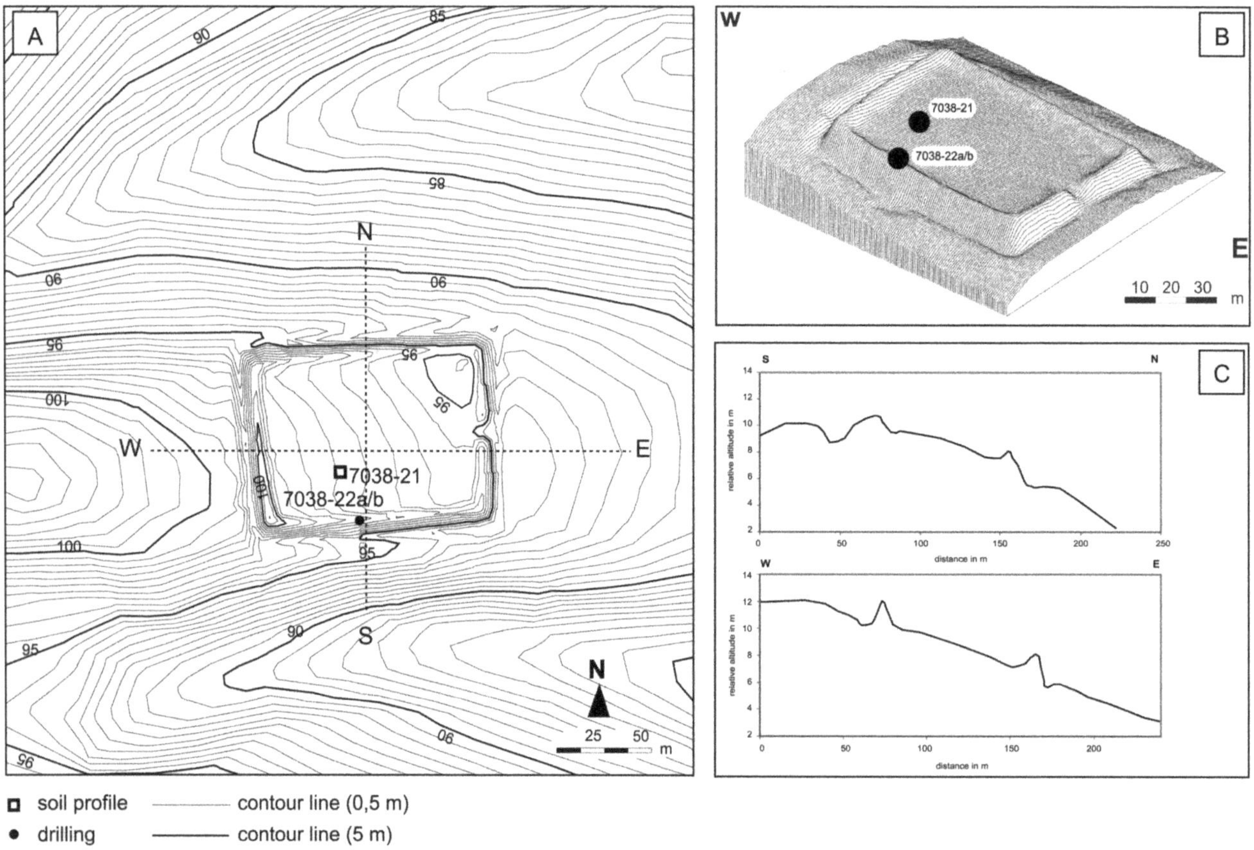


*Figure 2. A) Topographic map of the Viereckschanze Poign and its surrounding with the locations of the soil profiles and the lines of the cross sections (W-E, S-N) shown in C. Surveying was conducted in a local system. B) Digital Elevation Model of the Viereckschanze Poign with the location of the soil profiles 7038-21 and 7038-22a/b. C) Cross sections in South-North (S-N) and West-East (W-E) direction through the Viereckschanze Poign.*

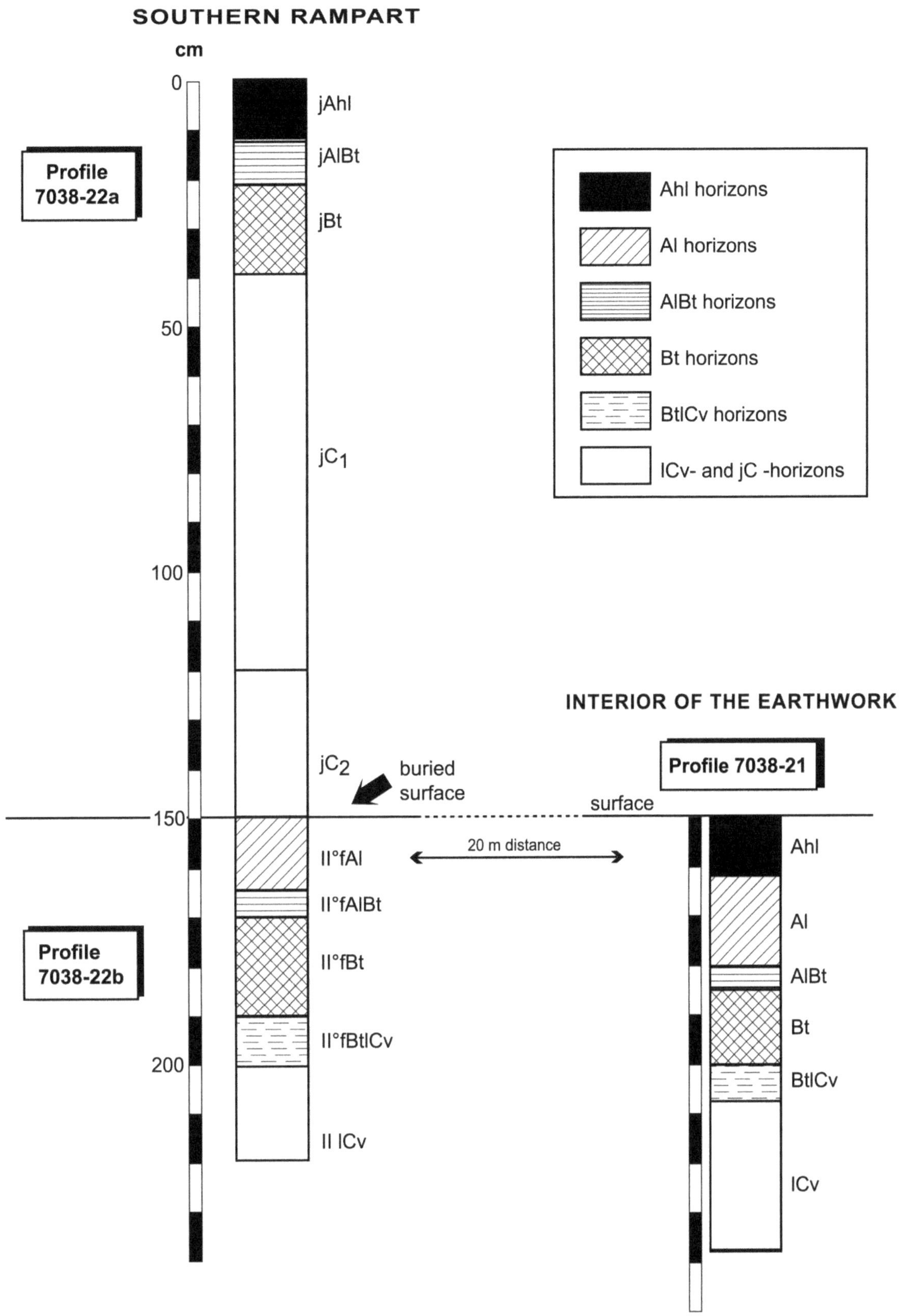


*Figure 3. Investigated soil profiles at the Viereckschanze Poign. Soil horizons of profile 7038-21 in the interior, of profile 7038-22a developed on top of the southern rampart of the earthwork and of profile 7038-22b buried under the earth wall.*

about 10 m deep, periglacially formed hollows. The ditch-earth rampart system with a surface area of ca. 60 x 45 m is well preserved and shows the typical characteristics of South German latène aged square enclosures (Fig. 2). The construction time of this Viereckschanze is also proven by archaeological findings like a snaffle-bit and several $^{14}$C-dates from soil profiles (Leopold and Völkel, 2005).

The present-day climate in the study area is continental with a mean annual air temperature of 8 - 9°C and an average annual precipitation between 640 and 710 mm (BayFORKLIM, 1996). The studied site lies at the eastern border of the Lower Bavarian Tertiary Hills and at the transition to the Dungau, which is a very old and continuously occupied settlement area since the Neolithic (Altsiedelland). At the earthwork, Miocene feldspar sands of the Upper Freshwater Molasse (*Obere Süßwassermolasse*) are overlain by metres thick Quaternary sediments comprising colluvium, loess and periglacial slope deposits. The latter were formed under former periglacial conditions and consist of carbonate-free loess-loam and feldspar sands. Under the Tertiary sediments, Mesozoic bedrock is present (Oschmann, 1958). The hilly topography in the study area was formed during the Pliocene and Quarternary, not at least under periglacial conditions. The typical soils of the study area developed as Luvisols or Calcic Luvisols (*Parabraunerde* according to Ad-hoc-AG Boden, 2005) on the loess and the periglacial slope deposits. Due to plowing and farming all over prehistoric and historic settlement periods these soils are often truncated at the top and upper slope positions. Consequently the soil development rejuvenated back to Calcic Regosols (*Pararendzina*). Locally the soil is completely eroded and loess can outcrop at the surface. Colluvial deposits, associated to soil erosion (Leopold and Völkel, 2007b) are present at lower slopes and depressions. On outcropping feldspar sands, Cambisols (Braunerden) occur. As documented by the Bavarian land register map (Bayerisches Landesvermessungsamt, 1942), the Celtic square enclosure and its closer surroundings are forested at least since 1862. In the interior of the earthwork, coniferous and deciduous trees (e.g., an 150 – 200 years old beech) are present; in the outer area, the coniferous forest is dominated by spruce. This forest cover, which lasts presumably since Celtic times, prevented the Viereckschanze of Poign from destruction.

## Methods

Soil profile 7038-21 was hand dug in the interior of the earthwork up to 88 cm depth. From 88 cm to 150 cm, the profile was drilled with a bucket auger. Profile description follows the German Soil classification (Ad-hoc-AG Boden, 2005). Soil sampling was carried out continuously in maximum 10 cm intervals, strongly orientated on soil horizon boundaries. The southern rampart was drilled through with a bucket auger, collecting samples in 10 cm intervals as well. From the top of the rampart to the final depth, the 220 cm long sediment sequence includes the soil on top of the rampart (profile 7038-22a), the sediments of the earth wall and the soil buried under the rampart (profile

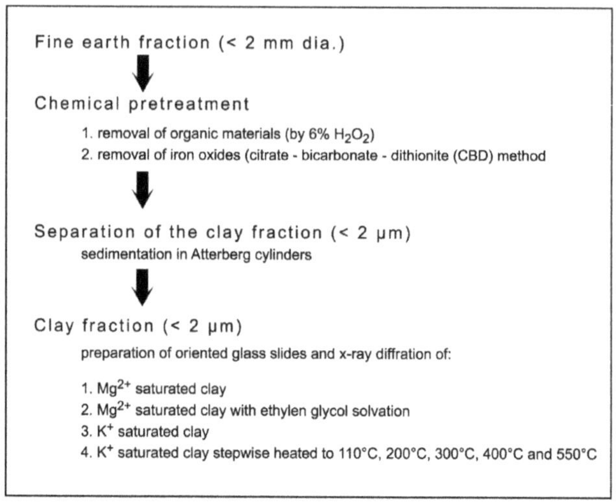


*Figure 4. Treatment of the fine fraction (< 2mm) and preparation steps for clay mineral analyses by X-ray diffraction (XRD).,*

7038-22b). Each sample was described in the field and bagged. The soil samples were dried in a drying furnace at 40°C. Afterwards, soil standard analyses were carried out on the fine fraction (< 2mm) comprising grain size analyses, pH($CaCl_2$)-values, organic carbon and $CaCO_3$-contents effective cation exchange capacity (CECeff.) and content of pedogenic oxides (Fe, Al, Mn) as described by Völkel (1995) and Völkel and Raab (1999). Selected parameters are presented. For clay mineralogy the fine fraction was chemically pre-treated (Fig. 4). Organic materials were removed with hydrogen peroxide (6% $H_2O_2$) and iron oxides were removed with the citrate-bicarbonate-dithionite (CBD) method. The separation of the clay fraction (< 2µm) was carried out by sedimentation in Atterberg cylinders. From the clay fractions, oriented glass slides were prepared and the following four treatment steps were measured with a Siemens D5000 X-ray diffractometer (XRD) using Cu kα radiation:

4. $Mg^{2+}$ saturated clay,
5. $Mg^{2+}$ saturated clay after ethylene glycol solvation,
6. $K^+$ saturated clay,
7. $K^+$ saturated clay stepwise heated to 110°, 200°, 300°, 400° und 550°C.

Sample treatment and preparation follows Völkel (1995). The evaluation was carried out with Siemens software (Siemens Diffrac AT).

***Results and Discussion***

<u>Soil profile in the interior of the earthwork (profile 7038-21)</u>

Studies on soil distribution showed that in the interior of the earthwork, the native Luvisol developed in loess is present, while outside the soils are truncated by erosion which is dated to the Latène age period (Völkel *et al.*, 2001). In the

| ***Sample*** | ***Horizon*** | ***Depth (cm)*** | ***Grain size distrubution (%)*** | | | ***pH ($CaCl_2$)*** | ***Effective Cation exchange capacity (CECeff.) (%)*** | | | | | | | ***Base saturation (%)*** |
|---|---|---|---|---|---|---|---|---|---|---|---|---|---|---|
| | | | sand | silt | clay | | K | Na | Mg | Ca | Al | Fe | Mn | |
| **Profile 7038-21 (interior of the earthwork)** | | | | | | | | | | | | | | |
| 1 | Ahl | 0-9/12 | 14.9 | 69.8 | 15.3 | 3.5 | 0.7 | 0.2 | 0.9 | 1.5 | 95.0 | 1.0 | 0.7 | 3.3 |
| 2 | Al | 12-21 | 13.1 | 68.1 | 18.9 | 3.8 | 0.8 | 0.0 | 0.5 | 0.8 | 97.1 | 0.1 | 0.7 | 2.1 |
| 3 | Al | 21-30 | 12.8 | 64.1 | 23.1 | 3.8 | 1.0 | 0.4 | 0.9 | 3.1 | 94.0 | 0.0 | 0.6 | 5.4 |
| 4 | AlBt | 30-36 | 10.8 | 57.8 | 31.4 | 3.9 | 1.4 | 0.2 | 2.4 | 5.0 | 90.4 | 0.0 | 0.6 | 9.0 |
| 5 | Bt | 36-43 | 16.6 | 52.4 | 37.0 | 3.9 | 1.7 | 0.2 | 7.5 | 18.7 | 71.4 | 0.0 | 0.6 | 28.1 |
| 6 | Bt | 43-50 | 12.6 | 51.5 | 35.9 | 4.2 | 1.5 | 0.2 | 11.7 | 28.6 | 57.8 | 0.0 | 0.2 | 42.0 |
| 7 | BtlCv | 50-58 | 15.7 | 47.9 | 36.4 | 4.2 | 1.7 | 0.3 | 16.0 | 40.0 | 41.3 | 0.5 | 0.2 | 58.0 |
| 8 | lCv | 58-68 | 20.3 | 46.7 | 33.0 | 4.4 | 1.8 | 0.4 | 18.7 | 53.5 | 25.1 | 0.4 | 0.1 | 74.4 |
| 9 | lCv | 68-78 | 23.0 | 44.2 | 32.8 | 4.5 | 1.9 | 0.5 | 20.8 | 61.9 | 14.8 | 0.0 | 0.1 | 85.1 |
| 10 | lCv | 78-88 | 24.0 | 45.7 | 30.3 | 4.6 | 1.9 | 2.9 | 18.8 | 64.4 | 11.4 | 0.5 | 0.1 | 88.0 |
| **Profile 7038-22a (top of the southern rampart)** | | | | | | | | | | | | | | |
| 1a | jAhl | 0-12 | 18.4 | 47.1 | 34.5 | 3.7 | 1.2 | 0.1 | 3.9 | 12.5 | 81.1 | 0.0 | 1.2 | 17.7 |
| 2a | jAlBt | 12-22 | 14.4 | 44.1 | 41.5 | 4.0 | 1.2 | 0.2 | 9.9 | 28.4 | 59.9 | 0.1 | 0.3 | 39.7 |
| 3a | jBt | 22-32 | 14.9 | 42.7 | 42.4 | 4.1 | 0.4 | 0.2 | 12.3 | 34.3 | 52.8 | 0.0 | 0.0 | 47.2 |
| 4a | jBt | 32-40 | 16.6 | 42.0 | 41.4 | 4.2 | 0.6 | 0.1 | 14.0 | 38.3 | 46.9 | 0.0 | 0.1 | 53.0 |
| 5 | $jC_1$ | 40-50 | 20.2 | 40.5 | 39.3 | 4.2 | 1.8 | 0.1 | 19.3 | 61.8 | 16.2 | 0.0 | 0.8 | 83.0 |
| **Profile 7038-22b (buried profile under the southern rampart)** | | | | | | | | | | | | | | |
| 17a | II°fAl | 150-155 | 17.6 | 61.7 | 20,7 | 5.3 | 1.5 | 0.6 | 15.0 | 82.7 | 0.0 | 0.0 | 0.2 | 99.8 |
| 17b | II°fAl | 155-165 | 14.5 | 66.2 | 19.3 | 5.5 | 1.9 | 1.0 | 15.5 | 81.2 | 0.0 | 0.0 | 0.4 | 99.8 |
| 18 | II°fAlBt | 165-170 | 14.5 | 56.6 | 28.9 | 5.4 | 1.7 | 0.9 | 16.6 | 80.6 | 0.0 | 0.0 | 0.2 | 99.8 |
| 19 | II°fBt | 170-180 | 15.2 | 42.8 | 42.0 | 5.5 | 2.0 | 0.9 | 16.8 | 80.2 | 0.0 | 0.0 | 0.1 | 99.9 |
| 20 | II°fBt | 180-190 | 19.7 | 44.2 | 36.1 | 5.7 | 2.2 | 0.9 | 17.6 | 79.2 | 0.0 | 0.0 | 0.1 | 99.9 |
| 21 | II°fBtlCv | 190-200 | 21.9 | 44.1 | 34.0 | 5.8 | 1.1 | 1.0 | 16.0 | 81.8 | 0.0 | 0.0 | 0.1 | 99.9 |
| 22 | II lCv | 200-210 | 25.6 | 45.4 | 29.0 | 5.9 | 1.1 | 1.1 | 15.9 | 81.8 | 0.0 | 0.0 | 0.1 | 99.9 |
| 23 | II lCv | 210-220 | 30.5 | 45.3 | 24.2 | 6.1 | 1.0 | 1.0 | 14.5 | 83.4 | 0.0 | 0.0 | 0.1 | 99.9 |

*Table 1. Soil standard parameters of profiles 7038-21, 7038-22a and 7038-22b.*

interior of the earthwork, neither admixtures of feldspar sands, nor components > 2 mm are found up to a profile depth of 150 cm. In contrast, west of the earthwork, on the other side of the ditch, loess loam is absent due to soil erosion and periglacial slope deposits outcrop instead. Only in the interior of the earthwork, the complete soil is present, including loess which was certainly $CaCO_3$-rich. On the south-exposed slope, outside of the earthwork, primary loess is present, in form of strips on middle and lower slope positions.

Profile 7038-21 in the interior of the earthwork is therefore the only soil profile, with a soil development spanning the whole Holocene and presumably reaches into the Late Glacial period. It is a well developed Luvisol. The most characteristic soil parameters are presented in Tab. 1. The Ahl-horizon has a pH value of 3.5, which is very acidic and also the lCv-horizon reaches pH values of 4.6 also indicating strong acidification. Accordingly, the CECeff. is low with percentages between 2.1 and 9.0% in the Ahl-horizon until the lower boundary of the AlBt-Horizon at a profile depth of 36 cm. $Al^{3+}$ dominates the CECeff. with > 90%. From the Bt-horizon until the base of the soil profile, the CECeff. and the base saturation increase, especially with a strong rise of calcium and magnesium percentages.

## Soil profile on the top of the earth rampart (profile 7038-22a)

The earth rampart is predominantly made up by Bt material excavated from the ditch. Sedimentological analyses show that the rampart material can be subdivided into two units $jC_1$ and $jC_2$ (Fig. 3). Since about 2000 years, a Luvisol developed on the top of the rampart (Fig. 3), evidencing that intensive clay translocation in terms of lessivation is possible in $CaCO_3$-free materials within a comparatively short time interval. In comparison to the interior of the earthwork (profile 7038-21), the soil on the top of the rampart (profile 7038-22a) is not as deeply acidified. This is a consequence of the erosive truncation of the jAhl-horizon on top of the wall. The topographic position on the culmination point on the rampart and the steep slopes, have enforced erosion processes. The comparison of the pH-values from the enclosure profile (7038-21) and from the rampart profile (7038-22a) shows no evidence for environments with different acidification. However, the depth plot of the CECeff. shows that the base cation supply in the rampart profile is much higher and that the base saturation reaches up to 83% ($jC_1$-horizon). The degree of acidification of the jAhl-horizon of the rampart profile, represented by the $Al^{3+}$ saturation, is a little bit lower (81%) than in the equivalent depth of the enclosure (Ahl-horizon with 95% $Al^{3+}$). The standard soil parameters are presented in Tab. 1.

## Buried soil under the earth wall (profile 7038-22b)

The development depth of the buried Luvisol is about 50 cm. Lessivation and clay accumulation horizons are distinctive features in profile 7038-22b. The analyses of the grain size distributions (Tab. 1) support the field results. The content of 42% clay in the Bt horizon is typical for Luvisols derived from $CaCO_3$-rich loess in the Regensburg region at the southern margin of the Danube valley (Völkel, 1995). Profile 7038-22b shows a moderate to slightly acidic soil reaction. The depth plot of the pH documents an even decrease of the soil acidification. The CECeff. percentages confirm this uniformity, except for Mn. On the contrary, the absolute values (CECeff. mval/100g soil) – not shown here – have the lowest calcium and magnesium contents in the buried II°fAl horizon.

## Clay mineral composition and pedogenic transformation of the studied soils

The clay mineral spectrum of the soils is composed of 1.4 nm minerals, 1.0 - 1.4 nm mixed-layered minerals, illite (1.0 nm and 0.5 nm), kaolinite (0.72 nm and 0.357 nm), muscovite (0.447 nm), quartz (0.427 nm and 0.333 nm) and the feldspars plagioclase (0.319 nm) and orthoclase (0.325 nm). Furthermore, there are reflections at 0.47 nm, which hint on the presence of chlorite according to Grim (1968) and Köster and Schwertmann (1993). On the same position in the diffractogram, also reflections of the 1.4 nm minerals vermiculite and smectite occur. In the majority of the clay mineral samples, there is a 0.354 nm peak which is partly present as shoulder on the 002-peak of kaolinite and partly developed as single peak, hinting on the presence of secondary chlorite. In addition to a 0.299 nm peak which belongs to K feldspar according to Brindley and Brown (1980), there is a 0.284 nm peak, pointing to the presence of non-expandable 1.4 nm minerals. After K saturation of the clays, this peak collapses, hence it should belong to high-charged vermiculites. The x-ray diffraction files are shown in Figures 5-7.

According to the estimation made by Niederbudde and Schwertmann (1980) neoformation and transformation of clay minerals in soils of Holocene age is rather low. This can also be observed in our study. Furthermore, on the basis of the x-ray diffraction patterns, there are two alterations concerning the 2:1 layer minerals. These alterations concern firstly, the degree of crystallinity of illites in the top soil and secondly, the degree of secondary chloritisation of 1.4 nm minerals.

In each of the three investigated profiles with different time intervals for soil development, there is a striking tendency of increasing illite contents as well as an increasing crystallinity of illites from deeper soil depths towards the top of the soil profiles. This could hint on an illitisation of 1.4 nm minerals in terms of Niederbudde (1975), an aspect which is discussed contrariwise (e.g., Tributh, 1974; Allen and Hajek, 1989). Subfractioning of the clay fraction and X-ray analyses could give further information (e.g., Tributh, 1990).

In addition, a secondary chloritisation of the 1.4 nm minerals occurs as well in profile 7038-21 (interior of

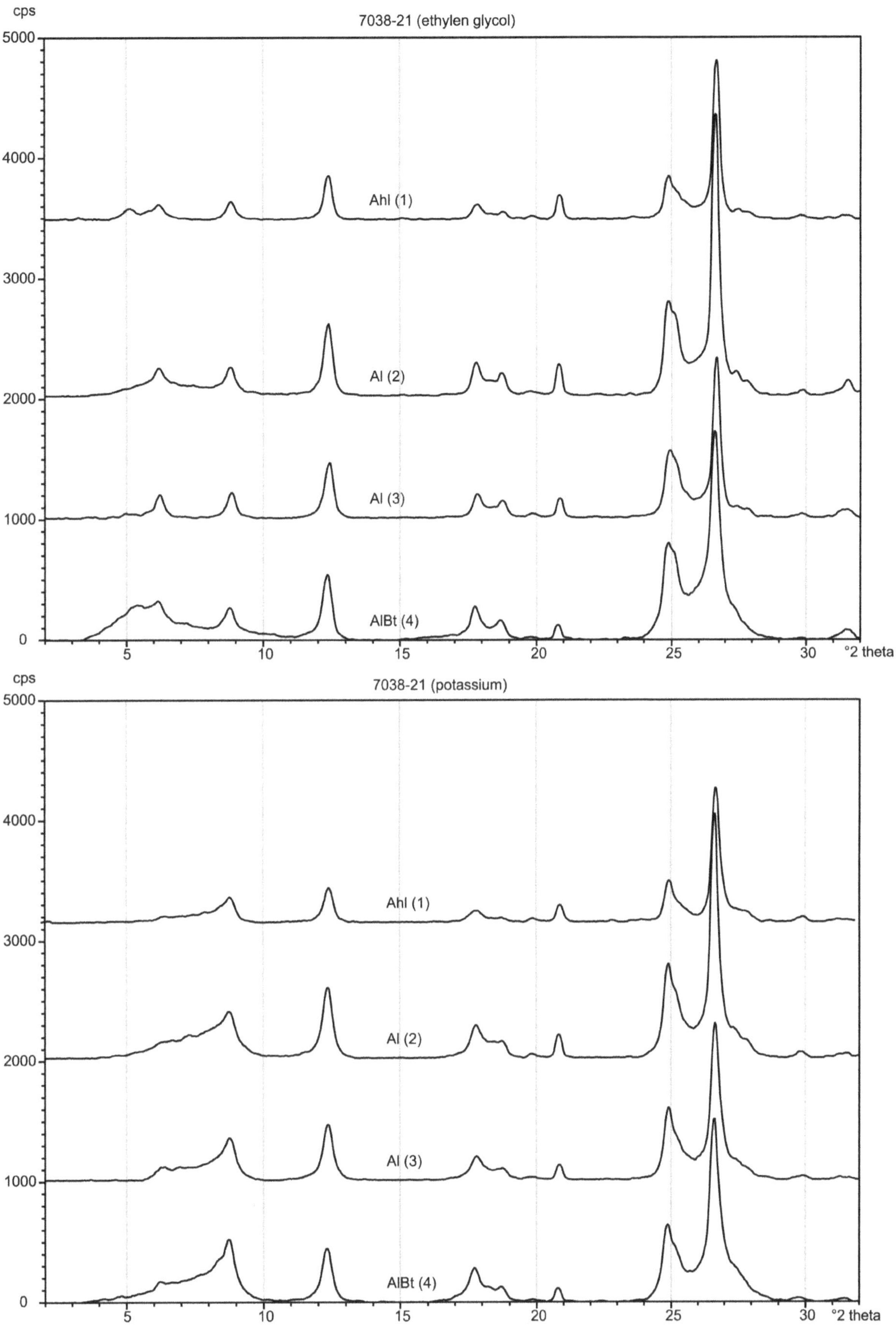


*Figure 5. X-ray diffractometer patterns (Cu kα) of the clay fraction derived from different top soil horizons of profile 7038-21 (Luvisol developed in loess, interior of the earthwork). Top: Mg saturated clay after ethylene glycol solvation. Bottom: Mg saturated clay after potassium treatment.*

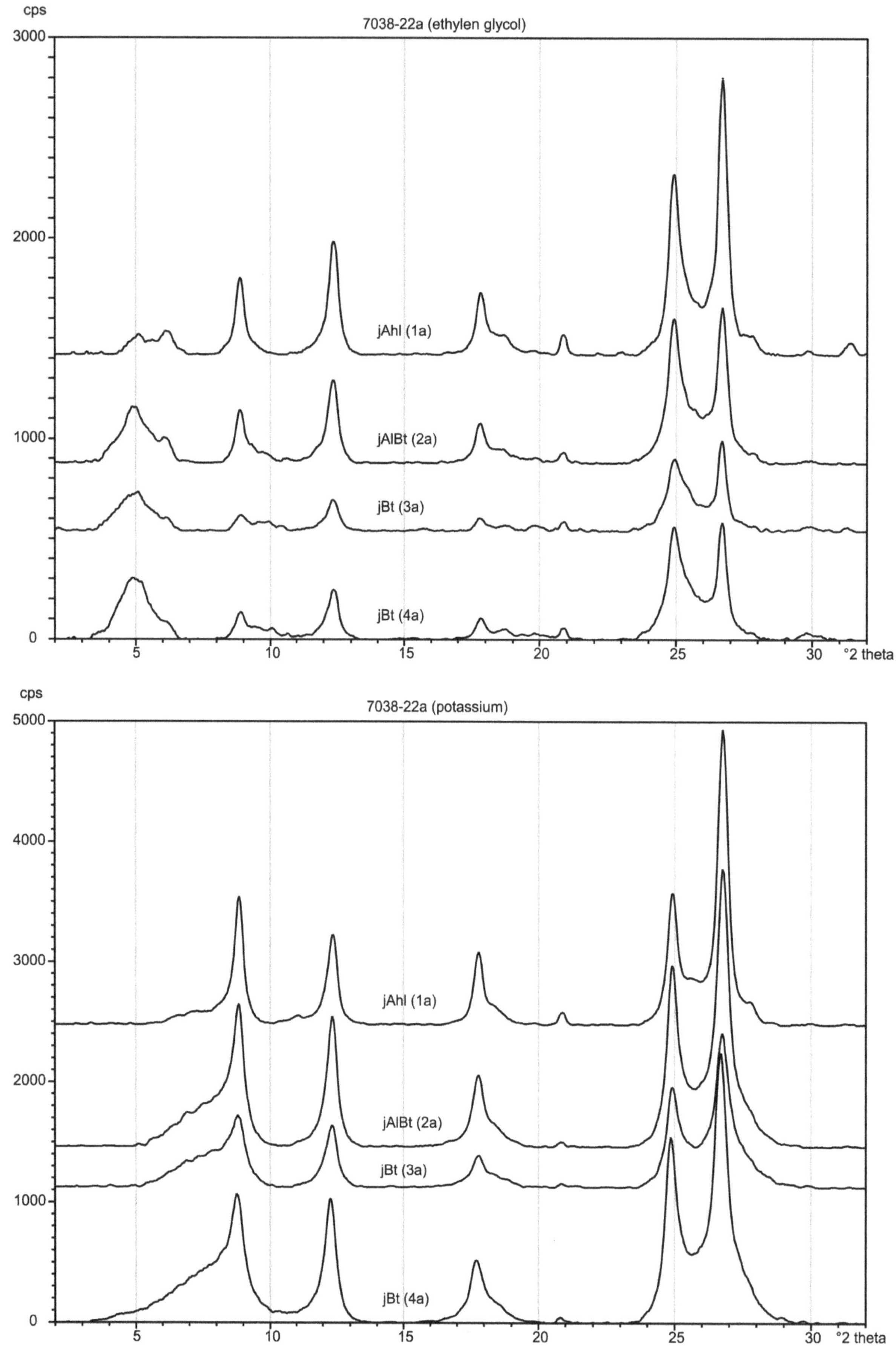


*Figure 6. X-ray diffractometer patterns (Cu kα) of the clay fraction derived from different top soil horizons of profile 7038-22a (Luvisol developed in the earth wall material, top of the rampart). Top: Mg saturated clay after ethylene glycol solvation. Bottom: Mg saturated clay after potassium treatment.*

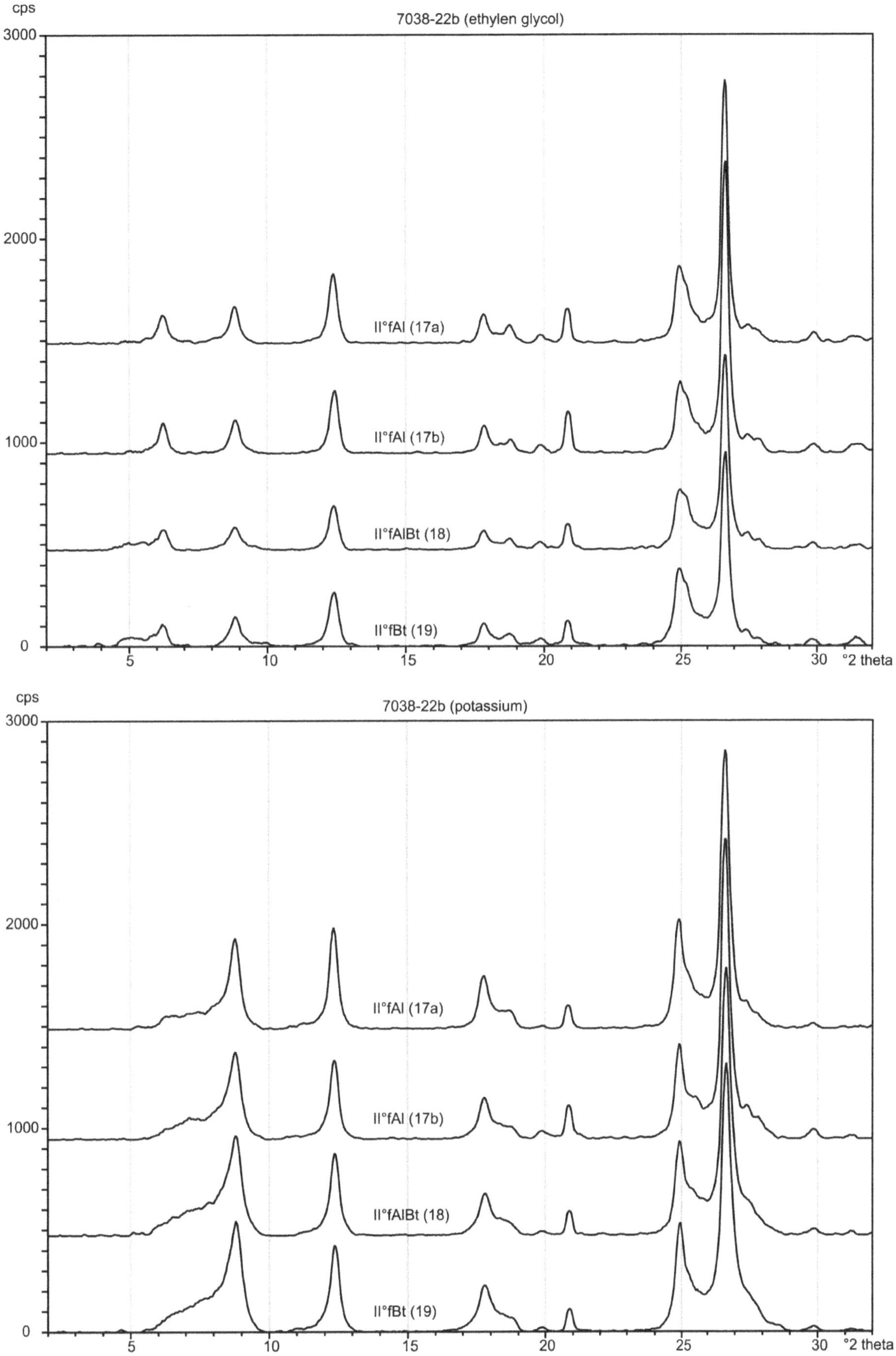


*Figure 7. X-ray diffractometer patterns (Cu kα) of the clay fraction derived from different top soil horizons of profile 7038-22b (Luvisol developed in loess, buried under the rampart). Top: Mg saturated clay after ethylene glycol solvation. Bottom: Mg saturated clay after potassium treatment.*

the earthwork) as in the anthropogenic fossilised profile 7038-22b under the earth rampart. The X-ray diffraction of the K saturated and stepwise heat-treated clay samples proves an instability of the non-contractable 1.4 nm peak from 200° C. It is assumed that the interlayering of Al-Fe-hydroxy cations in form of the so-called atoll structure in terms of Barnhisel and Bertsch (1989) collapse at comparatively low temperatures.

While pedochemical standard parameters do not give information about pedogenic differentiation of the anthropogenic fossilised Luvisol, the clay minerals preserved this information persistently. Hence, a slight secondary chloritisation of the surface horizons was already present about 2000 years ago. How far this aluminium chloritisation is reversible due to the high percentages of calcium (see CECeff.) is currently not known. On the contrary, the Luvisol on the top of the earth rampart (profile 7038-22a) shows no secondary chloritisation. We assume that first of all, the fresher and lesser acidified parent material consisting mainly of the Bt material derived from the ditch, prevented an Al chloritisation, though the aluminium percentages in the CECeff. in the surface soil (jAhl horizon) is also increased. Furthermore, the truncation of the soil as consequence of erosion at the exposed top position could also explain the missing secondary chloritisation of the soil.

## Conclusions

On the basis of our studies concerning Holocene soil development and pedogenic clay mineral transformation in Luvisols with different development ages at the Viereckschanze Poign, the following conclusions can be drawn:

- In the interior of the earthwork the complete and undisturbed soil is present (profile 7038-21). Soil formation started presumably during the Late Glacial and lasted at least the whole Holocene period. The soil is a well developed Luvisol showing strong acidification features.
- On the top of the rampart (profile 7038-22a) a Luvisol formed since 2000 years ago. This indicates that during a comparatively short time interval, intensive lessivation is possible in $CaCO_3$-free material. Concerning soil acidification, there is almost no difference in comparison to the soil profile in the interior of the earthwork (profile 7038-21).
- The soil under the rampart (profile 7038-22b) was buried about 2000 years ago. It is a well developed Luvisol, similar to profile 7038-21, showing a moderate to slightly acidic pH. Some modification occurred by leaching and redeposition of calcium and magnesium ions from the earth wall.
- Concerning pedogenic clay mineral transformation, two processes can be inferred from our investigations: A) The X-ray diffraction patterns of all three soil profiles hint on an illitisation in terms of Niederbudde (1975). Subfractioning of the clay and X-ray analyses could give further information. B) A secondary chloritisation can be observed in the X-ray patterns of profile 7038-21 and profile 7038-22b. The degree of secondary chloritisation is minor in profile 7038-22b. This shows that already 2000 years ago a slight secondary chloritisation was present among the natural soil forming processes. On the contrary the X-ray patterns of profile 7038-22a show no secondary chloritisation. This is most probably due to the fresher and lesser acidificated parent material preventing Al-chloritisation and the truncation of the soil profile on the exposed top position of the rampart.

## Synthesis

By using the buried soils of the Celtic square enclosure as a geoarchive, the hypothesis that acidification of soils in Central Europe is a relatively recent process during the Holocene and is mainly a man-induced phenomenon of the industrial age, can be redrawn. It could be clearly shown that pedogenic properties deriving from acidification (low base saturation, clay mineral transformation) developed during different periods of the Holocene. Acidification was a natural process already 2000 years ago. On the other hand lessivation can form Luvisols by sufficient translocation of clay within the last two millennia.

## References


Ad-hoc-AG Boden, 2005. *Bodenkundliche Kartieranleitung*, Hannover.

Allen, B.L. and Hajek, B.F., 1989. Mineral occurrence in soil environments.- In: Dixon, J.B. and Weed, S.B. (eds.), *Minerals in Soil Environments*: 199-278. (= Soil Science Society of America).

Barnhisel, R.I. and Bertsch, P.M., 1989. Chlorites and Hydroxy-Interlayered Vermiculite and Smectite. In: Dixon, J.B. and Weed, S.B. (eds.), *Minerals in Soil Environments*: 729-788 (= Soil Science Society of America).

Bayerisches Landesvermessungsamt (ed.), 1942. *Flurkarte Bayern von 1862*. Hofbibliothek Fürst von Thurn und Taxis.

BayFORKLIM, 1996. *Klimaatlas von Bayern*, München.

Birkeland, P.W., 1999. *Soils and geomorphology*. New York.

Brindley, G.W. and Brown, G., 1980. *Crystal structures of clay minerals and their identification*. (= Mineralogical Society, Monographs 5).

Grim, R.E., 1968. *Clay Mineralogy*, New York.

Köster, H.M. and Schwertmann, U., 1993. Beschreibung einzelner Tonminerale. In: Jasmund, K. and Lagaly, G. (eds.), *Tonminerale und Tone. Struktur, Eigenschaften, Anwendung und Einsatz in Industrie und Umwelt*, Darmstadt: 33-88.

Leopold, M. (this volume). *Buried soils and surfaces, a key archive for Geoarchaeology*.

Leopold, M. and Völkel, J., 2007a. Quantifiying prehistoric soil erosion – a review of soil loss methods and their application to a celtic square enclosure (Viereckschanze) in southern Germany, *Geoarchaeology*, 22/8: 873-898.

Leopold, M. and Völkel, J., 2007b. Colluvium: definition, differentiation and their possible suitability to reconstruct Holocene climate data, *Quaternary International*, 162-163: 133-140.

Leopold, M. and Völkel, J., 2005. Methodological approach and case study for the reconstruction of a (pre)historic land use model, *Zeitschrift für Geomorphologie,* Suppl. N.F., 139: 173-188.

Niederbudde, E.A., 1975. Veränderungen von Dreischicht-Tonmineralen durch natives Kalium in mitteldeutschen und niederbayerischen holozänen Lößgebieten, *Journal of Plant Nutrition and Soil Science,* 138: 214-217.

Niederbudde, E.A. and Schwertmann, U., 1980. Clay mineralogy of soils, *Geologisches Jahrbuch, Reihe D*, 39: 99-114.

Oschmann, F.,1958. *Erläuterungen zur Geologischen Karte von Bayern, 1:25 000, Blatt Nr. 7038 Bad Abbach*, München.

Raab, A., Leopold, M. and Völkel, J., 2005. Vegetation and land-use history in the surroundings of the Kirchenmoos (Central Bavaria, Germany) since the late Neolithic Period to the Middle Ages, *Zeitschrift für Geomorphologie*, Suppl. N.F., 139: 35-61.

Schaetzl, R.J. and Anderson, S., 2005. *Soils. Genesis and geomorphology*. Cambridge.

Torbrügge, W., 1984. Die Landschaften um Regensburg in vor- und frühgeschichtlicher Zeit. In: Rieckhoff-Pauli, S. and Torbrügge, W. (eds.): *Führer zu archäologischen Denkmälern in Deutschland* 5, Regensburg: 28-121.

Tributh, H., 1974. Die Umwandlung von Dreischichtmineralen in Böden aus Löß, *Mitteilungen der Deutschen Bodenkundlichen Gesellschaft*, 18: 413-418.

Tributh, H., 1990. Die Tonmineralentwicklung in Abhängigkeit von der Bodengenese, *Mitteilungen der Deutschen Bodenkundlichen Gesellschaft*, 62: 153-156.

Völkel, J., 1995. *Periglaziale Deckschichten im Bayerischen Wald und seinen Randgebieten als geogene Grundlagen landschaftsökologischer Forschung im Bereich naturnaher Waldstandorte.* (= Zeitschrift für Geomorphologie, Suppl., N.F., 98).

Völkel, J., 2006. Kulturlandschaften als Geoarchive – der paläoökosystemare Ansatz. In: Gaese, Sandholz, S. and Böhler, A. (eds.): *Denken in Räumen. Nachhaltiges Ressourcenmanagement als Identitätssicherung - Durch Veränderung der Rahmenbedingung veränderte Kulturlandschaften und das Problem ihrer Erhaltung. – Verhandlungsband zum Symposium des Instituts für Tropentechnologie (ITT) der FH Köln, der Deutschen Bundesstiftung Umwelt (BDU) und der Deutschen UNESCO-Kommission, Osnabrück 3.-5. November 2004*: 289-304.

Völkel, J. and Raab, T., 1999. Zur Differenzierung periglazialer, glazigener und kolluvialer Sedimente im Hangrelief von Mittel- und Hochgebirgen. *Zentralblatt für Geologie und Paläontologie*, I (5-6): 541-57.

Völkel, J., Leopold, M. and Weber, B., 2001. Neue Befunde zur Landschaftsentwicklung im niederbayerischen Donauraum während der Zeitenwende (Keltisches Oppidum von Manching / Viereckschanze von Poign bei Bad Abbach), *Zeitschrift für Geomorphologie,* Suppl. N.F.: 128: 47-66.

Völkel, J., Raab, A., Raab, T., Leopold, M. and Simm, H., 1998. Methoden zur Bilanzierung spätlatènezeitlicher Bodenerosion am Beispiel der Viereckschanze von Poign, Lkr. Regensburg. In: Küster, H., Lang, A. and Schauer, P. (eds.): *Archäologische Forschungen in urgeschichtlichen Siedlungslandschaften,* Regensburg: 541-558. (= Regensburger Beiträge zur prähistorischen Archäologie, 5.

# Chapter 7
# An Example of a Fruitful Discussion Between a Pedologist and an Archaeologist.
# A 1st-4th Century AD Agricultural Enclosure with a Stable and a Manure Pit at "Chapelle Saint-Nicolas" in Saint-Brice-sous-Forêt (Val-d'Oise, France)


*Vanessa Rouppert (1), Jean-Yves Dufour (1), Kai Fechner (2, 3)*

*(1) Inrap / Centre - Île-de-France (Paris), e-mail: vanessa.rouppert@inrap.fr*
*(2) Inrap Nord - Picardie (Villeneuve d'Ascq)*
*(3) formerly: RooTS and Université libre de Bruxelles (Brussels/ Belgium)*


## Introduction

North of Paris, in the *Plaine de France*, the increase in open area excavations of very large surfaces has led to the discovery and study of complete small Roman rural settlements (**fig. 1**). The sites are identified by the presence of one or more joined enclosures. Their surface area varies from 4000 $m^2$ to 1 ha.

The characterization of these settlements and of their relationship with the surrounding territory is rarely possible without the contribution of the natural sciences. When an archaeological model is proposed, the palaeo-environmental approach, as in this case pedology, greatly reduces the likelihood of an error of interpretation.

The site of the Chapelle Saint-Nicolas in Saint-Brice-sous-Forêt (Rouppert *et al.*, 2009), which covers approximately 6,000 $m^2$, was supposed, at the beginning of the study, to be one of these small settlements. It was occupied between the 1st and the 4th centuries AD. The field evaluation had shown the presence of a ditched enclosure surrounding a sector thought to be devoted to agriculture or stockholding and another for lodgings. The full excavation would confirm or correct this interpretation and perhaps establish the status of the site. The question was which methods were best adapted to the study of this site.

## The Pedological Study

Spatial analysis of occupation ecofacts and artefacts was chosen as the first line of attack. This mainly comprised the mapping of phosphate levels ($P_2O_5$) which could indicate activity within the enclosure. This is a proven method but has been little used in France (Zimmermann 2004:18-20; Fechner *et al.*, 2004; Van de Wetering and Wansleeben, 1988; Conway, 1983). The chemical soil element mapping covered the whole surface area of the enclosures as well as their immediate surroundings. Two methods were used so as to treat a large number of samples rapidly and cheaply:

- a test revealing the relative amount of phosphorus for all samples,
- a quantitative analysis of the precise amount of total phosphate, on selected samples chosen according to their position (**fig. 2**).

The latter method allows a certain control of the tests and comparisons with other sites. These quantitative data also distinguished between the amounts of organic and inorganic phosphate in the total phosphate count determined using the Ghent and Copenhagen methods (Eidt research protocol : Eidt, 1984).

But phosphate cartography does not work if the dynamic soil formations and transformation processes, including in this case post-Roman pollution, are not understood. So a complete pedological study was necessary. Such an approach requires the precise characterisation of both significant structures and finds, which is after all the basis of archaeological method. To this end, the natural sciences, and specifically pedology, contributed to a better understanding of the field stratification and site interpretation through soil analysis and micromorphology completed by parasitological and phytolithic analyses. From the numerous structures identified at the Chapelle Saint-Nicolas after the confrontation of archaeological and pedological approaches, an example related to stockholding has been chosen to illustrate the method. It concerns a stable and its manure pit.


[1] Translation: Peter MacIntyre, Inrap. Special thanks to Ph. Lenhardt (architect, Inrap) for his contribution to the field work and the illustrations. The work of Iliana Pasquier on the drawings was of inestimable help.

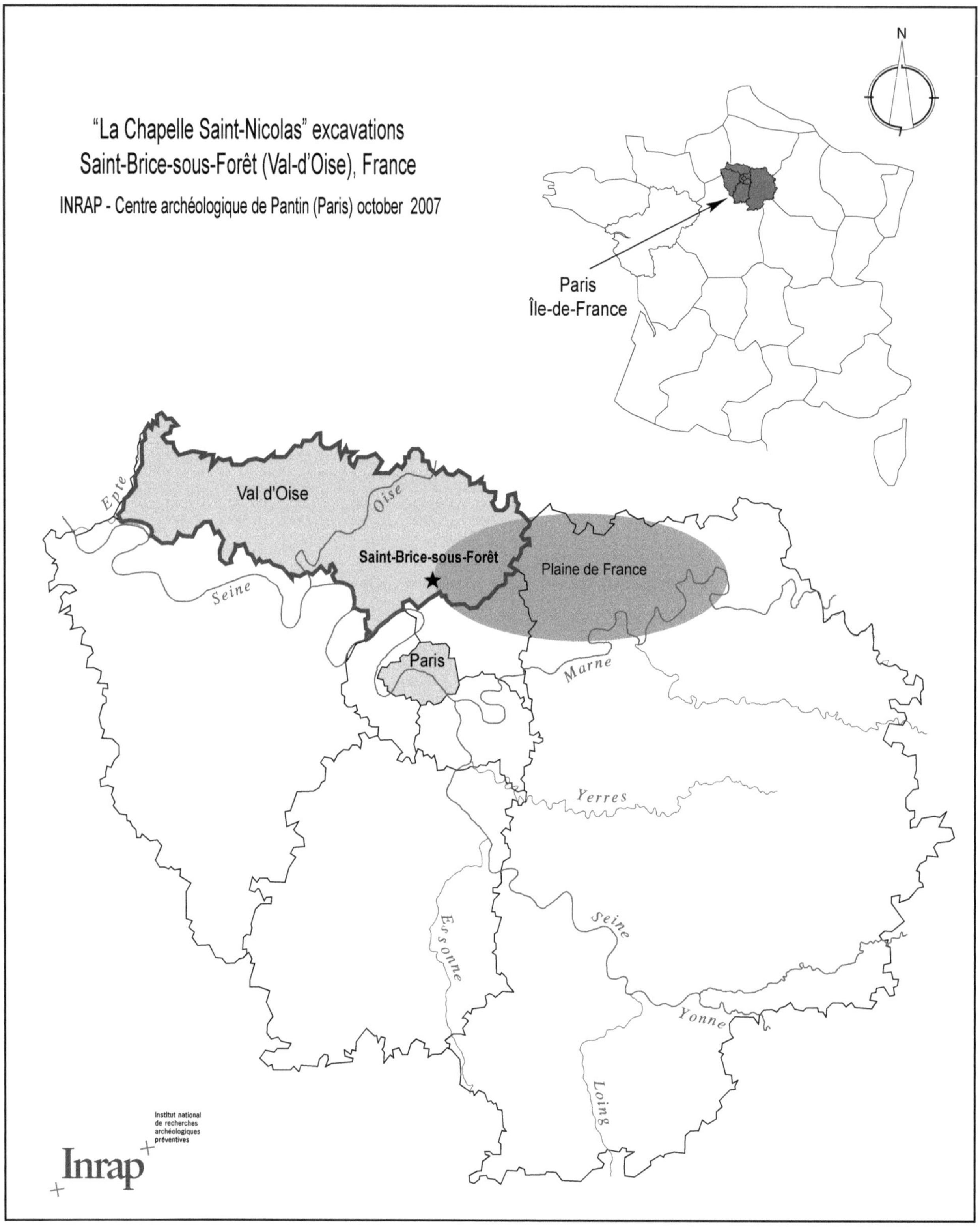


*Figure 1. Situation of the site.*

| Sample nr. | Total phosph. ppm | inorg. phosph. ppm | organic phosph. ppm |
|---|---|---|---|
| Extremely Low | | | |
| B3 | 257 | 54 | 203 |
| A14 | 268 | 104 | 164 |
| X17 | 272 | 69 | 203 |
| A16 | 280 | 97 | 183 |
| V20 | 293 | 107 | 186 |
| C3 | 305 | 98 | 207 |
| V25 | 318 | 110 | 208 |
| Very low | | | |
| Y18 | 333 | 143 | 190 |
| X10 | 334 | 191 | 143 |
| Y17 | 347 | 167 | 180 |
| W18 | 349 | 151 | 198 |
| C14 | 351 | 306 | 45 |
| X9 | 353 | 204 | 149 |
| C9 | 357 | 193 | 164 |
| X18 | 363 | 121 | 242 |
| W17 | 369 | 178 | 191 |
| V18 | 369 | 183 | 186 |
| Y11 | 374 | 222 | 152 |
| W9 | 385 | 197 | 188 |
| C17 | 388 | 101 | 287 |
| Z9 | 389 | 181 | 208 |
| Low | | | |
| X16 | 390 | 194 | 196 |
| X11 | 398 | 204 | 194 |
| X8 | 408 | 243 | 165 |
| C15 | 408 | 197 | 211 |
| D17 | 411 | 208 | 203 |
| Medium-low | | | |
| Z10 | 418 | 272 | 146 |
| X14 | 420 | 210 | 210 |
| C12 | 422 | 254 | 168 |
| H3 | 424 | 290 | 134 |
| Y16 | 427 | 246 | 181 |
| X15 | 439 | 235 | 204 |
| Medium-high | | | |
| G3 | 447 | 318 | 129 |
| V24 | 453 | 254 | 199 |
| D12 | 480 | 263 | 217 |
| E11 | 515 | 385 | 130 |
| V23 | 518 | 237 | 281 |
| D10 | 522 | 259 | 263 |
| Y15 | 523 | 342 | 181 |
| D15 | 527 | 286 | 241 |
| X12 | 559 | 379 | 180 |
| D11 | 595 | 337 | 258 |
| Y14 | 597 | 409 | 188 |
| High | | | |
| X13 | 621 | 414 | 207 |
| X22 | 680 | 488 | 192 |
| D9 | 763 | 521 | 242 |
| V19 | 836 | 598 | 238 |
| Y13 | 856 | 579 | 277 |
| K26 | 914 | 658 | 256 |
| Very high (in pits) | | | |
| Z32 | 1134 | 908 | 226 |
| Z30 | 1222 | 1038 | 184 |
| Particularly high (in pits) | | | |
| Z26 | 1834 | 1553 | 281 |

*Figure 2. Content of total, inorganic and organic Phosphates of the Chapelle Saint-Nicolas site*

## 1/ General mapping of the phosphate results

In many places, the limit of the enclosure correspond precisely with the limit of phosphate enrichment (**fig. 3 & 4**); the contrast between inside and outside being clearest between the sample series on the south and the east sides of the enclosure. A clear relationship between the limits of the strong pollution and the Roman structures was also observed near a Roman conducting water system separating the enclosure diagonally in two parts.

It is clear that the phosphate pollution and the Roman occupation are contemporary. Moreover, the ancient water conduct marks the limit between the low to medium pollution to the North and the high dosage to the South. The sector delimited by the water conduct to the North-West and the enclosure to the South-West suffered considerable phosphate contamination.

## 2/ Identification of a stable: how mapping chemical elements can confirm a hypothesis and contribute to the interpretation

An attempt to compare this long narrow building (35 m that is 1 actus or 120 Roman feet[2] long by 3.50 m or 12 feet broad) with similar constructions on other Roman sites in Northern France did not go very far. The problem is that interpretations vary and are often not well argued, usually because of the absence of a floor level related to the building or occupation phases and in structures preserved or identified in or near the building.

French archaeological literature mentions « agricultural buildings », « annexes», living quarters as well as cowsheds or sheep pens, but mostly without clear evidence. Generally, little has been done to verify these identifications.

At the Chapelle Saint-Nicolas site, it was decided to go further in the interpretation by using, among others approaches, chemical analysis (see also Fechner, this volume). Chemical soil analysis can provide solid arguments as to whether a building or a space was used for tethering or sheltering animals. The resultant natural pollution tends to be massive in scale and is difficult to miss when the appropriate method is used. This is true even when the circulation levels and specific structures

[2] A Roman foot equals 29.55 cm. An *actus* is about 35.5 m.

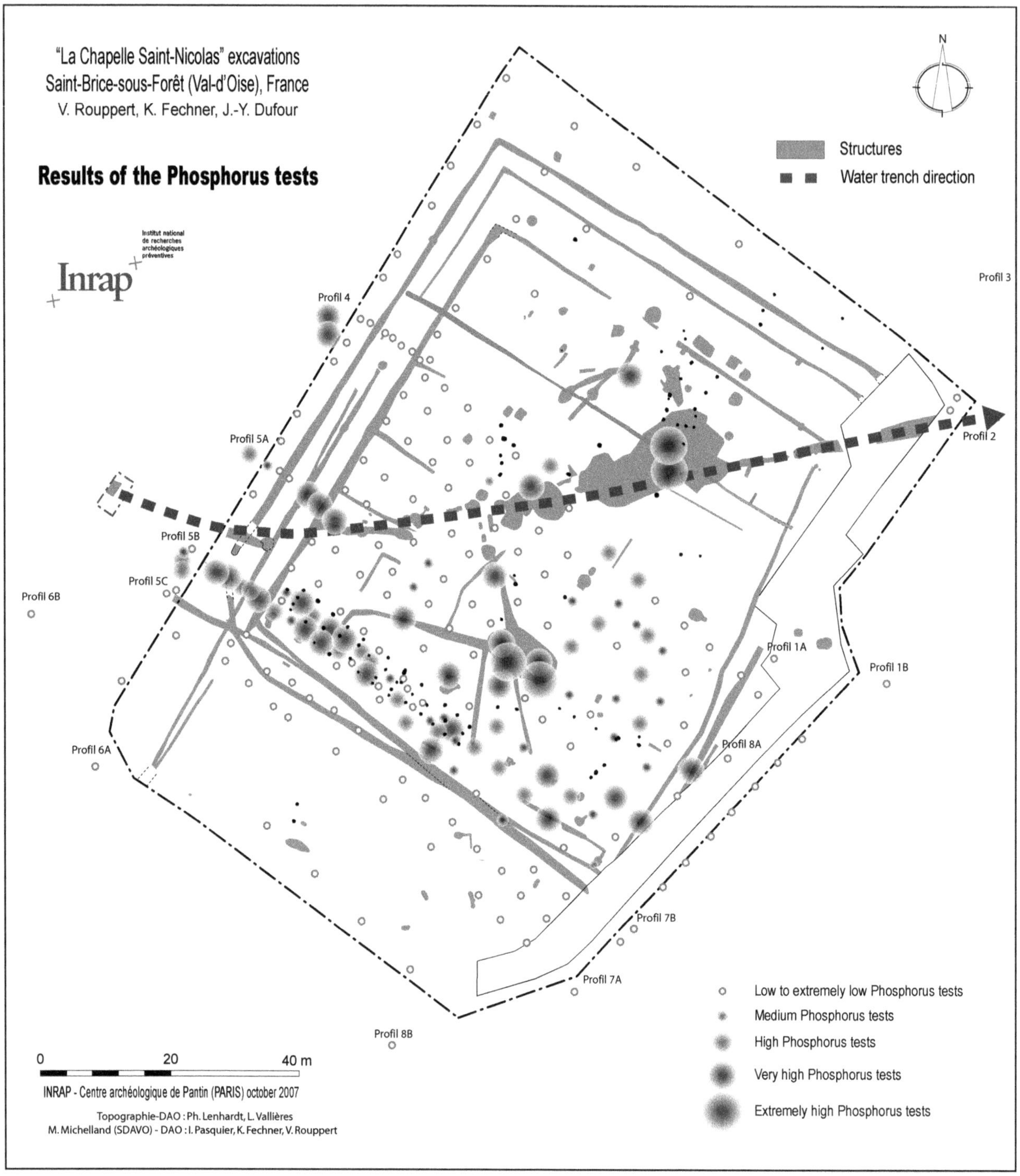


*Figure 3. Results of the Phosphorus tests.*

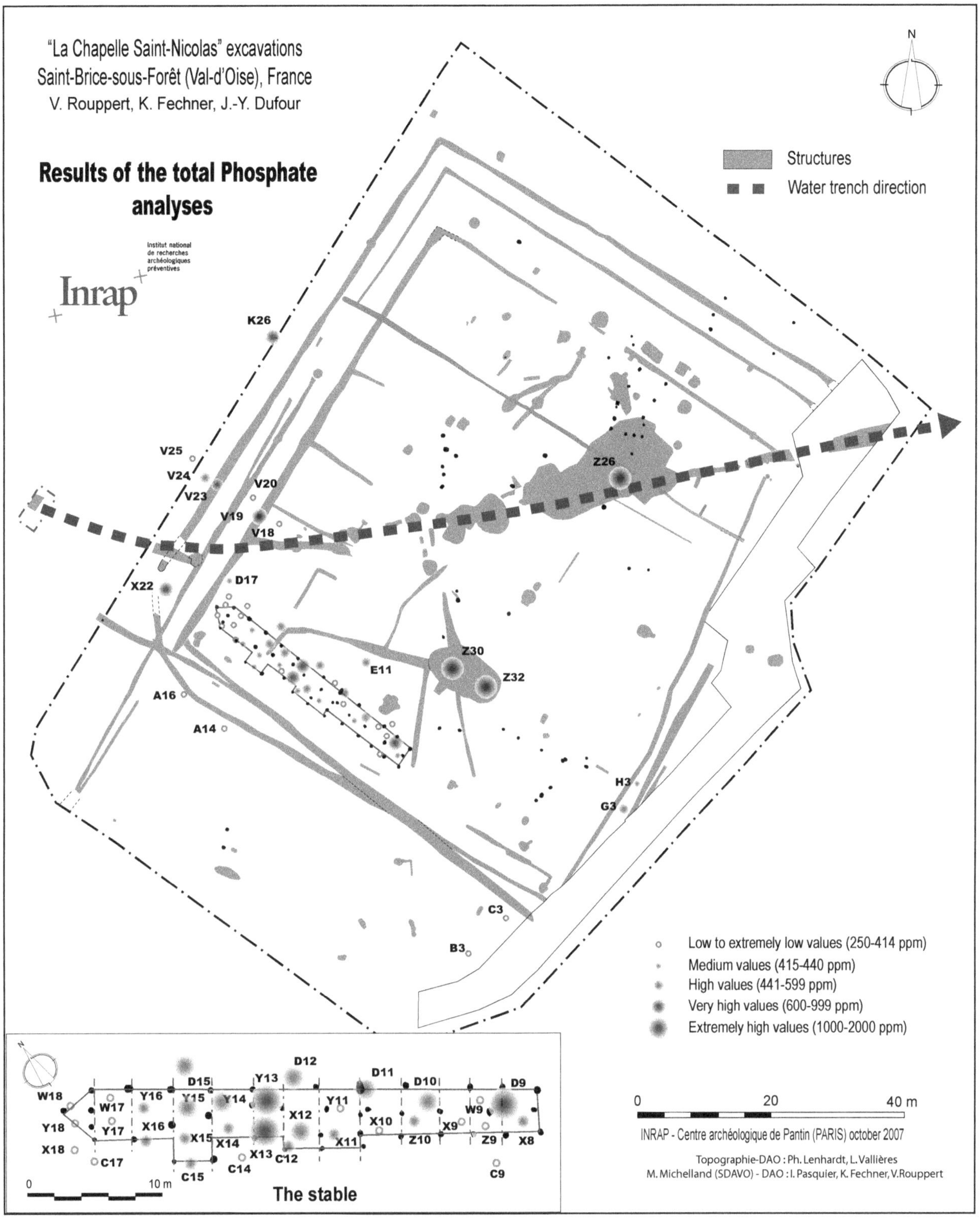


*Figure 4. Results of the total Phosphate analyses.*

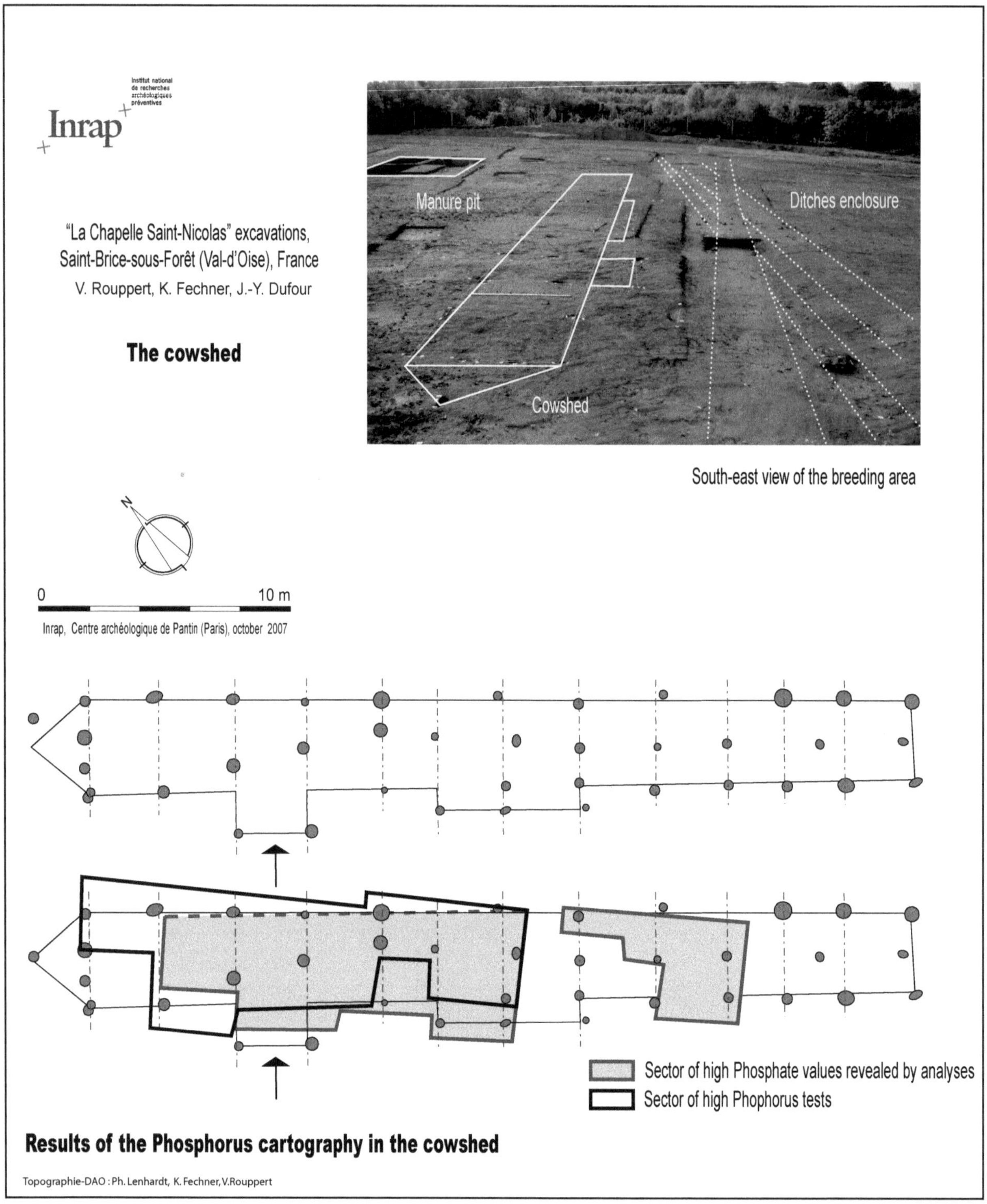


*Figure 5. South-est view of the breeding area and results of the Phosphorus cartography in the cowshed.*

are absent, since the pollution penetrates deeply, under the archaeological levels[3].

This is why the large scale analyses used on the entire Chapelle Saint-Nicolas site are complemented by detailed mapping in and around the building.

When the results of tests and quantitative analyses were compared, the limits of the building and of the strong pollution coincided exactly (**fig. 5**). The two methods concur in situating the strongest and continuous pollution in the north-west part of the building. Further, taking into account the particular soil development (Fechner, this volume), the observed phosphate amounts are high and are compatible with a pollution related to the presence of animals. Hence, the archaeological hypothesis of animal stabling has been supported by the soil chemical data and the complete building was interpreted as a stable, probably a cowshed (Rouppert *et al.*, 2009 ; Dufour and Roupert, in preparation).

## 3/ The stable and the double manure pit

### *3.1. Description of the double pit*

Two large quadrangular pits were found ten metres east of the building. One measured 6 by 7 metres and the other 4 by 6 metres. They were connected to the building by two ditches. Two other ditches were present, and, taken together, the four of them formed a star-shape around the pits.

The tops of both pits were covered and partially filled by the same continuous layer. At an early stage of the excavation, the most likely interpretation seemed to be a pond. After removal of the continuous layer it became clear that the pits overlapped to form one construction, vertical on one side and slightly sloped on the other.

### *3.2. Pedological field observations and chemical analyses in the double pit*

Pedological observations about pit 1 show a clear discontinuity between the cut and fill and a later rapid backfill. In contrast, pit 2 shows a local bioturbation under the fill, in one case separated by a small flat gravel. This initial bioturbation could possibly indicate a longer open phase. Nevertheless, the mere interpretation of the pedological field observations would suggest extraction of non-calcareous clayey silt (Bt-horizon of a Luvisol), rapidly followed by some deliberate back filling with a homogeneous sediment that was less clayey (E-horizon). However the archaeological and historical data did not confirm this.

### *3.3. Confrontation of archaeological, agronomical and environmental data : identification of a manure pit*

Ponds, drinking troughs, washing places, manure pits and extraction pits are five types of structures whose morphologies are close to that of the discovered double pit. Indeed, it can be supposed that all of these structures need a large opening on one complete side and an access with a more gradual slope.

But, in this case, the hypothesis of a simple clay extraction pit is not compatible with the organisation of the ditches and the pit. In particular, one ditch establishes the connection of the pit with the polluted north part of the nearby stable, early in the Roman occupation. Moreover, the mixed dating material contained in the fill implies that the backfilling of the pits differs by at least two centuries from the early date at which the pit was abandoned. As a result, the limit between the digging and the fill coincides with a hiatus due to the cleaning of the former basal pit fill. This interpretation explains the localized bioturbation observed at the base of pit 2. It is also compatible with selective backfilling, due to the scraping of surface and sub-surface horizons.

Once the hypothesis of a clay extraction pit is eliminated, the possibilities which remain are a pond, a watering place, a washing place or a manure pit. Agricultural manuals ascribe paved bases to all of these structures[4]. However, none have been found. Moreover, the size of the structures can vary. The different types have in common that they may produce some traces related to cleaning or clearing out, but the manure pit is the only one where a complete cleaning is fully justified before its abandonment, either for the use of the manure or for the cleaning of an environment that will be dedicated to other activities.

A pond with clean water, a watering place or a washing place would not deliver traces of strong or prolonged pollution. The combination of the archaeological stratigraphy and the chemical data show that one or more phases with some rather polluting activities intervened between the digging and the filling of this pit : the phosphate analysis below the pit appeared to be one of the richest in the site, as opposed to the weak signal in the overlying fill[5]. Finally, the hypothesis of a pond or other watering place with polluted water is improbable in the absence of traces of iron and manganese crusts that are typical of stagnation water containing some organic matter (see also discussion in Fechner, this volume).

[3] The depth depends on various factors and so can differ from one site to another. At the Chapelle Saint-Nicolas, where the profile P5B, was completely tested for phosphorus, there was the same medium enrichment as in the fill (-45 cm to -120 cm), under the less rich colluvium (-0 cm to -45 cm, including plough soil). After removing the colluvium, there were 75 cm of phosphorous enriched soil starting from the Roman level (between -45 cm et -120 cm). This is the horizon between the accumulated clay and the transition to the underlying impoverished clay. The P5A profile, which was also tested and analysed in the laboratory, confirms this information (medium phosphorous indicated in mauve, -35 cm under ground level, at the transition between horizon E et horizon Bt (32,5% clay) transition. Here the phosphate level was 318 ppm, of which the majority was P organic.

[4] The discussion about the manure pits is based on the contents of several manuals.

[5] The two total phosphate analyses in the ground under the double pit (F1117) indicate levels between 1134 and 1122 ppm (samples Z32 and Z30). Elsewhere on the site, the results are near 915, with the exception of 1834 ppm recorded in a dark layer (F1196, US 1410, sample Z26) which was interpreted as collapsed or colluviated ploughed horizon with bone fragments (extremely rich in phosphorus), possibly used as manure.

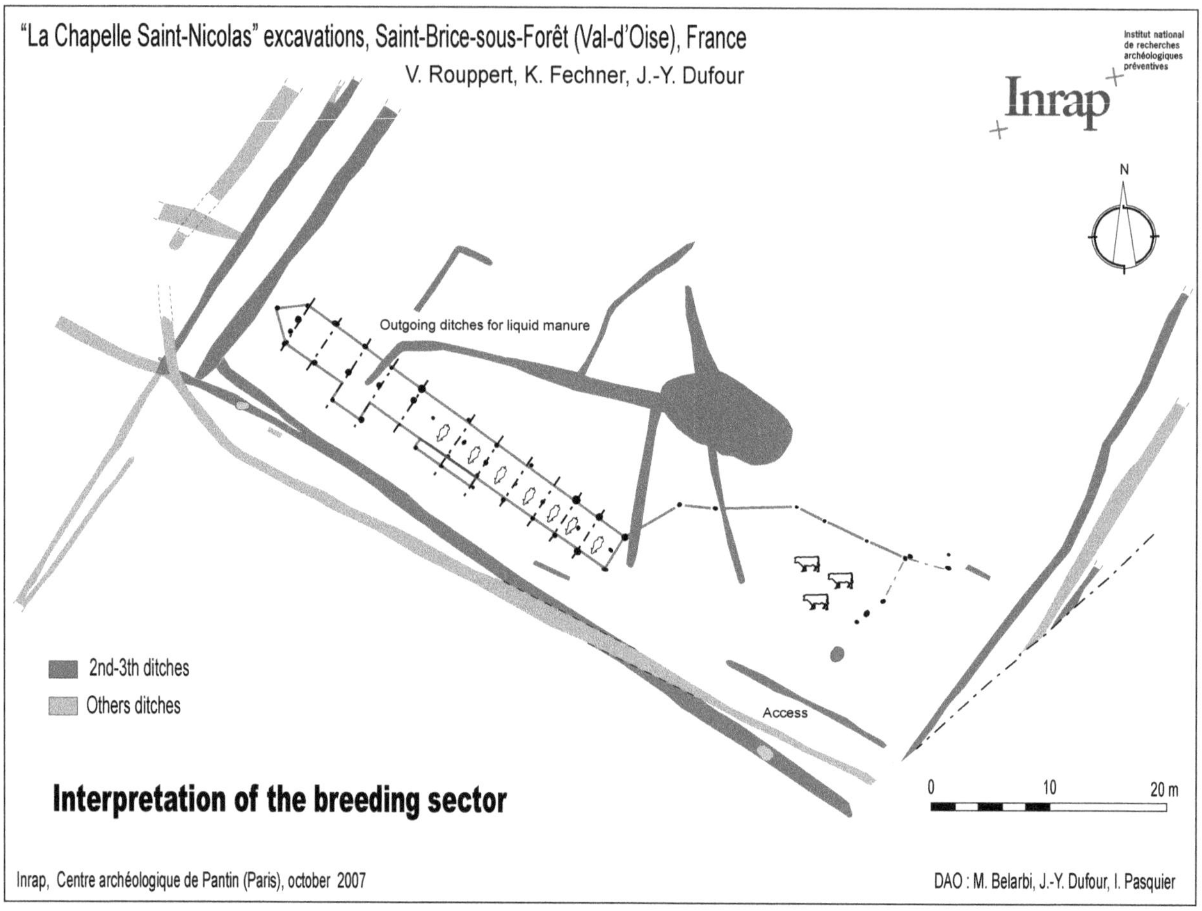


*Figure 6. Interpretation of the breeding sector.*

The most decisive argument for a manure pit is the fact that this phase is, as mentioned, linked with the most polluted part of the stable.

This study is a significant example of the complexity of such approaches and of the success of an ongoing interactive collaboration to cope with this complexity : starting from an initial archaeological hypothesis (a pond), pedological field observations suggested a first hypothesis (a clay extraction pit). The close study of the archaeological data showed that this was impossible and led to the re-evaluation of the pedological data and towards a common solution finally confirmed by a new chemical analysis. A double pit was fed with liquid manure via ditches which came from the neighbouring building. The base of the pits was cleaned to recover the manure until it was filled with subsurface soil material.

## What did the pedological study add to the interpretation of the site of the Chapelle Saint-Nicolas ?

The results of the field evaluation suggested the presence of what seemed to be a Gallo-Roman farm. The full excavation thus was supposed to find an area devoted to agriculture as well as to living quarters within the enclosure. Both of these, obviously, are basic requirements for a farm, whatever its size.

The results of the phosphate analysis in the building and in the associated structure, of the general pedological analysis and of the post excavation studies were compared with the information that can be found in old agricultural treatises and manuals[6] (Dufour and Rouppert, in preparation). This confrontation allows the recognition of the double pit as being part of a system for the evacuation of animal manure associated with the raising of cattle or horses. A cattle pen, next to the stable and itself connected to the manure pit by a ditch, might also be recognised by correlation of phosphate cartography and the excavation plan. It is represented by a row of higher phosphorus between ditches and the north-eastern side of the building. These various structures were all within the area bounded by the ditch relating pollution phenomena with archaeological delineations of Roman age. The structures that revealed the stock breeding activities date from the 1st and the 2nd centuries AD and they no longer exist in the 4th century (**fig. 6**). These activities were particularly intense during the 2nd and 3rd centuries in this dedicated sector, separated from the human living quarters situated in the north-north-eastern part of the

[6] Only twenty eight agricultural manuals out of the one hundred and fifty consulted deal with the question of cattle stabling.

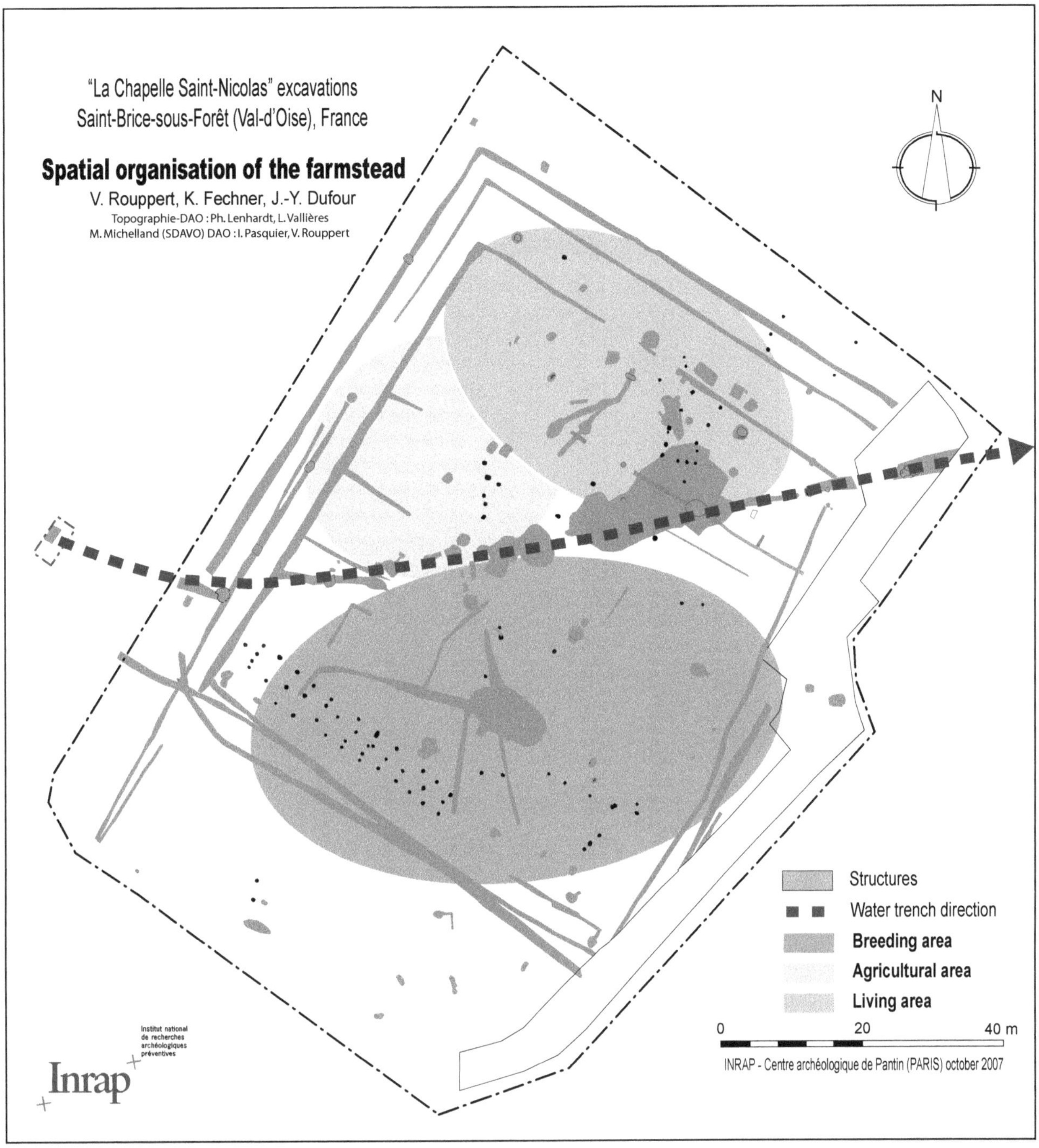


*Figure 7. Spatial organisation of the farmstead.*

enclosure. West of these quarters, the supposed presence of a vegetable garden and perhaps even an orchard (based on archaeological data, see Rouppert *et al.*, 2009) suggests that the separation of activities was more complex than that the bi-partition suggested by the field evaluation (**fig. 7**).

## Conclusion

The cartography of chemical soil elements of the enclosure and the study of the double pit of the site of Chapelle Saint-Nicolas are especially good illustrations of how archaeological and pedological approaches to the interpretation of structures can be fruitful. It is also suggested to reproduce on other sites the thorough preparation that was done here before any phosphorus cartography: the control of the influencing soil form factors (see Devos *et al.*, this volume), including past human activities reflected by old maps and documents on recent agricultural activities on the site.

On a broader note, this collaboration demonstrates that applied environmental studies should be a part of general field practice by archaeologists, which is today unfortunately not the case. It also confirms once more that when experts work together all along the study to produce a common synthesis the results go beyond the contents of their individual studies.

While pedological analysis has contributed to the interpretation of the activities undertaken and the spatial organisation of the Chapelle Saint-Nicolas site in Saint-Brice-sous-Forêt, other methods have to be used to go beyond this and to aid our appreciation of the status of this rural settlement.

## Bibliography

Conway, J.S., 1983. An Investigation of Soil Phosphorus Distribution within Occupation Deposits from a Romano-British Hut Group, *Journal of Archaeological Science*, 10: 117-128.

Devos, Y., Fechner, K. and Mikkelsen, J.H., this volume. The application of phosphorus cartography to archaeological structures: the development of a protocol. In: Fechner, K., Devos, Y., Leopold, M. and Völkel, J. (eds.), *Enclosed and buried surfaces as key sources in Archaeology and Pedology (papers of the EAA congress of Krakow)*, British Archaeological Reports. International Series.

Dufour, J.-Y. and Rouppert, V., in preparation. Analyses croisées des vestiges archéologiques et des manuels d'agriculture pour une étude des bâtiments agricoles : l'exemple d'étables à bovins en Île-de-France, *Revue Archéologique d'Île-de-France*, 17 p.

Eidt, R.C., 1984. *Advances in abandoned settlement analyses. Application to Prehistoric Anthrosols in Colombia, S. America*, Milwaukee.

Fechner, K., Blanquaert, G. and Langohr, R., 2004. L'utilité de l'archéopédologie : l'exemple du site du Marais de Dourges (Pas-de-Calais), *Archéopages*, 12: 10-13.

Fechner, K., this volume. Case studies of phosphorus mapping in Late La Tène to Roman enclosures and ponds between Rhine and Seine. In: Fechner, K., Devos, Y., Leopold, M. and Völkel, J. (eds.), *Enclosed and buried surfaces as key sources in Archaeology and Pedology (papers of the EAA congress of Krakow)*, British Archaeological Reports. International Series.

Rouppert, V., Dufour, J. L. and Fechner, K., 2009. *Saint-Brice-sous-Forêt "La Chapelle Saint-Nicolas" (95 Val d'Oise, Île-de-France), essai de caractérisation d'une exploitation agricole gallo-romaine, INRAP-CIF - DRAC-SRA IDF*, Pantin : INRAP CIF, 2009. 3 vol. (vol. 1 : texte, 253 p. ; vol.2 : illustrations, 200 p., vol. 3 : annexes, 530 p.).

Van de Wetering, H.T.J. and Wansleeben, M., 1988. Fosfaatgehalten binnen twee huisplattegronden in Oss-Ussen. In: Van den Broeke, P.W. and Van der Sanden, W.B. (eds.), *Getekend zand. Tien jaar archeologisch onderzoek in Oss-Ussen*, Waalre: 169-176. (= Bijdragen tot de Studie van het Brabants Heem, 31).

Zimmermann, W. H., 2004. Vom Pfostenloch zum Hausbefund, *Archäologie in Deutschland*, März-April, 2004-2: 18-20.

# Chapter 8
# Buried Anthropic Soils in the Centre of Brussels (Belgium): Looking for Fields in a (Proto-) urban Context

*Yannick Devos (1), Luc Vrydaghs (2), Kai Fechner (3), Christine Laurent (1), Ann Degraeve (4) & Sylvianne Modrie (4) (in collaboration with Deligne F. (1))*

(1) Centre de Recherches Archéologiques, Université Libre de Bruxelles. Avenue F.D. Roosevelt, 50, CP850, B-1050 Brussels, Belgium
(2) Research Team in Archaeo- and Palaeo_Sciences. Rue Belliard, 197 bte. 3, B-1040 Brussels, Belgium
(3) INRAP Nord-Picardie. Rue des Champs, 11, F- 59650 Villeneuve d'Ascq, formerly: 1 and 2
(4) Direction of Monument and Sites – Ministry of the Brussels Capital Region. Rue du Progrès, 80 bte 1, B-1035 Brussels, Belgium

*Abstract: The last ten years, at several occasions, dark coloured, humiferous horizons with an abrupt lower boundary have been uncovered during excavations in the centre of Brussels. The C14-datings and the archaeological material point to the 11th-13th century, the transition period when historians suspect the first urban development of Brussels.*

*The dark coloured horizons have been interpreted as being the remains of old agricultural land. The aim of this article is to synthesise the available data on these particular horizons, and to evaluate to what extent they can contribute to the reconstruction of ancient (pre-) urban space organisation in the 11th-13th century in the actual town centre.*

*Keywords:* Agriculture, urban development, archaeopedology, micromorphology, phytoliths, anthracology, C-14, carpology, micro-archaeology.

## 1. Introduction

Cultivation of the soil means human interference on the soil, either directly by forest clearance, hoeing, ploughing or adding fertilizers on the soil, or indirectly by causing erosion and colluviation. *In situ* archaeological evidences of ancient farming can sometimes hardly be differentiated from other human activities such as pasturing, human induced phenomena like colluvium, or purely natural events. Occasionally, the field observation of soil features, such as plough marks, ard marks, spade marks, terraces, ridges and lynchets, allow the direct identification of ancient agricultural practices (Thomas, 1990; Bryant and Davidson, 1996; Macphail *et al.*, 1990; Langohr, 2001).

In an attempt to identify ancient agricultural practices when such soil features are not observed, several methods have been applied: palynology, malacology, carpology and micromorphology (Davidson and Carter, 1998; French, 2003; Courty et al., 1989).

All these methods are based on the comparison with actual plough horizons. The major drawback of this concept is that processes that cause diagnostic features in actual soils may not occur in ancient soils or may be confused with unknown ancient processes (Davidson and Carter, 1998; Usai, 2001; Macphail and Cruise, 2001). Recent research has therefore focussed on the establishment of long term controlled field experiments on the one hand and the study of recently abandoned 'traditional' agricultural systems on the other hand (Davidson and Carter, 1998; Gebhardt, 1995; Gebhardt, 2000). A second drawback relies on the preservation of the agricultural horizons and the related features. Near-surface horizons can be heavily transformed through biological and/or human activities in a time span of only a few years (Davidson, 2002).

As to what concerns fertilizers, their use is poorly documented. Although, testimonies for the application of ash (Bieleman, 1995), travertine (Fechner and Sartieaux, 2000), shells (Bieleman, 1995), seaweed (Simpson, 1997; Pieters, 1990; Davidson and Carter, 1998) and turfs (Simpson et al., 1998; Langohr, 2001) exist.

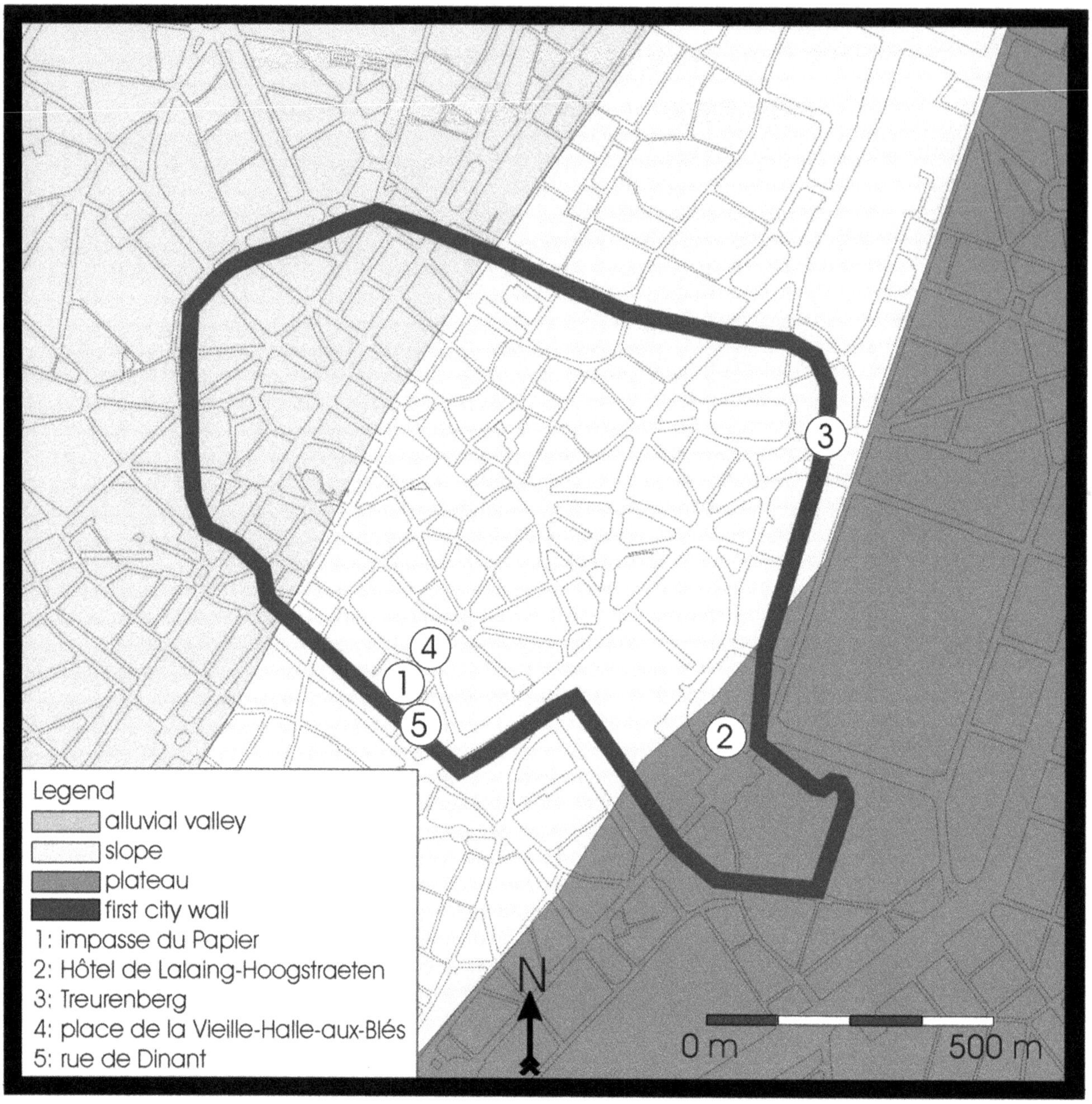


*Figure 1. Map of the Brussels town centre showing the location of the different archaeological sites (after Cabuy and Demeter, 1995: fig.15).*

The present study focuses on an alternative strategy, the study of ancient plough horizons in modern urban contexts. Almost no research on the presence of agricultural activities within such contexts has been performed. The main advantage of this strategy is that in actual urban regions ancient plough horizons are often sealed. Consequently they may not only inform about the fertility of these ancient fields, but also inform about the application of fertilizers (Bakels, 1997).

## 2. Geographical setting

The town centre of Brussels (Belgium) is characterised by its particular relief. It can be divided in a valley basin, a steep slope and a smoothly undulating plateau (figure 1). In the upper part of the centre of Brussels, we are dealing with sandy to loamy sandy soils, developed on the tertiary

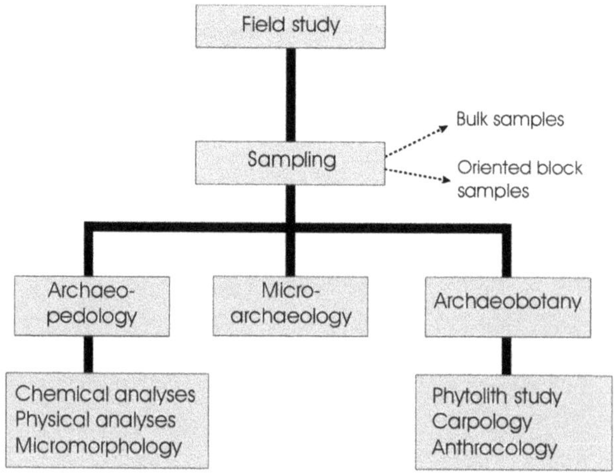


*Figure 2. Protocol of analyses.*

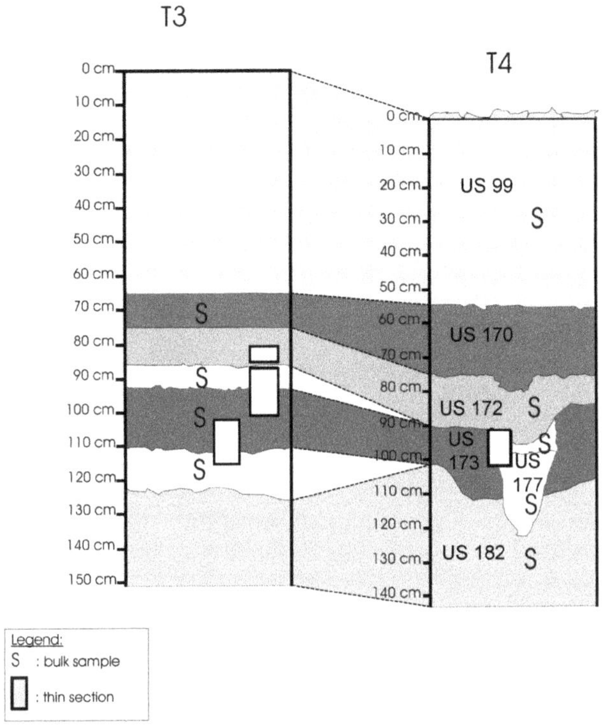


*Figure 3. (A)*

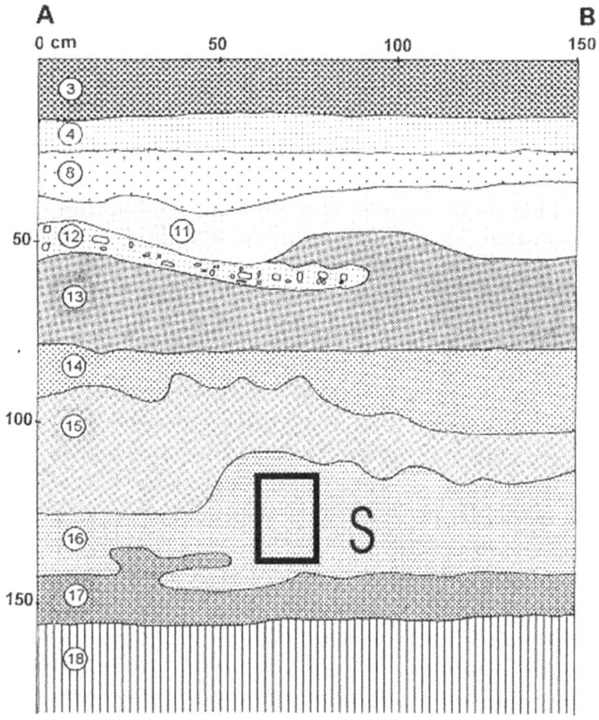


*Figure 3. (B)*

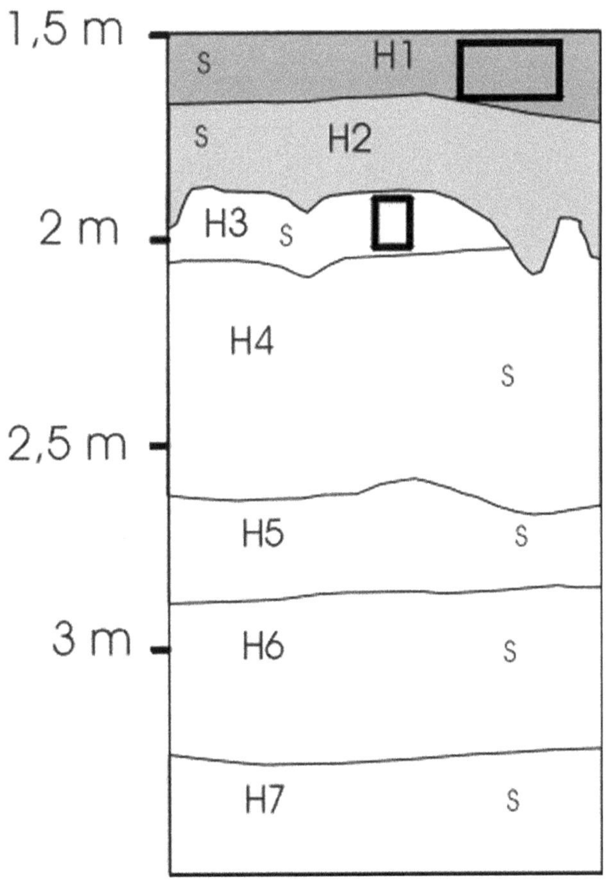


*Figure 3. (C)*

*Figure 3. Location of the samples on the Treurenberg site (A), the site of Vieille-Halle-aux-Blés (B) and Impasse du Papier (C).*

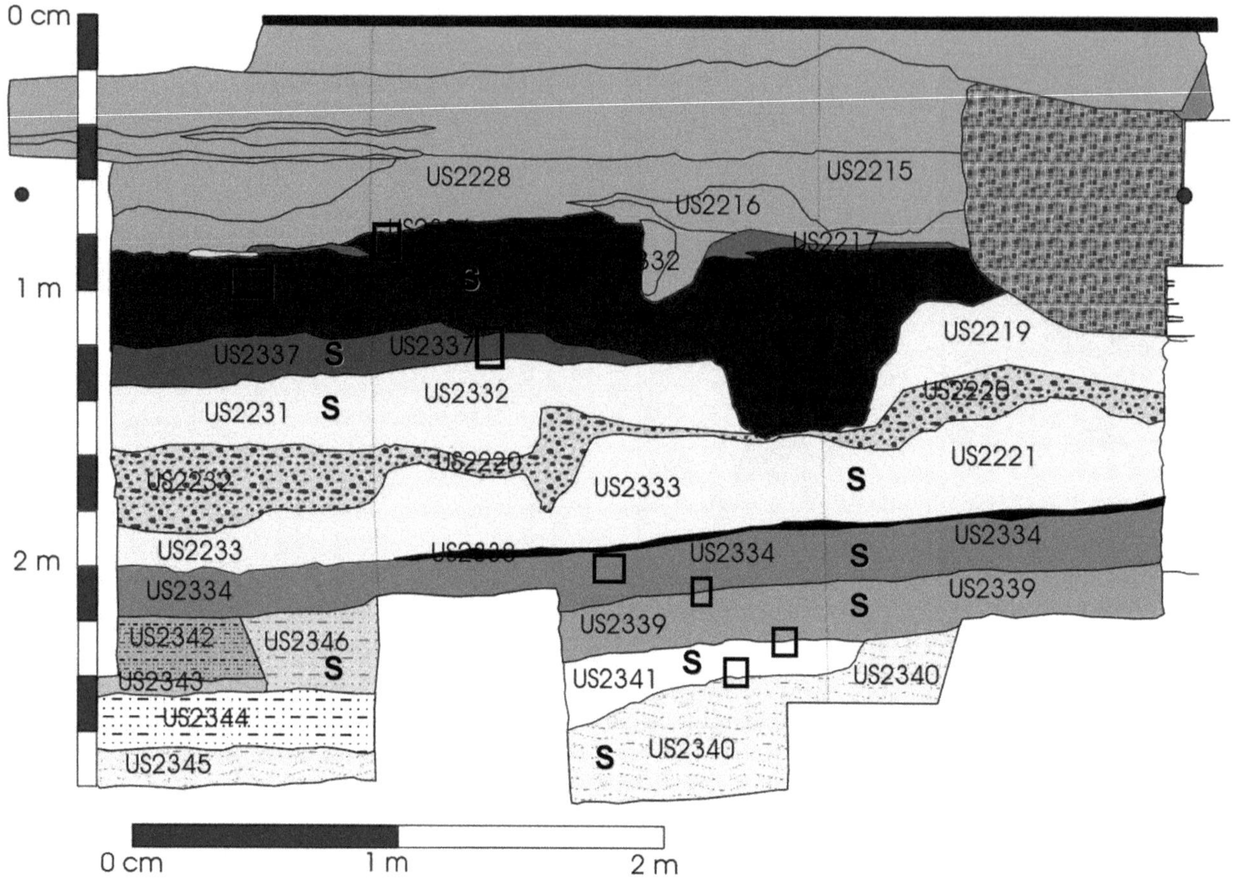


*Figure 4.A Location of the undisturbed block samples taken on the site of Hôtel de Lalaing-Hoogstraeten (A: Tr43/44-56).*

Brusselian sands or locally on a mixture of Weichsel silty (loess)deposits and tertiary sands. On top of these deposits, the remains of human occupations accumulated through time, sometimes sealing the successive occupation surfaces.

## 4. Material and methods

So far, four sites have been systematically investigated: Impasse du Papier, Hôtel de Lalaing- Hoogstraten, Treurenberg and Vieille-Halle-aux-Blés. A fifth site, Rue de Dinant, is currently under study (figure 1).

Field descriptions have been conducted according to the "Comprehensive Field Data Bases" (Langohr, 1994) and the "Guidelines for soil profile description" (F.A.O., 1990). Adapted checklists have been applied for the description of associated soil traces (Fechner et al., 2004).

In order to check the field working hypothesis, a protocol of analyses has been set up (figure 2).

Bulk samples and undisturbed oriented blocks were taken from the available soil profiles. Special attention has been taken to avoid contaminated areas. On the sites of Hôtel de Lalaing-Hoogstraeten, Treurenberg and Impasse du Papier, bulk samples have been taken on each recognised horizon. On the site of Vieille-Halle-aux-Blés only one bulk sample has been taken in the dark horizon. (figures 3 and 4).

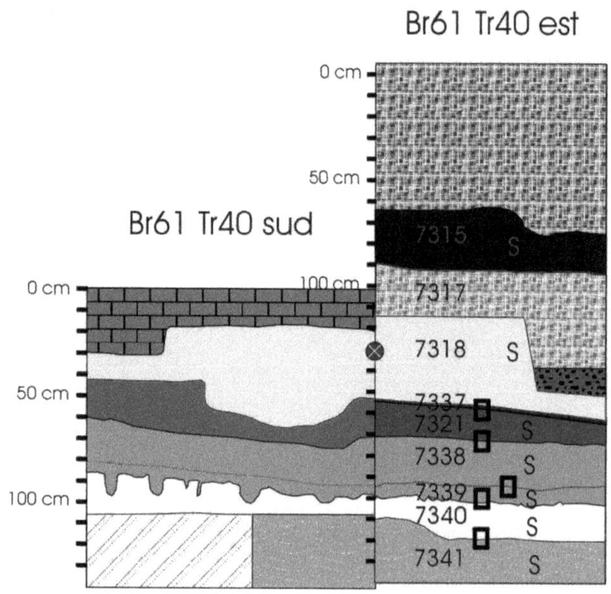


*Figure 4.B Location of the undisturbed block samples taken on the site of Hôtel de Lalaing-Hoogstraeten (B: Tr40).*

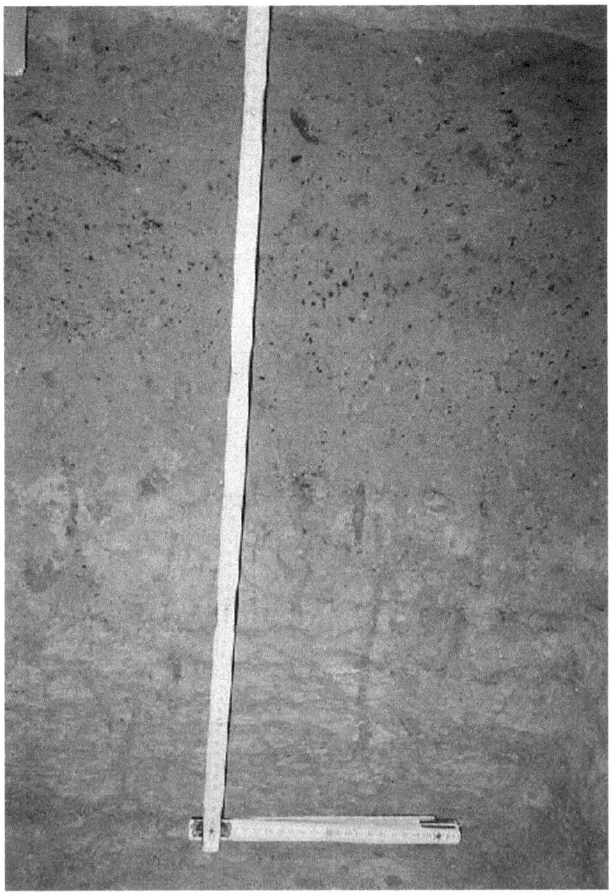

*Figure 5. Dark coloured horizons on the site of Hôtel de Lalaing-Hoogstraeten. Tape subunits are 20cm long.*

After gentle crushing, most of the bulk samples have been sieved at a 2 mm sieve.

Soil thin sections were prepared by Thomas Beckmann (Germany), using standard procedures. The thin sections were described following the terminology of Stoops (2003). The descriptions were completed by limited semi-numerical area counting to provide better estimates of the abundance of several features (Macphail and Cruise, 2001).

Particle size distribution of the <2 mm fraction has been performed by the laboratory of Arras following AFNOR NF-X 31-107 (Baize, 2000). Similarity indices have been calculated following Langohr, Scoppa and Van Wambeke (1976).

The laboratory of Arras also performed several chemical analyses (Baize, 2000):

- pH ($H_2O$) following NF ISO 10390.
- organic carbon according to NF ISO 10694.
- nitrogen according to NF ISO 13878.
- CEC according to Metson, AFNOR NF X 31-130.
- exchangeable cations: $Ca^{++}$, $Mg^{++}$, $K^{+}$, $Na^{+}$ according to AFNOR NF X 31-108.

Phosphate analysis has been performed at the laboratory of Soil Science at the University of Ghent (Mikkelsen, 1997).

Phytolith analysis has been undertaken on bulk samples (Treurenberg and Impasse du Papier) and on thin sections (all sites). Sediments were processed according to Bowdery (1999). Material was mixed with the dispersing agent Calgon and put into an ultrasonic bath. Descriptions were made under the light microscopy in suspension in demineralised water and conducted at magnifications x250, x400 and x1000.

The phytolith analysis of the thin sections has been conducted according to Vrydaghs et al. (in press). All the observations are classified following Runge's inventory (Runge, 1999) with a substantial review of the G category under which fall all morphologies deriving from the Graminae and Cyperaceae.

Naming of the phytolith was conducted according to the ICPN 1.0 (Madella et al., 2005).

Microarchaeological, carpological (Laurent, 1997b) and anthracological (Deligne, F., 2003) analysis of the samples have also been performed. Bulk samples for microarchaeological studies were wet sieved through a set of four wire-mesh screens (2, 1, 0.5 and 0.2 mm).

Sieving residues were sorted under binocular microscope (magnification maximum : 40x) to establish their « ID » card. Micro-elements were subdivided in four categories: anthropic, plant, animal and mineral.

Carpological and anthracological remains discovered were identified according a modern reference collection and bibliographical references.

C-14 dating on charcoal fragments of the plough horizons has been realised at the Beta Analytic Radiocarbon Dating Laboratory (Miami, U.S.A.). To make the dates as reliable as possible, charcoals were identified previously to their datation.

## 5. Results

### *5.1 The field evidences*

The major characteristic of all the sites is the almost systematic occurrence of one or several dark coloured, homogeneous humiferous horizons (Fechner, 1997; Degryse and Fechner, 2001; Devos, 2004a; Vrydaghs et al., in press). All these horizons display the following evidences (table 1):

- humiferous as reflected by a homogeneous dark colour with Munsell Colours (Anonymous, 1975) of hue 5YR to 5Y, value 4 to 6; and chroma 1 to 2 and locally up to 4;
- the base is abrupt, more or less horizontal and straight;
- high textural and structural similarity between the dark

| *field checklist* | | | | | | | | |
|---|---|---|---|---|---|---|---|---|
| ***site*** | ***Treurenberg*** | ***Lalaing*** | ***Lalaing*** | ***Lalaing*** | ***Lalaing*** | ***Lalaing*** | ***Impasse du Papier*** | ***Halle-aux-Blés*** |
| US | 173 | 2339 | 2334 | 7321 | 2337 | 2331 | H1 | 16 |
| Indices | | | | | | | | |
| homogeneous | P | P | P | A | P | P | P | P |
| humiferous | P | P | P | P | P | P | P | P |
| abrupt base | P | P | P | P | P | P | P | P |
| horizontal base | P | P | P | P | P | P | P | P |
| straight base | P | P | P | P | P | P | P | P |
| textural similarity with horizon underneath | P | P | P | - | P | P | - | - |
| structural similarity with horizon underneath | P | P | P | - | P | P | - | - |
| splash | A | A | A | A | A | A | A | A |
| sheet | - | - | - | - | - | - | - | - |
| rill | A | A | A | A | A | A | A | A |
| gully | A | A | A | A | A | A | A | A |
| slumping tunneling | A | A | A | A | A | A | A | A |
| microlamination | A | A | A | A | A | A | A | A |
| ripening cracks | A | A | A | A | A | A | A | A |
| subsidence | A | A | A | A | A | A | A | A |
| Fe/Mn-depl | - | - | - | - | - | - | - | - |
| Fe/Mn-red | - | - | - | - | - | - | - | - |
| Fe/Mn-acc | - | - | - | - | - | - | - | - |
| P-accum | ? | ? | ? | ? | A | A | ? | ? |
| Ca-depl | A | A | A | A | A | A | - | - |
| Ca-acc | - | - | - | - | - | - | - | - |
| clay-accum | A | A | A | A | A | A | - | - |
| clay-depl | A | A | A | A | A | A | - | - |
| humus-depl | - | - | - | - | - | - | - | - |
| humus-acc | - | - | - | - | - | - | - | - |
| spade marks | A | A | A | A | A | A | A | A |
| ard marks | A | A | A | A | A | A | A | A |
| plough marks | A | A | A | A | A | A | A | A |
| trampling | A | A | A | A | A | A | - | - |
| puddling | A | A | A | A | A | A | - | - |
| traffic pan | A | A | A | A | A | A | A | - |
| floral bioturbation | ? | P | P | P | P | P | P | P |
| faunal bioturbation | A | P | P | ? | P | P | P | P |
| burned | P | A | A | P | P | P | P | - |
| ceramics | P | P | P | P | P | P | P | - |
| scoria | A | A | A | A | P | - | P | A |
| bones | P | P | P | P | P | P | P | - |
| charcoal | P | P | P | P | P | P | P | P |
| stones | P | P | P | P | P | P | P | P |

*Table 1. Summary of the results of the archaeopedological field study (A = absent, P = present, ? = suspicion of presence, - = not checked).*

horizon and the horizon underneath; these observations indicate that the soil matrix of both horizons is very similar;

- absence of stratification or microlamination within the dark coloured horizons;
- presence of olive coloured mottles indicative for possible phosphate accumulations (more precise identification requires supplementary thin section analysis);
- presence of bone fragments, little ceramics and charcoal in the dark coloured horizons; all of these show no particular orientation.

This set of field characteristics identifies these dark coloured horizons as old plough horizons.

### *5.2 Micromorphological evidences*

In all the studied thin sections quartz grains largely dominate the mineral coarse material. Minor presence of glauconite, mica and chlorite is attested. The thin sections of the dark horizons are enriched with fine organic material.

On three sites (Vieille Halle aux Blés, Impasse du Papier and Hôtel de Lalaing-Hoogstraeten) the presence of soil fragments is attested (figure 6). They seem to originate from the Bt-horizon underneath.

The absence of traces of lamination of the sediments is also noted.

Some randomly distributed and unoriented anthropogenic materials are observed: bone fragments, charcoal, ash, limestone fragments and ceramics (figure 6; table 2).

Dusty clay coatings are noted in the dark horizons, as well as in the horizon underneath. Yellow phosphatic (apatite) nodules, radial ironphosphate crystal intergrowths and coprolites are also observed (figure 6).

The semi-quantitive analyses reveals following patterns:

- on the site of Vieille-Halle-aux-Blés, a diminishing of the void space (Devos, 2004c) with depth, No such clear pattern is observed for the site of Hôtel de Lalaing-Hoogstraeten where the whole sequence is investigated (Devos, 2004a), nor for the site of Impasse du Papier (Devos, 2004b).
- on three sites (Hôtel de Lalaing-Hoogstraeten, Impasse du Papier and Treurenberg) a clear decrease in number of the charcoal fragments with depth.

### *5.3 Physical evidences*

The treatment of the granulometrical data by calculating the Similarity Indices between the dark coloured horizon and the horizon underneath highlights the fact that both have *grosso modo* the same granulometry, indicating probably the same origin. An increase in the clay and silt fraction is however recorded for the dark coloured horizons.

### *5.4 Chemical evidences*

On all the investigated sites there is a higher concentration of organic matter in the dark horizon and a C/N ratio around 10, except for the site of Impasse du Papier and at the top of the Ap of the Treurenberg where the C/N ratio is slightly lower. This low C/N ratio is indicative for a high biological activity with a rapid decomposition of the organic matter (Nørnberg and Dalsgaard, 1982). Higher phosphate levels are also observed for the dark horizons.

### *5.5 Carpological evidences*

No remains of cereals are found, except for a few carbonised grains form the Impasse du Papier (Table 5) (Laurent, 1997a; Laurent, 2001; Laurent, sd.; Laurent and Deligne, sd.).

### *5.6 The opal component of biological origin*

The recorded opal component of biological origin includes phytoliths, diatoms, chrysophycean cysts and sponge spicules. All the thin sections present an association of phytoliths, diatoms and chrysophycean cysts (table 6). As to the sponge spicules, their occurrence pattern is irregular (table 6).

### *5.7 The phytolith analysis*

According to Runge (1999), all the recorded phytoliths of the dark coloured horizons fall under the A, F and G morphological categories. The latter category identifies one grass subfamily, the Pooideae as well as several crops genus, Avena sp., Triticum sp. and, Hordeum sp. (table 6). For where it could be studied (Rue de Dinant), phytoliths are not recorded for the Bt horizon.

Except for part of the material observed at Hôtel de Lalaing-Hoogstraeten, the distribution of the phytoliths within the dark coloured horizons is isolated (figure 6) (for definition see Vrydaghs et al., in press), unreferred and unrelated (Stoops, 2003) (table 7). At the site of Hôtel de Lalaing-Hoogstraeten, various grass phytoliths (G) appear in pockets or are contiguous. Such distribution might be related to the occurrence of excrements.

Differences in the preservation are observed between the dendritic phytoliths deriving from the cultivated crops (Vrydaghs, sd.a; Vrydaghs, sd.b), and the Pooideae-phytoliths. Frequently, the latter display surface features that are not relevant of the surface texture or ornamentation, despite pH conditions favouring their preservation. Such features are systematically not observed on the phytoliths from cereals (Vrydaghs et al., in press). This indicates that all the phytoliths do not share the same history.

### *5.8 Anthracological evidences & C-14 dating*

The results of the C14-datings show us that all the dark coloured horizons date from the 11th-13th centuries AD

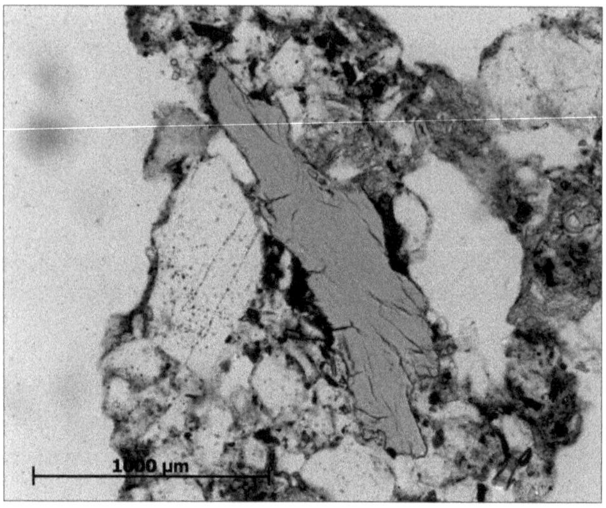


*Figure 6. (A)*

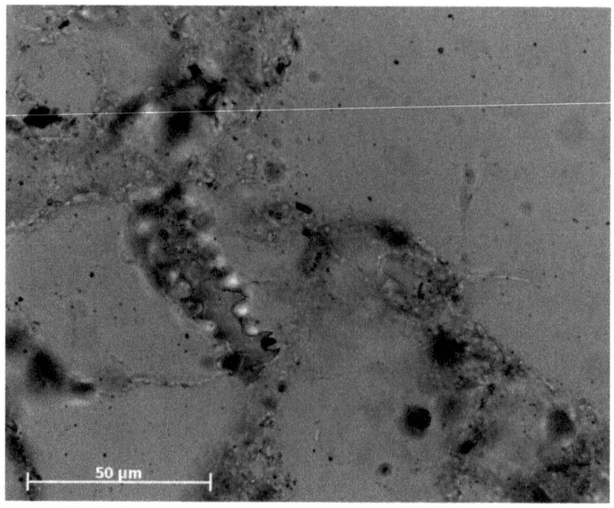


*Figure 6. (D)*

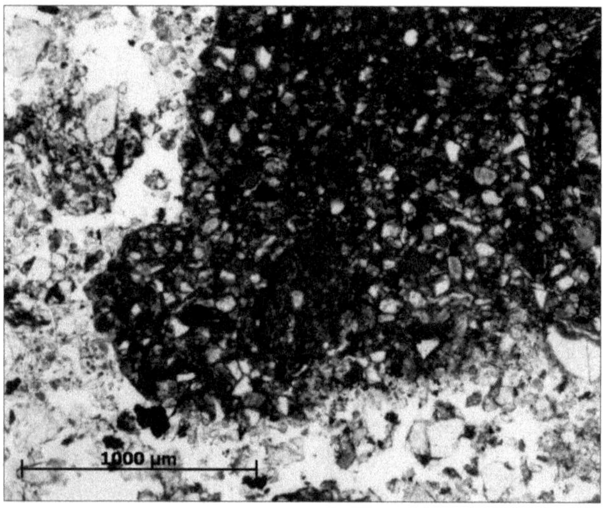


*Figure 6. (B)*

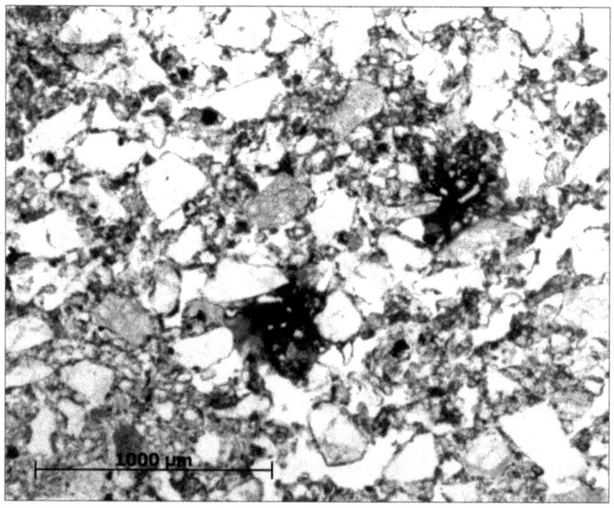


*Figure 6. (E)*

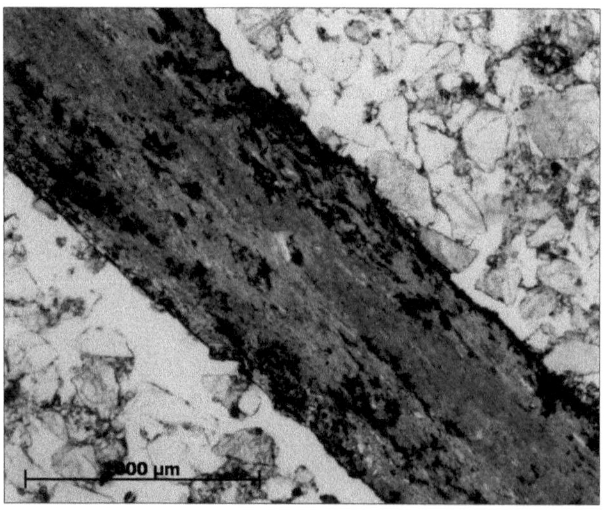

*Figure 6. (C)*

*Figure 6. (F)*

*Figure 6. Illustrations of microscopic anomalies pointing to cultivation and the application of fertilizers. A. Bone fragment (Hôtel de Lalaing-Hoogstraeten, US7321, plain polarized light); B. Ceramic fragment (Hôtel de Lalaing-Hoogstraeten, US 7321, plain polarized light); C. Burnt bone (Treurenberg, K34, plain polarized light); D. Isolated phytolith (Hôtel de Lalaing-Hoogstraeten, plain polarized light); E. Ironphosphate crystal intergrowth (Impasse du Papier, plain polarized light); F. Eggshell (Hôtel de Lalaing-Hoogstraeten, US7321, plain polarized light).*

| site | US / | microstructure | voids | gravel | coarse mineral material | | | | anthropogenic elements | | | | | | | groundmass | pedofeatures | | |
|---|---|---|---|---|---|---|---|---|---|---|---|---|---|---|---|---|---|---|---|
| | horizon | | | | (>10 µm & <2mm) | | | | | | | | | | | | | | |
| | | | | | quartz | glauconite | mica | chlorite | charcoal | bone | ash | CaCO3 | phytoliths | ceramics | phosphates | colour | clay coatings | | Fe/Mn nodules |
| | | | | | | | | | | | | | | | | | limpid | dusty | |
| Treurenberg | US173 | Subangular blocky | fff | f | fffff | ff | | | f | f | f | f | f | | f | orange | | | f |
| | US173 | Complex | fff | f | fffff | ff | f | | ff | f | | f | f | | f | grey-brown | f | f | f |
| | US173 | Complex | fff | f | fffff | ff | f | f | f | | | | f | | | orange to grey | f | f | f |
| Vieille-Halle-aux-Blés | 16 | Channel | fff | f | fffff | ff | | f | ff | f | f | | f | | f | orange to grey | f | f | ff |
| Impasse du papier | H1 | Granular | fff | fff | fffff | f | f | f | f | | f | | f | | f | light brown to grey | | f | f |
| | H1 | Granular | fff | ff | fffff | ff | f | f | f | | f | | f | | f | light brown | | f | f |
| Lalaing | 7337 | Massif | ff | ff | fffff | ff | f | f | f | f | | | f | | f | brown | f | ff | f |
| | 7321 | Massif | fff | f | fffff | ff | f | f | f | f | f | f | f | f | f | grey-yellow | f | f | f |
| | 7338 | Complex | fff | f | fffff | ff | f | | f | f | f | f | f | f | | yellow to brown | | f | f |
| | 7339 | Complex | fff | f | fffff | ff | f | f | f | | f | | f | | | orange | | f | f |

**legend**
ffffff very dominant (>70%)
fffff dominant (50-70%)
ffff frequent (30-50%)
fff common (15-30%)
ff few (5-15%)
f very few (<5%)

*Table 2. Summary of the results of the micromorphological study.*

| site | horizon | stratigraphic | clay | fine silt | coarse silt | fine sand | | coarse sand | | |
|---|---|---|---|---|---|---|---|---|---|---|
| | | unit | < 2µm | 2/20µm | 20/50µm | 50/100µm | 100/200µm | 200/500µm | 500/1000µm | 1-2mm |
| | | | % | % | % | % | % | % | % | % |
| Lalaing | Ap1 | 2334 | 5,9 | 6,2 | 16,1 | 9,4 | 27,3 | 33,9 | 0,7 | 0,5 |
| Lalaing | Ap1 | 7321 | 6,3 | 6,8 | 19,1 | 13,8 | 20,7 | 32,4 | 0,7 | 0,2 |
| Lalaing | Ap2 | 7338 | 5,6 | 5,8 | 14,9 | 9,6 | 28,8 | 33,9 | 0,7 | 0,7 |
| Lalaing | Ap2 | 2339 | 5,6 | 6,4 | 15,2 | 8,5 | 27,1 | 36,0 | 0,7 | 0,5 |
| Lalaing | Ap | 2341 | 6,1 | 6,9 | 17,1 | 9,8 | 26,8 | 32,4 | 0,5 | 0,4 |
| Lalaing | E | 7340 | 4,5 | 7,3 | 19,8 | 10,0 | 25,0 | 32,1 | 0,7 | 0,6 |
| Lalaing | Bt | 7341 | 13,0 | 11,4 | 31,0 | 10,3 | 15,4 | 18,6 | 0,2 | 0,1 |
| Papier | Ap | H1 | 7,2 | 11,2 | 22,6 | 9,1 | 16,3 | 32,4 | 0,7 | 0,5 |
| Papier | Abbi | H2 | 6,6 | 11,1 | 21,3 | 7,8 | 17,2 | 34,4 | 1,1 | 0,5 |
| Papier | Bbi | H3 | 6,6 | 14,8 | 31,1 | 8,9 | 11,8 | 25,7 | 1,1 | 0,0 |
| Papier | IIBt | H4 | 22,8 | 21,1 | 37,9 | 7,5 | 3,9 | 6,7 | 0,1 | 0,0 |
| Papier | IIBt/C | H5 | 22,3 | 18,6 | 39,9 | 6,3 | 5,1 | 7,5 | 0,1 | 0,2 |
| Papier | IIC | H6 | 21,4 | 20,1 | 38,4 | 6,2 | 5,4 | 8,4 | 0,1 | 0,0 |
| Papier | IIIC | H7 | 10,5 | 2,6 | 3,6 | 2,1 | 27,4 | 53,0 | 0,4 | 0,4 |
| Treurenberg | Ap | US173 | 5,0 | 4,2 | 9,2 | 5,7 | 20,8 | 53,9 | 0,7 | 0,5 |
| Treurenberg | | US182 | 3,2 | 2,6 | 4,9 | 4,3 | 27,6 | 54,9 | 1,3 | 1,2 |

*Table 3. Results of the physical analysis performed on the different plough horizons and their parent material (bi = important biological activity).*

| | 7318 | 7340 (E) | 2334 (Ap1) | 7321 (Ap1) | 7338 (Ap2) | 2341 (Ap) | 2339 (Ap2) |
|---|---|---|---|---|---|---|---|
| 7318 | 100% | 77% | 73% | 78% | 72% | 75% | 72% |
| 7340 (E) | | 100% | 95% | 94% | 93% | 96% | 93% |
| 2334 (Ap1) | | | 100% | 92% | 98% | 98% | 98% |
| 7321 (Ap1) | | | | 100% | 90% | 94% | 90% |
| 7338 (Ap2) | | | | | 100% | 96% | 97% |
| 2341 (Ap) | | | | | | 100% | 96% |
| 2339 (Ap2) | | | | | | | 100% |

*Figure 7. Results of the SI-index calculation on the site of Hôtel de Lalaing-Hoogstraeten.*

| | US99 | US170 | US172 | US173 | US182 |
|---|---|---|---|---|---|
| US99 | 100% | 76% | 51% | 76% | 77% |
| US170 | | 100% | 73% | 97% | 90% |
| US172 | | | 100% | 72% | 74% |
| US173 | | | | 100% | 91% |
| US182 | | | | | 100% |

*Figure 8. Results of the SI-index calculation on the Treurenberg site.*

| | H1 | H2 | H3 | H4 |
|---|---|---|---|---|
| H1 | 100% | 97% | 88% | 59% |
| H2 | | 100% | 85% | 57% |
| H3 | | | 100% | 71% |
| H4 | | | | 100% |

*Figure 9. Results of the SI-index calculation on the site of Impasse du Papier.*

(table 8), and are related to the older medieval occupation phases of Brussels. Artefacts found in the dark coloured horizons of the sites of Impasse du Papier (Massart, 2001) and Place de la Vieille-Halle-aux-Blés (Diekmann, 1997) point also to this period.

## 6. Discussion

Once the identification of the dark coloured horizons as old plough land is established by confronting all the available data and all alternative hypotheses like wasteland, waste deposits, alluvium, colluvium, forest soils, prairies or gardens are excluded (figure 10), the horizons can be further characterised and the landscape reconstructed.

### *6.1 The identification of ploughing*

First, confirmation is needed about the presence of ancient soil surface horizons. One type of evidence often quoted in literature is the presence of dusty clay coatings, typical for such horizons with a bare surface (Courty et al., 1989). However, this kind of feature does not necessarily point to cultivation of the soil surface (Usai, 2001).

Ploughing a soil means mixing of the upper part of the soil. Evidences of such type of soil mixing are the presence of soil fragments coming from the horizon underneath and the absence of any traces of lamination within the dark horizons. Unless new sediments are added (alluvium, colluvium, waste deposits turfs), as for example with plaggen soils (Dercon et al., 2003; Langohr, 2001), one can estimate that the plough horizon and the horizon underneath often will have grosso modo the same granulometry (e.g. a very high Similarity Index), as observed on the studied dark horizons. Finally, the random distribution of the anthropogenic material also points to the direction of mixing of the soil.

The basic assumption about the taphonomy of phytoliths is that they tend to be liberated through in situ decay mechanisms (Vrydaghs et al., in press). Without any post-depositional disturbance, the phytoliths should appear as contiguous features. But most of the phytoliths are isolated and randomly distributed and oriented, as are most of the other observed features. This is also in agreement with the mixing of the topsoil. Additional observations, of which some are reported by previous publications on archaeological ploughs lands, are:

| site | horizon | US | N | C | C/N | OM | pH | Ca | Mg | K | Na | CEC | Porg | Pinorg | Ptot |
|---|---|---|---|---|---|---|---|---|---|---|---|---|---|---|---|
| | | | g/kg | g/kg | | g/kg | (H2O) | cmol+/kg | cmol+/kg | cmol+/kg | cmol/kg | cmol+/kg | ppm | ppm | ppm |
| Lalaing | Ap | 2341 | 0,241 | 2,28 | 9,46 | 3,94 | 7,64 | 5,040 | 0,359 | 0,330 | 0,081 | 3,720 | 38 | 1396 | 1434 |
| Lalaing | Ap1 | 2334 | 0,272 | 2,82 | 10,37 | 4,88 | 7,83 | | | | | | 81 | 1233 | 1314 |
| Lalaing | Ap2 | 2339 | 0,237 | 2,18 | 9,20 | 3,77 | 7,59 | | | | | | 61 | 972 | 1033 |
| Lalaing | Ap1 | 7321 | 0,263 | 3,19 | 12,13 | 5,51 | 8,15 | 6,610 | 0,369 | 0,554 | 0,071 | 5,330 | 16 | 1659 | 1675 |
| Lalaing | Ap2 | 7338 | 0,231 | 2,07 | 8,96 | 3,59 | 7,94 | 4,460 | 0,232 | 0,438 | 0,048 | 3,360 | 69 | 911 | 980 |
| Lalaing | E | 7340 | 0,151 | 1,15 | 7,62 | 1,98 | 8,11 | | | | | | 95 | 438 | 533 |
| Lalaing | Bt | 7341 | 0,244 | 1,13 | 4,63 | 1,95 | 7,95 | | | | | | 2 | 795 | 797 |
| Papier | Ap | H1* | | 3,20 | | 5,50 | 7,00 | | | | | | | | |
| Papier | Ap | H1 | 0,540 | 3,90 | 7,22 | 6,70 | 7,10 | 6,700 | 0,770 | 2,187 | 0,215 | 7,600 | 1724 | 4787 | 6511 |
| Papier | Abbi | H2 | 0,430 | 3,30 | 7,67 | 5,70 | 7,00 | 5,700 | 0,720 | 2,223 | 0,254 | 6,800 | | | |
| Papier | Bbi | H3 | 0,290 | 1,90 | 6,55 | 3,30 | 6,80 | 3,300 | 0,540 | 2,008 | 0,233 | 5,400 | | | |
| Treurenberg | Ap (top) | US173 | 0,220 | 1,54 | 7,00 | 2,60 | 8,50 | | | | | | | | |
| Treurenberg | Ap | US173 | 0,210 | 2,36 | 11,24 | 4,10 | 7,90 | 3,87 | 0,292 | 0,339 | 0,0718 | 3,18 | 404 | 1160 | 1564 |
| Treurenberg | | US182 | 0,060 | 0,67 | 11,17 | 1,20 | 7,70 | | | | | | 73 | 1007 | 1079 |
| VHB | Ap | 16 | 0,405 | 3,71 | 9,16 | 6,42 | 7,57 | 9,550 | 0,717 | 0,706 | 0,180 | 7,150 | | | |

*Table 4. Summary of the results of the chemical analysis performed on the different plough horizons and their parent material.*

| *Site* | *Samples/US* | | *c* | *nc* |
|---|---|---|---|---|
| Impasse du Papier | 12 (Tr.G, phase Ib) | Cereal, non identified fragments | X | |
| | 13 (Tr.G, phase Ib) | Galium cfr. palustre | X | |
| | | Sambucus nigra | | X |
| | | Cereal, non identified fragments | X | |
| Lalaing | 7321 | Sambucus sp. | | X |
| | | Triticum sp. | X | |
| | 7338 | Sambucus sp. | | X |
| | | Hordeum sp. | X | |
| Treurenberg | 173a | Sambucus nigra | | X |
| | | Sambucus sp. | | X |
| | | Vicia sp. | | X |
| | | Secale cereale | X | |
| | | Triticum sp. | X | |
| | | Cereal, non identified fragments | X | |
| | 173b | Juncus sp. | | X |
| | | Pisum sativum | X | |
| | | Rubus fruticosus | | X |
| | | Sambucus nigra | | X |
| | | Sambucus sp. | | X |
| | | Vicia sp. | X | |
| | | Avena sativa | X | |
| | | Secale cereale | X | |
| | | Cereal, non identified fragments | X | |
| | 173c | Pisum sativum | X | |
| | | Sambucus nigra | | X |
| | | Sambucus sp. | | X |
| | | Secale cereale | X | |
| | | Triticum sp. | X | |
| | | Cereal, non identified fragments | X | |

*Table 5. Summary of the results of the carpological studies performed on the different plough horizons. (c : carbonised; nc: non-carbonised).*

| *Site* | *US* | *Opal components* | | | |
|---|---|---|---|---|---|
| | | *phyto.* | *diato.* | *chryso.* | *spicu.* |
| Treurenberg | 173 | X | X | X | X |
| Lalaing-Hoogstraeten | 7321 | X | X | X | X |
| | 7338 | X | X | X | X |
| Vieille-Halle-aux-Blés | 16 | X | X | X | |
| Impasse du Papier | H1 | X | X | X | |

*Table 6. Opal components from biological origin and organic components of the black earth horizons. Opal components: phyto.: phytolith; diato.: diatoms; chryso.: chrysophycean cysts; spicu.: sponge spicules.*

| ***Site*** | ***US*** | ***G*** | | | | | |
|---|---|---|---|---|---|---|---|
| | | ***Tabular*** | ***Hair*** | ***Rondle*** | ***Saddle*** | ***Trapeziform*** | ***Lobate*** |
| Treurenberg | 173 | X | X | X | | X | |
| Hôtel de Lalaing-Hoogstraeten | 7321 | X | X | X | | X | |
| | 7338 | X | X | X | | X | |
| Vieille Halle aux Blés | 16 | X | X | X | | X | |
| Impasse Papier | H1 | X | X | X | | X | |

| ***Site*** | ***US*** | ***Dendriform*** | | | |
|---|---|---|---|---|---|
| | | ***Triticum*** | ***Hordeum*** | ***Avena*** | ***Lolium*** |
| Treurenberg | 173 | X | X | X | |
| Hôtel de Lalaing-Hoogstraeten | 7321 | ? | ? | X | |
| | 7338 | X | X | X | |
| Vieille Halle aux Blés | 16 | ? | ? | X | |
| Impasse Papier | H1 | ? | ? | X | |

*Table 6.a: Tables 6. a & b: Summary of the results of the phytolith study. Tables 6. a & b sum up all the observations from bulk samples and soil thin sections.*

| ***thin section*** | ***A/P*** | ***short elements*** | | | ***long elements*** | | | ***identified cereals*** |
|---|---|---|---|---|---|---|---|---|
| | | ***MI*** | ***C*** | ***D*** | ***MI*** | ***C*** | ***D*** | |
| K34 (top) | 2% | rondel | good to sufficient | i,u,r,c | tabular | good | i,u,r,c | Avena sp |
| | | trapeziform | | | dendritic | | | |
| | | unciform hair cell | | | serrated | | | |
| | | trapeziform sinuate | | | | | | |
| | | square | | | | | | |
| K34 (base) | 2% | rondel | good to sufficient | i,u,r,c | tabular | good | i,u,r,c | Avena sp & Hordeum sp |
| | | trapeziform | | | dendritic | | | |
| | | trapeziform sinuate | | | serrated | | | |
| | | unciform hair cell | | | | | | |
| K47 (top) | < 2% | rondel | good to sufficient | i,u,r,c | tabular | good | i,u,r,c | Avena sp & Hordeum sp |
| | | polyhedral | | | trapeziform | | | |
| | | | | | dendritic | | | |
| K47 (base) | 2% | rondel | good to sufficient | i,u,r,c | tabular | good | i,u,r,c | Avena sp |
| | | trapeziform | | | dendritic | | | |
| | | square | | | | | | |

*Table 7. Summary of the phytolith analysis of thin sections from Treurenberg. A/P presence of the phytoliths. MI: Morphotype identification. C: conservation state. D: distribution of the phytoliths. i: isolate; u: unrelated; r: unreferrred; c: clustered.*

| *Site* | *Sample* | *Method* | *Measured* | *13C/12C ratio* | *Conventional* | *2 sigma calibration* |
|---|---|---|---|---|---|---|
| | | | *radiocarbon age* | | *radiocarbon age* | |
| Treurenberg | US170 | AMS | 910 +/- 40 BP | -25.8 ‰ | 900 +/- 40 BP | Cal AD 1030 to 1230 |
| | US173 | AMS | 840 +/- 40 BP | -25.8 ‰ | 830 +/- 40 BP | Cal AD 1160 to 1270 |
| Hôtel de Lalaing- | US2334 | AMS | 1000 +/- 40 BP | -27.8 ‰ | 950 +/- 40 BP | Cal AD 1010 to 1180 |
| Hoogstraeten | US2339 | AMS | 1040 +/- 40 BP | -27.4 ‰ | 1000 +/- 40 BP | Cal AD 980 to 1060 & Cal AD 1080 to 1150 |

*Table 8. C-14 data obtained for the different dark horizons.*

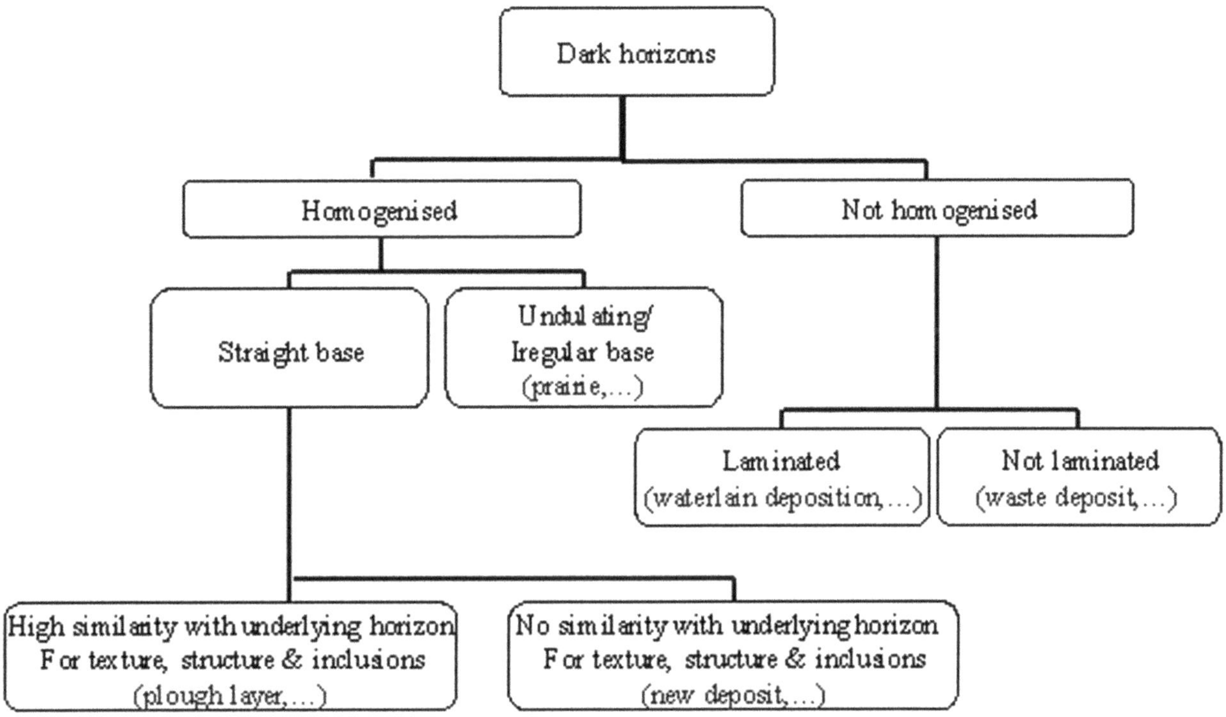


*Figure 10. Scheme summarizing steps towards the identification of ancient plough layers.*

- the diminishing of the void space recorded at the base of the plough layer (Macphail et al., 1990; Bryant and Davidson, 1996);
- an increase in clay and silt fractions. According to Mikkelsen, Langohr and Macphail (sd.) that is also typical for old plough horizons;

The absence of non-carbonized cereal grains is not in disagreement with the ploughing hypothesis as the oxidising environment in these well drained sandy to sandy loamy soils do not favour their conservation. Careful removal of the plant remains such as leaves, grains and fruits during harvesting might be another explanation. Nevertheless, the phyotlith study shows the presence of direct markers of cultivated crops (*Triticum* sp., *Avena* sp. and *Hordeum* sp.) proposing to relate all these processes to crop cultivation.

The main restriction to cultivate these initially acid soils was probably the limited nutrient supply. In order to bring them under cultivation, soil fertility has to be enhanced by liming and manuring. A series of evidences for the application of fertilizers have been observed.

The evidence for liming includes:

- charcoal and ash. The conservation of ash points to a rapid sealing of the horizon as it is easily dissolved;
- bone fragments. Some of these are transformed into apatite pseudomorphs (figure 6);
- on the Treurenberg site: small fragments of partially dissolved limestone.

The evidence for manuring comprises:

- enhanced phosphate levels. Phosphorus has often been used to trace human activities like manuring (Sjöberg, 1976; Lillios, 1992; Leonardi et al., 1999; Macphail et al., 2000). The micromorphological study demonstrates the presence of yellow phosphatic nodules, ironphosphate crystal intergrowths and occasional vivianite crystal intergrowths;
- enhanced organic carbon levels. Cultivation without application of organic fertilizers is believed to reduce organic carbon levels in the topsoil by harvesting the crops and therefore part of the organic carbon.. Enhanced levels are therefore an indication of an extra input of organic material (manure).
- coprolites;
- ceramics. The presence of ceramics is viewed as an indirect evidence for the application of manure, as it is often part of the rubbish thrown on the field (Pieters, 1990). It needs to be recalled that the carpological study reveals a picture of domestic residues which might have been thrown on the fields.

Putting together the macro-, meso and microscopic evidences with the results of the chemical analyses, it can be concluded that fertilizers have been added. Due to the better chemical conditions of the soils, biological activity increased, and the soil structure improved.

Thanks to the sudden sealing contexts, an estimation of the soil fertility at the time of burial can be made. According to the Belgian texture triangle (Bah et al., 2005) the soils can be classified as loamy sands (S) and light loamy sands (P). These light textured soils offer good cultivation possibilities. The granular structure of the dark coloured surface horizons, as result of the intense biological activity, is also favourable for farming.

Soil pH values are rather high, compared to actual cultivated loamy-sandy and light loamy sands. The C/N ratio indicates high biological activity and a rapid decomposition of the organic matter (Nørnberg and Dalsgaard, 1982). The humus content is rather low.

According to the actual chemical nutrient status of Belgian soils (Verloo, 1992), calcium content is normal to very high, magnesium content normal to high, potassium content high to very high, sodium is low to very high.

The plough horizon of the Treurenberg appears to be poorer in nutrients, compared to the sandy-loamy plough horizons of the other investigated sites. For the site of Hôtel de Lalaing-Hoogstraeten a net increase of fertility is marked for the younger plough layers (Ap1).

All these data support the hypothesis that it was possible to successfully cultivate oats, wheat and barley on these particular soils.

### *6.2 Importance for the archaeological study of Brussels*

The 10th-13th century is a crucial period in the development of Brussels. Unfortunately this period is historically very poorly documented (Billen, 2000; Deligne, C., 2003; Despy, 1979; Despy, 1997).

According to the actual results, the crop fields are not only present on the richer sandy-loamy soils; they are also attested on more sandy soils, which need a more intensive application of fertilizer and on non-calcareous quarry fills. As these more marginal soils were also becoming part of the cultivated area, an intensification of the agricultural activities might be advanced. This is in agreement with the general picture shown by historians (Duby 1977; Mazoyer and Roudart 2002).

These old plough horizons indicate a rather rural environment. As some fields seem to continue underneath the first city wall, there are between the 11th and 13th centuries no obvious separations between what will become afterwards the inner and the outer part of the city. Therefore the construction of the wall seems to be an important marker of change in the landscape organisation.

## 7. Conclusions

The development of an integrative protocol permits not only to identify plough layers; it also allows gaining important information about the agricultural activities between the 11-13th centuries. These include the mixing up of the surface soil, whether by spading, hoeing or ploughing in order to cultivate crops. So far the presence of '*Triticum*','*Hordeum*', and *Avena*' is demonstrated.

The integrated approach made it also possible to detect the application of liming and manuring.

Whereas medieval plough horizons in the rural environment around Brussels are not always well preserved, in Brussels a high concentration of traces of old agricultural activities are detected. As these sites are often sealed rather fast after their occupation, they become an important source to study old agricultural practices.

## Acknowledgements

The authors wish to thank the Brussels Capital Region for financing this research.

They also want to thank Roger Langohr of the University of Ghent, dr. Richard Macphail of the University College London and Stephane Demeter of the Direction of Monuments and Sites of the Brussels Capital Region for their valuable advice.

## Bibliography

Anonymous, 1975. *Munsell Soil Color Charts*, Baltimore.

Bah, B., Engels, P. and Colinet, G. 2005. *Légende de la Carte Numérique des Sols de Wallonie (Belgique) sur base de la légende originale de la carte des Sols de la Belgique de l'IRSIA à 1/20.000*, Gembloux.

http://www.fsagx.ac.be/gp/L%C3%A9gende%20carte%20num%C3%A9rique.pdf

Baize, D., 2000. *Guide des analyses en pédologie, choix – expression – présentation – interprétation*, Paris. (=Techniques et pratiques).

Bakels, C.C., 1997. The beginnings of manuring in western Europe, *Antiquity*, 71: 442-445.

Bieleman, J., 1995. Van traditionele naar technologische vruchtbaarheid en verder... Het mestprobleem in de Nederlandse landbouw in historisch perspectief, *Tijdschrift voor ecologische geschiedenis*, 1995/2: 2-8.

Billen, C., 2000. Espace et société. In: Billen, C. & Duvosquel, J.-M. (eds.), *Bruxelles*, Antwerpen: 36-141. (=L'esprit des villes d'Europe).

Bowdery, D., 1999. Phytoliths from tropical sediments: reports from southeast Asia and Papua New Guinea, *Bulletin of the Indo-Pacific Prehistory Association,* 18: 159-168.

Bryant, R.G. and Davidson, D.A., 1996. The Use of Image Analysis in the Micromorphological Study of Old Cultivated Soils: an Evaluation Based on Soil from the Island of Papa Stour, Shetland, *Journal of Archaeological Science*, 23: 811-822.

Cabuy, Y. and Demeter, S., 1995. *Atlas van de archeologische ondergrond van het gewest Brussel, 10.1: Brussel vijfhoek, archeologisch potentieel*, Brussel.

Courty, M.A., Goldberg, P. and Macphail, R., 1989. *Soils and micromorphology in archaeology*, Cambridge. (=Cambridge Manuals in Archaeology).

Davidson, D.A., 2002. Bioturbation in Old Arable Soils: Quantitative Evidence from Soil Micromorphology, *Journal of Archaeological Science*, 29: 1247-1253.

Davidson, D.A. and Carter S.P., 1998. Micromorphological Evidence of past Agricultural Practices in Cultivated Soils: The Impact of a Traditional Agricultural System on Soils in Papa Stour, Shetland, *Journal of Archaeological Science*, 25: 827-838.

Degryse, H. and Fechner, K., 2001. Etude archéopédologique dans l'impasse du Papier. In: Blanquart, P., Demeter, S., De Poorter, A., Massart, C., Modrie, S., Nachtergael, I. and Siebrand, M. (eds.), *Autour de la première enceinte*, Bruxelles: 300-311. (=Archéologie à Bruxelles, 4).

Deligne, C., 2003. *Bruxelles et sa rivière. Genèse d'un territoire urbain (12ᵉ-18ᵉ siècle)*, Turnhout. (=Studies in European Urban History (1100-1800), 1).

Deligne, F., 2003. *Bruxelles-Treurenberg: résultats anthracologiques*, unedited GiEP report.

Dercon, G., Davidson, D.A., Simpson, I.A., Dalsgaard, K. and Spek, T. 2003. The nature and formation of Anthrosols in NW Europe: a comparison between 3 countries. In: Boschian, G. (ed.), *Second International Conference on Soils and Archaeology, pisa, 12th-15th May, 2003. Extended abstracts*, Pisa: 22-24.

Despy, G., 1979. La genèse d'une ville, In: Stengers, J. (ed.), *Bruxelles, Croissance d'une capitale*, Anvers: 28-40.

Despy, G., 1997. Un dossier mystérieux: les origines de Bruxelles, *Bulletin de la Classe de Lettres*, 8 (1-6): 241-303.

Devos, Y., 2004a. *Une perspective diachronique de l'environnement à Bruxelles et les nouvelles méthodes pour valoriser le patrimoine archéologique: étude archéopédologique du site de l'Hôtel de Lalaing-Hoogstraeten, rapport intermédiaire*, unedited GiEP report.

Devos, Y., 2004b. *Une perspective diachronique de l'environnement à Bruxelles et les nouvelles méthodes pour valoriser le patrimoine archéologique: synthèse de l'étude archéopédologique du site de l'impasse du Papier*, unedited GiEP report.

Devos, Y., 2004c. *Une perspective diachronique de l'environnement à Bruxelles et les nouvelles méthodes pour valoriser le patrimoine archéologique: Micromorphological Study of the Site of 'place de la Vieille-Halle-aux-Blés' (Brussels)*, unedited GiEP report.

Diekmann, A., 1997. Recherche archéologique sur le site de la place de la Vieille-Halle-aux-Blés. In: Diekmann, A. (ed.), *Artisanat médiéval et habitat urbain, rue d'Une Personne et place de la Vieille-Halle-aux-Blés*, Bruxelles: 103-130. (= Archéologie à Bruxelles, 3).

Duby, G., 1977. *L'économie rurale et la vie des campagnes dans l'Occident médiéval (France, Angleterre, Empire, IXe - XVe siècles). Essai de synthèse et perspectives de recherches*, Champs - Flammarion.

F.A.O. 1990, *Guidelines for Soil Description*, Rome.

Fechner, K., 1997. Etude archéo-pédologique du site de la Place de la Vieille-Halle-aux-Blés. In: Diekmann, A. (ed.), *Artisanat médiéval et habitat urbain, rue d'Une Personne et place de la Vieille-Halle-aux-Blés*, Bruxelles: 135-139. (=Archéologie à Bruxelles, 3).

Fechner, K., Langohr, R. and Devos, Y., 2004. Archaeopedological checklists. Proposal for a simplified version for the routine archaeological record in Holocene rural and urban sites of Nord-Western Europe. In: Carver, G. (ed.), *Digging in the dirt: Excavations in a new millennium*, Oxford: 240-256. (=British Archaeological Reports, International Series, S1256).

Fechner, K. and Sartieaux, P., 2000. Reconstituer le paléoenvironnement à l'échelle humaine. Essai de visualisation des résultats de l'archéopédologie sur le tracé occidental du Train à Grande Vitesse en Wallonie (Belgique). In: Louwagie, G. and Langohr R. (ed.), *Thematic Day. Soils and Palaeoenvironment Reconstruction. Applications in Geo- and Archaeopedology, 10 November 1999*, Leuven: 35-56. (=Pedologie-Themata, 8).

French, C., 2003. *Geoarchaeology in action. Studies in soil micromorphology and landscape evolution*, London & New York.

Gebhardt, A., 1995. Soil micromorphological data from traditional and experimental agriculture. In: Barham, A.J. and Macphail, R.I. (eds.), *Archaeological Sediments and Soils: Analysis, Interpretation and Management*, London: 25-40.

Gebhardt, A., 2000. Anthropisation, agricultures anciennes et formation des paysages ruraux: le rôle de la micromorphologie des sols. In: Louwagie, G. and Langohr R. (eds.), *Thematic Day. Soils and*

*Palaeoenvironment Reconstruction. Applications in Geo- and Archaeopedology, 10 November 1999*, Leuven: 77-83. (=Pedologie-Themata, N° 8).

Langohr, R., 1994. Directives and rationale for adequate and comprehensive field soil data bases. In: *New Waves in Soil Science. Refresher Course for Alumni of the International Training Centre for Post-Graduate Soil Scientists of the Ghent University, Harare 1994*, Gent: 176-191. (=ITC-Gent, Publications series, n°5).

Langohr, R., 2001. L'anthropisation du paysage pédologique agricole de la Belgique depuis le Néolithique ancien – Apports de l'archéopédologie, *Etude et Gestion des Sols*, 8, 2: 103-118.

Langohr, R., Scoppa, C.O. and Van Wambeke, A., 1976. The use of a comparative particle size distribution index for the numerical classification of soil parent materials: application to Mollisols of the Argentinean pampa, *Geoderma*: 305-312.

Laurent, C., 1997a. Etude micro-archéologique et archéopédologique du site de la place de la Vieille-Halle-aux-Blés. In: Diekmann, A. (ed.), *Artisanat médiéval et habitat urbain, rue d'Une Personne et place de la Vieille-Halle-aux-Blés*, Bruxelles: 131-134. (=Archéologie à Bruxelles, 3).

Laurent, C., 1997b. La micro-archéologie et la carpologie en contexte urbain-médiéval, méthode et résultats: les sites archéologiques de Bruxelles, de 1992 à 1996. In: De Boe G. (ed.), *Environment and subsistence in Medieval Europe-Papers of the Medieval Europe Brugge 1997 Conference, Volume 09*, Zellik: 169-173.

Laurent, C., 2001. Etudes microarchéologique et carpologique. In: Blanquart, P., Demeter, S., De Poorter, A., Massart, C., Modrie, S., Nachtergael, I. and Siebrand, M. (eds.), *Autour de la première enceinte*, Bruxelles: 311-315. (=Archéologie à Bruxelles, 4).

Laurent, C., sd. *Bruxelles, site du Treurenberg (Br59): résultats des études micro-archéologiques et carpologiques*, unedited report.

Laurent, C. and Deligne, F., sd. *Bruxelles, Hôtel Hoogstraeten-Lalaing: études micro-archéologiques et carpologiques*, unedited report.

Leonardi, G., Miglavacca, M. and Nardi, S., 1999. Soil Phosphorus Analysis as an Integrative Tool for Recognizing Buried Ancient Ploughsoils, *Journal of Archaeological Science*, 26: 343-352.

Lillios, K.T., 1992. Phosphate fractionation of soils at Agroal, Portugal, *American Antiquity*, 57, 3: 495-506.

Macphail, R.I., Courty, M.A. and Gebhardt, A., 1990. Soil micromorphological evidence of early agriculture in north-west Europe, *World Archaeology*, 22, 1: 52-69.

Macphail, R.I. and Cruise, J., 2001. The Soil Micromorphologist as Team Player. A Multianalytical Approach to the Study of European Microstratigraphy. In: Goldberg, P., Holliday, V.T. and Ferring, C.R. (eds.), *Earth Sciences and Archaeology*, New York: 241-267.

Macphail, R.I., Cruise, G., Engelmark, R. and Linderholm, J., 2000. Integrating Soil Micromorphology and Rapid Chemical Survey methods: new developments in reconstructing past rural settlement and landscape organisation. In: Roskams, S. (ed.), *Interpreting Stratigraphy. Site evaluation, recording procedures and stratigraphic analysis. Papers presented to the interpreting Stratigraphy Conferences 1993-1997*, Oxford: 71-80. (=British Archaeological Reports, international Series, 910).

Madella, M., Alexandre, A. and Ball, T., 2005. Intenational Code for Phytolith Nomenclature 1.0, *Annals of botany*, 96: 253-260

Massart, C., 2001. Etude archéologique de l'impasse du Papier (1996). In: Blanquart, P., Demeter, S., De Poorter, A., Massart, C., Modrie, S., Nachtergael, I. and Siebrand, M. (eds.), *Autour de la première enceinte*, Bruxelles: 262-299. (=Archéologie à Bruxelles, 4).

Mazoyer, M. and Roudart, L., 2002. *Histoire des agricultures du monde. Du néolithique à la crise contemporaine*, Paris.

Mikkelsen, J.H.,1997. Laboratory method for determination of organic, inorganic and total phosphate. In: *Round Table I, Micromorphology and Phosphate. Report on workshop at institute of archaeology/department of geology, university college London, Monday 24th March to Wednesday 26th March 1997.* (~http://www.gre.ac.uk/~at05/micro/soilmain/intro1.htm accessed in 2001)

Mikkelsen, J.H., Langohr, R. and Macphail, R.I., sd. Land Use and Environmental Degradation in Bronze Age Settlements, Eastern Jutland, Denmark. In: Engelmark, R. & Linderholm J. (eds.), *Proceedings from the VIII Nordic conference on the applications of Scientific Methods in Archaeology Umea 2001*, Umea: 81-92. (= Archaeology and Environment, 21).

Nørnberg, P. and Dalsgaard, K., 1982. A comparative studt of present and ancient dark-coloured topsoil for the purpose of evaluating soil fertility in the past. In: *Second Nordic Conference on the Application of Scientific Methods in Archaeology, Helsingør (Elsinore), Denmark, 17-19 August 1981*, Strasbourg: 247-256. (= PACT, 7).

Pieters, M., 1990. Bemesting en aanverwante grondverbeteringspraktijken archeologisch bekeken, *Archeologie*, 1990: 5-9.

Runge, F., 1999. The opal phytolith inventory of soils in central Africa - quantities, shapes, classification , and spectra, *Review of Palaeobotany and Palynology*, 107: 23-53.

Simpson, I., 1997. Relict Properties of Anthropogenic Deep Top Soils as Indicators of Infield Management in Marwick, West Mainland, Orkney, *Journal of Archaeological Science*, 24: 365-380.

Simpson, I.A., Dockrill S.J., Bull, I.D. and Evershed, R.P., 1998. Early Anthropogenic Soil Formation at Tofts Ness, Sanday, Orkney, *Journal of Archaeological Science*, 25: 729-746.

Sjöberg, A., 1976. Phosphate Analysis of Anthropic Soils, *Journal of Field Archaeology*, 3: 447-454.

Stoops, G., 2003. *Guidelines for Analysis and Description of Soil and Regolith Thin Sections*, Madison.

Thomas, K., 1990. Aspects of soils and early agriculture, *World Archaeology*, 22, 1: 7-13.

Usai, R., 2001. Textural Pedofeatures and Pre-Hadrian's Wall Ploughed Palaeosols at Stanwix, Carlisle, Cumbria, U.K., *Journal of Archaeological Science*, 28: 541-553.

Verloo, M. 1992. *Agrochemie. Cursus*, Gent, Faculteit van de Landbouwwetenschappen.

Vrydaghs, L. sd.a. *Rapport d'analyses phytolitariennes de lames minces de sol, Vieille-Halle-aux-Blés*, unedited report.

Vrydaghs, L. sd.b. *Rapport d'analyses phytolitariennes de lames minces de sol, impasse du Papier*, unedited report.

Vrydaghs, L., Devos, Y., Fechner, K. and Degraeve, A., in press. Phytolith analysis of ploughed land thin sections. Contribution to the early medieval town development of Brussels (Treurenberg site, Belgium). In: *New perspectives in Phytolith research: climate, environment and archaeology. Proceedings of the 4th International Meeting on Phytolith Research, Cambridge (UK). 28th-31st August 2002. McDonald Institute for Archaeological Research. University of Cambridge.*

www.ingramcontent.com/pod-product-compliance
Lightning Source LLC
LaVergne TN
LVHW071214210926
848482LV00015B/807